Springer Tracts in Civil Engineering

Springer Tracts in Civil Engineering (STCE) publishes the latest developments in Civil Engineering - quickly, informally and in top quality. The series scope includes monographs, professional books, graduate textbooks and edited volumes, as well as outstanding PhD theses. Its goal is to cover all the main branches of civil engineering, both theoretical and applied, including:

- Construction and Structural Mechanics
- Building Materials
- Concrete, Steel and Timber Structures
- Geotechnical Engineering
- Earthquake Engineering
- Coastal Engineering; Ocean and Offshore Engineering
- Hydraulics, Hydrology and Water Resources Engineering
- Environmental Engineering and Sustainability
- Structural Health and Monitoring
- Surveying and Geographical Information Systems
- Heating, Ventilation and Air Conditioning (HVAC)
- Transportation and Traffic
- Risk Analysis
- Safety and Security

Indexed by Scopus

To submit a proposal or request further information, please contact:
Pierpaolo Riva at Pierpaolo.Riva@springer.com (Europe and Americas) Wayne Hu at wayne.hu@springer.com (China)

Honghua Zhao · Gangqiang Kong · Wanghua Sui

Transparent Soil Modelling Technique and Its Application

Science Press
Beijing

Springer

Honghua Zhao
State Key Laboratory of Structural Analysis for Industrial Equipment
Department of Engineering Mechanics
Dalian University of Technology
Dalian, Liaoning, China

Gangqiang Kong
School of Civil and Transportation Engineering
Hohai University
Nanjing, Jiangsu, China

Wanghua Sui
School of Resources and Geosciences
China University of Mining and Technology
Xuzhou, Jiangsu, China

ISSN 2366-259X ISSN 2366-2603 (electronic)
Springer Tracts in Civil Engineering
ISBN 978-981-19-6824-2 ISBN 978-981-19-6825-9 (eBook)
https://doi.org/10.1007/978-981-19-6825-9

Jointly published with Science Press
The print edition is not for sale in China (Mainland). Customers from China (Mainland) please order the print book from: Science Press.

This Springer imprint is published by the registered company Springer Nature Singapore Pte Ltd.
The registered company address is: 152 Beach Road, #21-01/04 Gateway East, Singapore 189721, Singapore

Preface

It gives me immense pleasure to present this book on *Transparent Soil Modelling Technique and Its Application* for graduate students and professional researchers in geotechnical engineering and granular mechanics. This book presented the latest development in transparent soil modelling technique and more application examples and hence can provide better reference and guidance for the further usage of this technique.

Aim

During my Ph.D. study, I came across the concept of "transparent soil", which triggered my great interests on this topic. It is well known that natural soil is opaque, for the geotechnical people like me, it is difficult to guess what really happened inside the soil. What we observed commonly is only the phenomena that happened on the surface or the boundaries of soil mass. Through embedded transducers, discrete information can be obtained for what happened inside the soil mass, but it is not a whole field information. There are some advanced techniques such as Nuclear Magnetic Resonance Imaging (NMRI), Computed Tomography (CT) and X-Ray Computed Tomography (X-CT) that can capture the full image inside the soil mass. However, these techniques are sophisticated and of high cost to install and operate for daily research activities, on the other hand, also limited in the model size. Therefore, I devoted myself into the transparent soil modelling technique in my Ph.D. research life from 2002 to 2007. At that time, Dr. Iskander at New York University has done a pioneer work to advance the transparent soil modelling technique. They have systematically investigated several materials that can make transparent soil, compared their geotechnical properties with the natural soil and applied digital image correlation (DIC) to obtain the displacement field. They applied this technique to investigate some geotechnical problems such as shallow foundation and 2D multiphase flow in transparent soil. I have started to investigate the dynamic properties of transparent soils of silica gel and applied it to investigate the dynamic soil-structure interaction problem. These earlier works were compiled into the first book on the transparent soil modelling by Dr. Iskander in 2010.

Since then, more and more research works in geotechnical engineering and granular mechanics were conducted with the help of transparent soil modelling technique. Many new types of transparent soil materials have emerged. Three-dimensional transparent soil modelling and imaging were also advanced. In China, the advance and the application of the transparent modelling technique have achieved great success. My group at Dalian University of Technology, the research team led by Dr. Hanlong Liu at Hohai University and the group by Dr. Wanghua Sui at China University of Mining and Technology have conducted research on the transparent soil modelling technique for many years and got their research results published. Now we feel that it is a good time to summarize these works and compile them into a new book, which is to introduce the new development of transparent soil modelling technique to the interested readers and enhance further applications on the related research topics in the geotechnical engineering and granular mechanics.

Salient Features

The salient features of the book are the following:

(1) The newest transparent materials reported;
(2) The principle of digital image cross-correlation and 3D deformation field reconstruction were covered;
(3) Detailed application cases of transparent soil modelling in pile foundation were presented;
(4) Detailed application cases of transparent soil modelling in chemical grouting were provided.

Organization

This book is divided into three parts.

Part I of the book deals with the subject of transparent materials, which is focusing on the transparent soils developed till now and their geotechnical properties. It includes six chapters. Chapter 1 introduces available different types of transparent soils. Chapter 2 introduces the geotechnical properties of transparent soil made of silica gel. Chapter 3 describes the geotechnical properties of transparent soil of fused quartz. Chapter 4 describes the geotechnical properties of transparent clay of Carbopol U10. Chapter 5 describes transparent rock, and Chap. 6 describes different types of pore fluids to make transparent soil.

Part II of the book mainly deals with the digital image processing and 3D deformation measurement reconstruction. It includes five chapters. Chapter 7 introduces the laser speckle effect in transparent soil modelling. Chapter 8 describes the 2D transparent soil imaging and digital image cross-correlation methods. Chapter 9 describes the camera calibration through neural network method. Chapter 10 introduces the 3D transparent soil modelling and deformation reconstruction.

Part III of the book provides cases on the application of transparent soil modelling technique. It includes three chapters. Chapter 11 described the application of transparent soil modelling technique in pile foundations, which covers the soil plugging in pipe pile, soil-pile-cap interaction, the penetration mechanisms of pile foundation and the geosynthetics reinforced stone columns. Chapter 12 described the application of transparent soil modelling in chemical grouting. Chapter 13 describes the application of transparent soil modelling of cylindrical object penetration into granular materials.

Dalian, China — Honghua Zhao
Nanjing, China — Gangqiang Kong
Xuzhou, China — Wanghua Sui

Acknowledgements

On preparation and completion of the book *Transparent Soil Modelling Technique and Its Application*, I express my heartiest gratitude to my students, past and present, I also sincerely thank my Ph.D. advisor Dr. Louis Ge who supported and encouraged me to study transparent soil modelling topic in my Ph.D. research.

I also sincerely thank Dr. Gangqiang Kong who inspired me to write this book and shared their research works on this topic. I also thank Dr. Wanghua Sui who made contributions to this book. I appreciated the support from Dr. Gangqiang Kong and Dr. Wanghua Sui who made this book possible. I also owe an enormous debt of gratitude to my colleagues and students who helped me directly or indirectly in preparation of this treatise. Research funds (Nos. 11672066, 51308091, 52027812, 51478165, 5100811, 41072237, 41472268) from National Natural Science Foundation of China are greatly acknowledged.

Contents

About the Authors

Dr. Honghua Zhao received Ph.D. from the University of Missouri-Rolla (now Missouri University of Science and Technology), USA. She is currently an associate professor at Department of Engineering Mechanics, Dalian University of Technology, China. In her Ph.D. studies, she has studied the transparent soil modelling technique. She developed a scanning three-dimensional imaging system for the 3D deformation measurement and reconstruction in transparent soil modelling. She has published over 40 technical papers. In addition, she has co-authored three books and co-edited one conference proceedings. In 2016, she received Endeavour Australia Cheung Kong Research Fellowship awards. She has presided over three National Natural Science Foundation projects.

Dr. Gangqiang Kong received a Ph.D. degree from Dalian University of Technology, China. He is currently a professor at College of Civil and Transportation Engineering, Hohai University, China. He is also a young and middle-aged academic leader of the Qinglan Project in Jiangsu University and the deputy director of the Key Laboratory of Ministry of Education for Geomechanics and Embankment Engineering. His research focuses on energy geotechnics. He presided over five National Natural Science Foundation projects and participated in the National Natural Science Foundation Major/Key Projects and 111 Project. He received the second prize of National Technical Invention (rank 4/6) and the first prize of Liaoning Province and Hubei Province scientific and technological progress (rank 3/11, 4/15, respectively). He is concurrently serving as a member of the Energy Geotechnics Committee of the International Society for Soil Mechanics and Geotechnical Engineering (TC308), the vice chairman of the Energy Geotechnics Committee of the Chinese Society for Rock Mechanics & Engineering and the editorial board members for *Geomechanics for Energy and the Environment* and *Chinese Journal of Geotechnical Engineering*.

Dr. Wanghua Sui received his Ph.D. from China University of Mining and Technology in 1993. He is currently a professor in Mine Hydrogeology and Engineering geology at School of Resources and Geosciences, China University of Mining and

Technology. His research interests include mine engineering geology, mine hydrogeology, geological hazards prevention and grouting. He has published more than 100 technical papers and six books. He received the 2001 National Teaching Achievement Award, 2008 Scientific Awards from the Ministry of Education, P. R. China, etc.

Part I
Transparent Materials

Chapter 1
Introduction

Abstract The opaque properties of natural soils make the visual observation of soil-structure interaction and measurement of soil deformation a great challenge. Transparent soils made of transparent solids and matching refractive index pore fluid is an innovative technique to overcome this difficulty, with the digital image processing technique, the full field deformation inside the transparent soil can be successfully obtained. This book presents the state art of the transparent soil modelling techniques and its applications in various field.

Natural soil is opaque which poses difficulty for people to observe what really happened inside the soil mass when studying soil deformation under external loading, soil-structure interaction, slop failure etc. Even though there have been methods such as computer tomography (CT), X-ray tomography, magnetic resonance imaging (MRT) available to make the observation of the interior deformation of soils possible, these techniques are limited in laboratory research because of its high costs and small model scale level. Searching a suitable transparent material to substitute natural soil is a continuing research topic for many researchers.

An important aspect of transparent soil modeling is to find the appropriate transparent solids to simulate the behavior of the investigated soil/rock. Many types of different materials have been tried more than decades. Crushed glass [1], resin [2] were used to substitute the soils/rock. However, crushed glass was found too angular, stiff and brittle to substitute natural soils (sand). With the continuing effort of numerous trials and tests, a few types of materials were found which can make transparent soils to substitute sand, clay, rock etc. [3, 4] found that silica gel with refractive index matching pore fluid can be used to manufacture the transparent soil with similar behavior as sand, while amorphous silica can be used to substitute clay. Aqua beads were used to model marine deposit [5]. Further, fused quartz is found to manufacture transparent soil with more close behavior with natural sand [6] which is frequently used now. Currently more new types of transparent soils are emerging, such as Laponite_RD, Carpobol U10 etc., which were used to modelling clay [7, 8]. In this chapter, two types of typical transparent solids to make transparent soils are presented.

H. Zhao et al., *Transparent Soil Modelling Technique and Its Application*, Springer Tracts in Civil Engineering, https://doi.org/10.1007/978-981-19-6825-9_1

References

1. Konagai K, Tamura C, Rangelow P, Matsushima T (1992) Laser-aided tomography: a tool for visualization of changes in the fabric of granular assemblage. Proc of JSCE No.455 I-21 Struct Eng/Earthquake Eng 9(3):193–201
2. Geng J, An Z (1978) Thin section preparation for loose rock and soil materials using epoxy resin. Geochimica 3:194–196
3. Iskander MG, Sadek S, Liu J (2002) Optical measurement of deformation using transparent silica gel to model sand. Int J Phys Model Geotech 2(4):13–26
4. Iskander MG, Liu J, Sadek S (2002) Transparent amorphous silica to model clay. J Geotech Geoenviron Eng 128(3):262–273
5. Tabe K, Iskander M, Honma S (2011) Transparent aquabeads to visualize flow in porous material. Adv Mater Res 239–242:2602–2605
6. Ezzein FM, Bathurst RJ (2011) A transparent sand for geotechnical laboratory modeling. Geotech Test J 34(6):1–12
7. Wallace J, Rutherford C (2015) Geotechnical properties of laponite RD. Geotech Test J 38(5):1–14
8. Kong G, Zhou Y, Yang Q, Wang C, Chen Y (2020) Geotechnical properties of a new transparent clay. Int J Geomech ASCE 20(12):06020032

Chapter 2
Transparent Sand of Silica Gel

Abstract Silica gel is a porous, amorphous form silica (SiO2), which closely resembles natural sand. It is composed of a vast network of interconnected microscopic pores. It is used to make transparent soil modelling sand. When mixed with a pore fluid having a matched refractive index. The physical, geotechnical, and dynamic properties of some silica gels are discussed and compared with the natural sand in this chapter.

Silica gel is a porous, amorphous form silica (SiO_2), which closely resembles natural sand [1]. It is composed of a vast network of interconnected microscopic pores with diameters ranging from 5×10^{-10} m to 300×10^{-10} m. It has either granular or beaded shape and is made by partial dehydration of metasilicic acid and is commonly used as a drying agent, catalysts, and for purifying various substances. Silica gel specimens usually exhibit a high apparent total void ratio e due to the internal porosity of the transparent aggregates [2]. Figure 2.1 shows the image of the typical silica gel particles.

The interaggregate void ratio e_i is usually used to describe the effective void space between silica gel aggregates. The interaggregate void ratio can be defined as follows.

$$e_i = \frac{V_v - V_{vi}}{V_s + V_{vi}} = \frac{e - a\gamma_s}{1 + a\gamma_s} \tag{2.1}$$

where V_v is the total volume of voids; V_{vi} is the volume of voids inside solid aggregates; V_s is the volume of solids; γ_s is the unit weight of solids; and a is the adsorption factor. A typical value of $a = 0.3$ ml/g for silica gel is usually used. Iskander et al. [3] first developed transparent soil with silica gel and refractive index matching pore fluid. This is the first type of transparent soil with the similar behavior as natural sand. It has been used to study spread footing [4], flow near PVDs [5], pile penetration [6], shallow foundations [7], and to study soil deformation due to tunneling [8, 9].

H. Zhao et al., *Transparent Soil Modelling Technique and Its Application*, Springer Tracts in Civil Engineering, https://doi.org/10.1007/978-981-19-6825-9_2

Fig. 2.1 Silica gel particles

2.1 Static Properties of Silica Gel

Iskander et al [3] have conducted comprehensive investigations on the static properties of silica gel. From a series of tests, they compared the static properties of silica gel with the sand, and found the similar behavior. The two silica gels (shown in Fig. 2.1) were studied, the first type is angular, 0.5–1.5 mm in size (SG1). The second is round, 2–5 mm in size (SG2). These can be used for modelling medium to coarse sand. The specific gravity of silica gel is 2.2, which is approximately 80% of that natural sand. The dry unit weight is 6–9 kN/m^3. The saturated unit weight is 11–14 kN/m^3 for mineral oil and solvent pore fluid which may vary depending on the pore fluid used. Figure 2.2 shows the particle size distribution curve of silica gels tested for static geotechnical properties.

Consolidated drained triaxial compression tests were performed on dry and saturated specimens (0.5–1.5 mm size, angular). The obtained internal friction angle for SG1 is varying from 31° to 42°, listed in Table 2.1; and the internal friction angle for

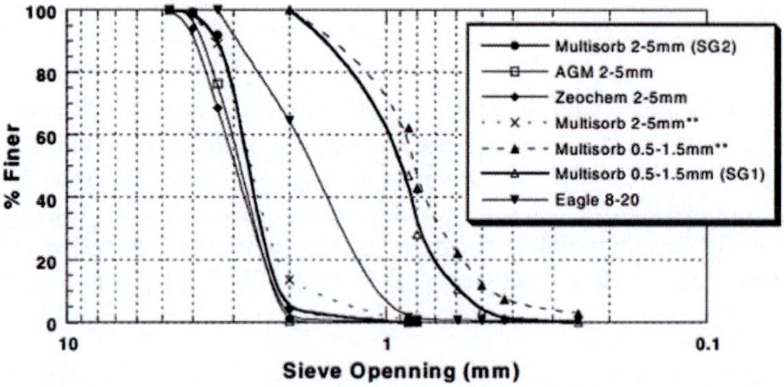

Fig. 2.2 Particle size distribution of silica gels (Solid lines and dashed lines are for samples before and after triaxial testing, [7])

Table 2.1 Triaxial test results for SG1 (0.5–1.5 mm size)

Testing condition	Average unit weight (kN/m^3)	Angle of internal friction (°)
Saturated loose	7.25	31°–36°
Sat. dense	8.53	35°–42°
Dry loose	6.68	36°–39°
Dry dense	7.08	31°–40°
Sat. med. dense	8.18	33°–40°

Table 2.2 Triaxial test results for SG2 (2–5 mm size)

Testing condition	Average unit weight (kN/m^3)	Angle of internal friction (°)
Saturated loose	7.24	29°–32°
Sat. dense	7.79	33°–36°
Dry loose	7.74	31°–37°
Dry dense	8.12	33°–35°

SG2 is varying from 29° to 37°, listed in Table 2.2. These values fall in the range of those for typical natural sand.

The stress–strain curves of silica gel display a strain hardening behavior at loose state, but a strain softening behavior at dense state. Typical stress–strain curves of silica gel are shown in Fig. 2.3.

Direct sear tests were conducted on the dry silica gel. The stress–strain curves by direct shear test were consistent with those natural sands. Angles of internal friction ranged between 34° and 50°. The angles of friction of the Mohr–Coulomb failure envelopes were 36–40° for SG1 and 38–42° for SG2 envelopes, which is consistent with most natural sands and the triaxial test results.

From triaxial test conducted, E_{50} (Young's modulus at 50% of failure strain) were found to be 15–22 MPa (2–3 ksi) for loose samples and 26–32 MPa (3–5 ksi) for dense samples of SG1. For SG2, E_{50} was found to be 24–52 MPa (3–8 ksi) and 36–84 MPa (5–12 ksi) for loose and dense respectively. These values are similar to the reported values for sand [1].

Consolidation tests on the silica gel indicates that the compressibility of silica gel is similar with that of calcareous sand because of its internal porous structure, but more compressible than the natural sand. The compression curve of silica gel is similar to that of natural sand.

Confined compression odometer tests were conducted on silica gel specimens according to ASTM test method D2439-96. They found that the silica gel is much more compressible compared to Ottawa sand. The "virgin" compressibility of SG1 is larger than that of calcareous sands, but when the compression pressure is greater than 100 kPa, the compressibility of 2–5 mm silica gel is similar to that of calcareous sands. A significant particle breakage was observed for the silica gel in consolidation tests.

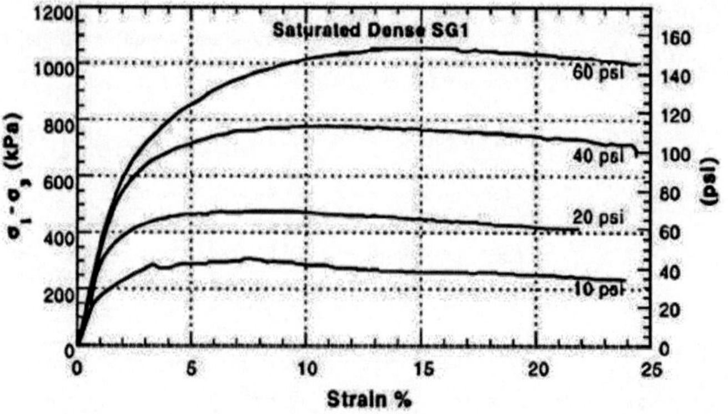

(a) Saturated dense SG1

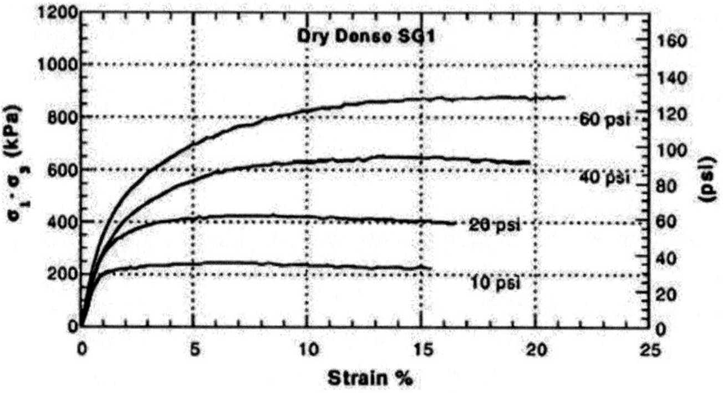

(b) Dry dense SG1

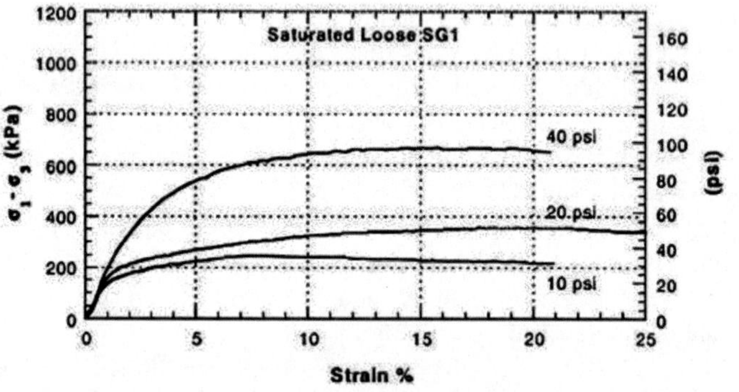

(c) Saturated loose SG1

Fig. 2.3 Typical stress–strain curves of silica gel in CD triaxial test

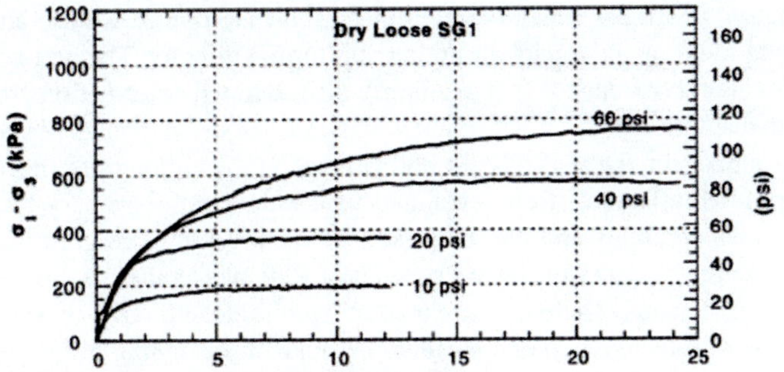

(d) Dry loose SG1

Fig. 2.3 (continued)

Flexible wall permeability tests were performed on the silica gels according to ASTM test standard D5084. The hydraulic conductivities obtained are 1.5×10^{-4} cm/sec for SG1 and 7×10^{-4} cm/sec for SG2 with pore fluid mineral oil & solvent mixture.

2.2 Dynamic Properties of Silica Gel

To apply the transparent soil modeling technique in dynamic geotechnical research field, it is necessary to know the dynamic properties of silica gel. Here a detailed investigation on the shear modulus and damping ratio of dry silica gel was presented.

2.2.1 Resonant Column Tests and Sample Preparation

Resonant column tests have been widely used for determining dynamic properties of soil. The basic principle of the resonant column test is to excite a cylindrical specimen of soil in its fundamental mode of torsional vibration. Once the fundamental mode of vibration is established, resonant frequency and amplitude of vibration are measured. Shear modulus and shear strain amplitude can be obtained from these data along with system constant and specimen properties. Material damping is determined from the free-vibration decay curve, which is recorded after turning off power at resonance. A torsional-resonant column apparatus from Geotechnical Consulting and Testing Systems (GCTS) was used for the tests. The GCTS Torsional Shear/Resonant Column system design features a fixed base with the top of the specimen able to rotate freely with the application of torsional force (as shown in Fig. 2.4). The torsional force is applied by a torsional drive at first-mode resonance at frequencies up to 250 Hz. A non-contacting rotational sensor (accelerometer) and a torque transducer attached to the top plate monitor the torsional motion and torque directly. The non-contacting sensor is located on extended arms to amplify the rotational deformation providing a shear strain resolution of 10^{-6}. The input voltage was adjusted to maintain consistent strain values between the tests. Hardin and Drnevich [13] summarized that the shear strain amplitude, effective stress level, and void ratio are the primary factors that affect the shear modulus and damping ratio of clean sands. The grain size, grain shape, and degree of saturation have a secondary influence on the shear modulus and damping ratio of sands. Therefore, shear strain amplitude, effective stress level, and void ratio were selected as the controlled variables in designing the experimental program. Also, the lowest shear strain value of 10 − 5 was achieved. The ASTM D 4015 Specification was followed as the operating and data reduction procedure. Summaries of the tests conducted are given in Table 2.3. Silica gel used in this study is manufactured by Qingdao Haiyang Chemical Co., LTD, China. Its specific gravity is 2.21. Four gradations of silica gels were tested within the ranges of 2–5 mm (SG1), 1–3 mm (SG2), 0.5–1 mm (SG3), and 0.5–5 mm (SG4). Figure 2.5 shows the gradation curves for all the tested silica gel samples. For each gradation, silica gel specimens at two different void ratios were prepared, representing the dense and loose states (e.g. SG1-L and SG1-G), respectively. The obtained void ratios from 1.48 to 1.91 were much higher than the uniform sized sands which typically range from 0.45 to 0.7. This is due to the special structure of the silica gel, which has both voids between the aggregates and inside the aggregate gates. The confining pressure for each specimen varied from 50–100, 200, 300, and 400 kPa. The testing shear strain level was ranging in the order of 10 − 5 to 10 − 3. The air pluviation technique was used to prepare the loose specimens, where silica gel was poured from a nozzle while maintaining a constant drop-height (4 cm) through the preparation. The confining pressure was applied to the specimen and was raised to the desired level. For dense specimens, the entire specimen was evenly divided into five layers, and each layer was compacted by a miniature tamper. The mass of silica gel for each layer was determined by trial and error. The average specimen diameter and height were 70 and 158 mm, respectively. Silica gels were received from the manufacturer, dry, sealed. Once they were tested, they were open to the atmosphere. More detailed information on the testing procedure and specimen preparation can be found in Zhao and Ge [25]. Liquid bridge effect was considered minimal in each test, as majority of the tested silica gel particles are larger than 1 mm in diameter. Also, it does not have a significant influence on the macroscopic engineering properties under the tested confining pressures.

2.2.2 Shear Modulus of Silica Gel

The maximum shear modulus G_{max}, at a strain level of 10^{-6}, under five different confining pressures (50, 100, 200, 300, and 400 kPa) for testing series SG1, SG2, SG3 and SG4, is presented in Fig. 2.6. Since the lowest strain level of the silica gel

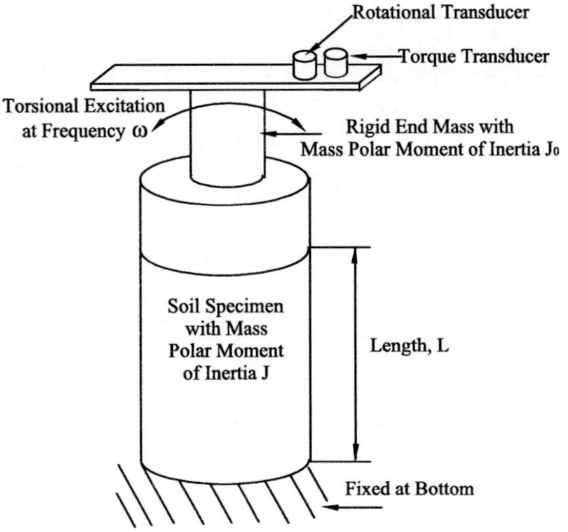

Fig. 2.4 The schematic illustration of GCTS resonant column apparatus

Table 2.3 Summary of experimental testing program

Test series	Void ratio	Grain size (mm)	Confining pressure (kPa)
SG1-L	1.885	2–5	50, 100, 200, 300, 400
SG1-D	1.797	2–5	50, 100, 200, 300, 400
SG2-L	1.882	1–3	50, 100, 200, 300, 400
SG2-D	1.791	1–3	50, 100, 200, 300, 400
SG3-L	1.912	0.5–1	50, 100, 200, 300, 400
SG3-D	1.831	0.5–1	50, 100, 200, 300, 400
SG4-L	1.536	0.5–5	50, 100, 200, 300, 400
SG4-D	1.481	0.5–5	50, 100, 200, 300, 400

from resonant column tests was 10^{-5}, the maximum shear modulus was determined through regression analyses. Two approached were adopted in this study. The first approach was given by Stokoe et al. [21, 22] as shown in Eq. (2.2) where G/G_{max} is called normalized shear modulus (no units), γ_r is a reference strain, and α is a curve-fitting variable called curvature parameter. The reference strain is described by the following equation.

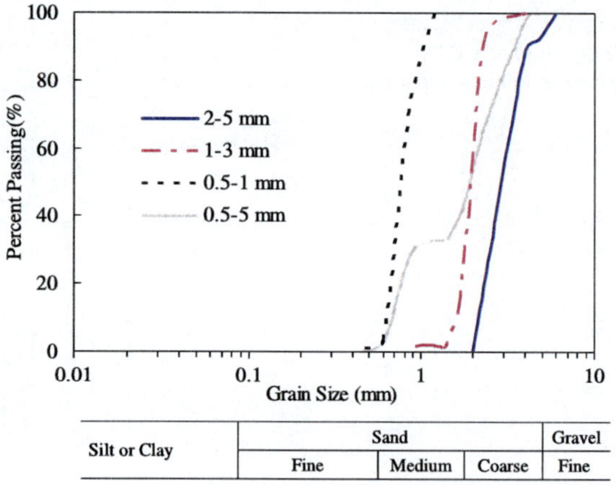

Fig. 2.5 The gradation curves for all the testing silica gel samples

$$\gamma_r = \gamma_{r1}\left(\frac{\sigma'_m}{p_a}\right)^k \tag{2.2}$$

where γ_{r1} is the reference strain at the effective mean stress of 100 kPa, p_a is the reference stress of 100 kPa, σ'_m is effective mean stress and k is a stress correction exponent.

The second approach was given by Borden et al. [10].

$$\frac{G}{G_{max}} = \frac{1}{(1 + m\gamma^n)^p} \tag{2.3}$$

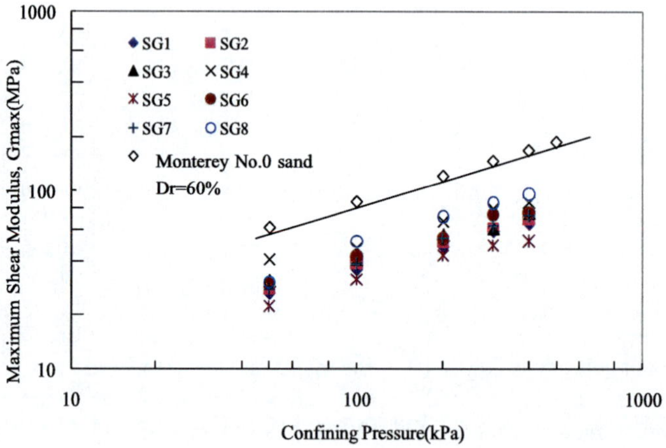

Fig. 2.6 Comparison of maximum shear modulus for Monterey No.0 sand and silica gel

where, γ is the shear strain and *m, n,* and *p* are model parameters. Both Eqs. (2.2) and (2.3) give the normalized shear modulus value as a function of shear strain. In the current study, it is found that both approaches lead to the same values of maximum shear modulus for silica gel, which are almost identical to the shear modulus at a shear strain level of 10^{-5}. Figure 2.6 also shows a general trend toward increasing G_{max} with confining pressures for each gradation of silica gel. For the same test series, shear modulus of loose specimen is greater than those of the dense specimen, which is obvious from Fig. 2.6. Grain size also plays an influence on the maximum shear modulus under the same confining pressure. An empirical equation proposed by Eq. (2.4) was used to fit the testing data between the maximum shear modulus (G_{max}) and confining pressure (σ_0).

$$G_{\max} = \frac{S \cdot OCR^k}{(0.3 + 0.7e^2)} \sigma_0^n p_a^{(1-n)} \tag{2.4}$$

where *S* is a non-dimensional stiffness coefficient, *OCR* is the overconsolidation ratio taken as 1 in the paper, *k* is a parameter depending on the plasticity of the soil ($k = 0$) for cohesionless soils), P_a is the atmosphere pressure, *e* is the void ratio, *n* is the power of the stress. Both parameters *S* and *n* are function of all particle contacts as described by Rowe [18]. Table 2.2 summarized parameters of *S* and *n* for all the test series. It can be seen that the parameter *S* is ranging between 751 and 1282, while the parameter *n* is ranging between 0.38 and 0.50. Chung et al. [11] reported an *S* value of 523 and an *n* value of 0.48 for Monterey No.0 sand with $D_r = 60\%$. When compared with the fitting data for Monterey No.0 sand (as shown in Fig. 2.6), the maximum shear modulus of the silica gel is slightly lower than that of sand because of their high apparent void ratios. For the natural soils, Hardin [14] has reported typical *S* value of 600 for the natural sand. The difference for the value of *n* between the silica gel and the natural soils shows that the shear modulus of silica gel is much lower than those of natural soils under the same confining pressure (Table 2.4).

The shear modulus as a function of shear strain amplitude as determined from resonant column tests for confining pressures of 50, 100, 200, 300, and 400 kPa is given in Fig. 2.5 for the loose and dense silica gel with a grain size of 2–5 mm (SG1). It is evident that the shear modulus is confining stress dependent, the higher confining stress and the higher shear modulus (e.g. [12, 15, 17, 22, 24]).

The dense specimen has a higher shear modulus than the loose specimen. This trend was also observed for the other tested samples (SG2, SG3 and SG4). From Fig. 2.7, it is also observed that the elastic shear strain threshold for the dense silica

Table 2.4 Parameters of *S* and *n*

Specimen	*S*	*n*	Specimen	*S*	*n*
SG1-L	989	0.44	SG3-L	890	0.39
SG1-D	978	0.42	SG3-D	1100	0.44
SG2-L	1137	0.41	SG4-L	891	0.50
SG2-D	1282	0.38	SG4-D	751	0.45

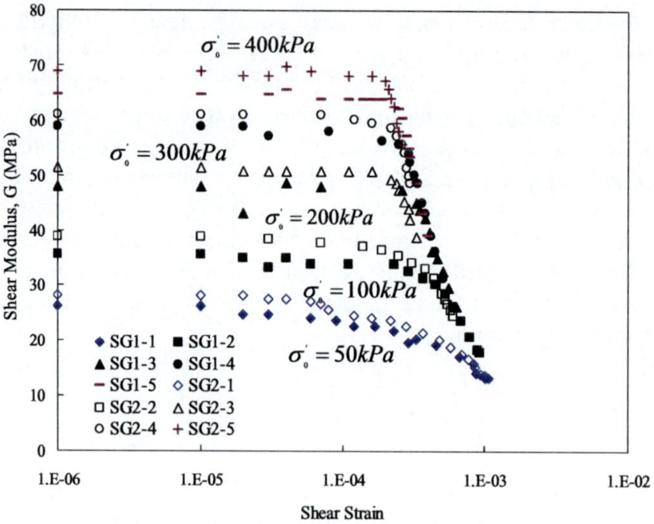

Fig. 2.7 The influence of confining pressure on the shear modulus of silica gel

gel is less than that for the loose silica gel, which indicates that the dense silica gel enters into the plastic state earlier than the loose silica gel.

The normalized shear modulus versus shear strain of all tested silica gel is shown through Figs. 2.8, 2.9, 2.10, 2.11. Figure 2.8 displayed the normalized shear modulus versus shear strain for SG1 series. The normalized shear strain for all the samples with the same grain size distribution tends to fall into a line when the shear strain becomes larger. The similar trend can be observed from the testing data for the samples of other grain size (SG2, SG3, and SG4), as can be seen from Figs. 2.9 (1–3 mm), 2.10 (0.5–1 mm) and 2.11 (0.5–5 mm). For each gradation at a given initial void ratio, the normalized shear moduli fall into a narrow band. Also, the elastic threshold strain level is found to be very consistent for all four gradations and is about 2×10^{-4}.

The upper and lower bounds of G/G_{max} as a function of shear strain for sand reported by Seed and Idriss (1970) with different void ratios (relative densities from 30 to 90%) for an effective vertical stress of 144 kPa and $k_0 = 0.5$ (coefficient of lateral earth pressure at rest), those for gravels reported by Rollins et al. [17] from loose to dense to very dense state (of 15 studies, most of confining pressures between 100 and 400 kPa, relative densities from 40 to 100%), and those for four gradations and is about 2×10^{-4}.

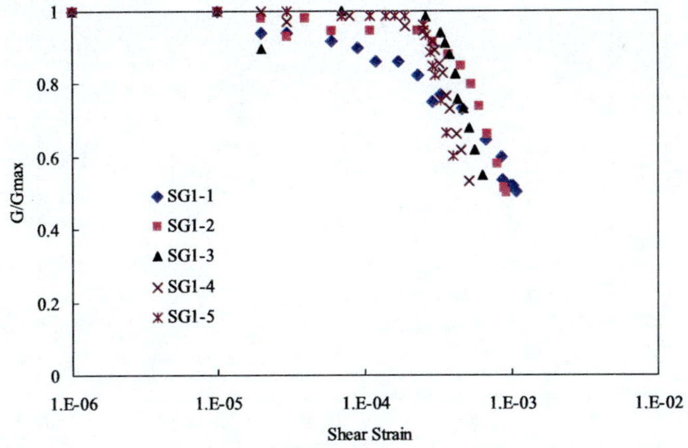

Fig. 2.8 The normalized shear modulus versus shear strain for the silica gel with a grain size of 2–5 mm (SG1)

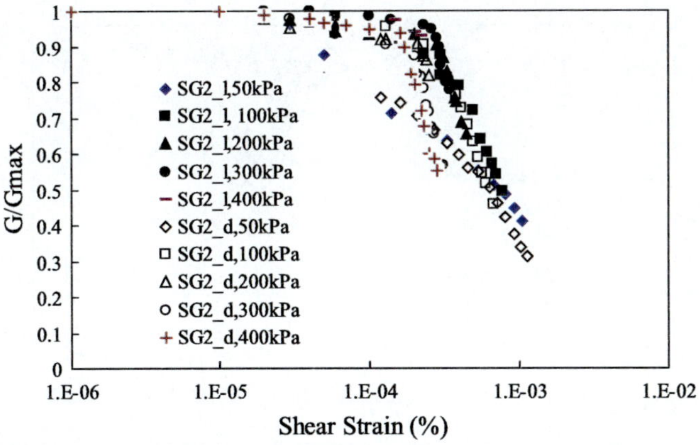

Fig. 2.9 The normalized shear modulus versus shear strain for the silica gel with a grain size of 1–3 mm (SG2)

2.2.3 Comparison with Shear Modulus of Clay, Sand and Gravel

The upper and lower bounds of G/G_{max} as a function of shear strain for sand reported by Seed and Idriss [19] with different void ratios(relative densities from 30 to 90%) for an effective vertical stress of 144 kPa and $k_O = 0.5$ (coefficient of lateral earth pressure at rest), those for gravels reported by Rollins et al. [17] from loose to dense to very dense state (of 15 studies, most of confining pressures between 100 and 400 kPa, relative densities from 40 to 100%), and those for clay reported by Vucetic

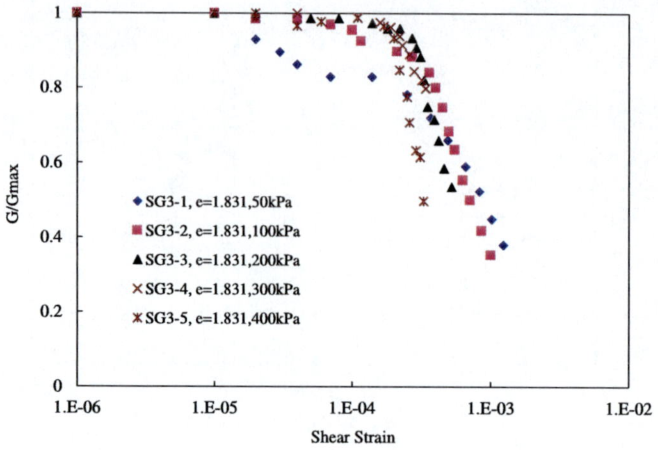

Fig. 2.10 The normalized shear modulus versus shear strain for the silica gel with a grain size of 0.5–1 mm (SG3)

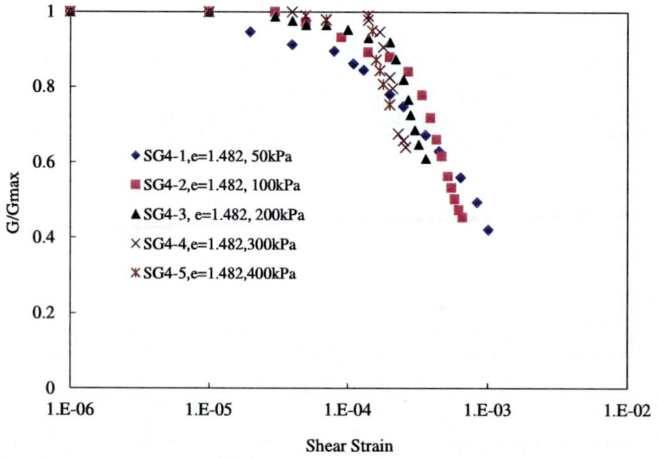

Fig. 2.11 The normalized shear modulus versus shear strain for the silica gel with a grain size of 0.5–5 mm (SG4)

and Dorbry [23] with a plastic index (PI) value from 0 to 200 (OCR from 1 to 15) are also plotted in Fig. 2.12. The normalized shear modulus for sand and gravel has the similar distribution range. While the normalized shear modulus for clay is located above that for sand and gravel. The normalized shear modulus of silica gel is shifted above the upper bound values for sand and gravel, falling within the region for clay. In the elastic region, most of the normalized shear modulus of silica gel is located above that of clay. Hence, the shear modulus response of silica gel behaves more like clay in that regard. The elastic threshold shear strain for the silica gel is ranging

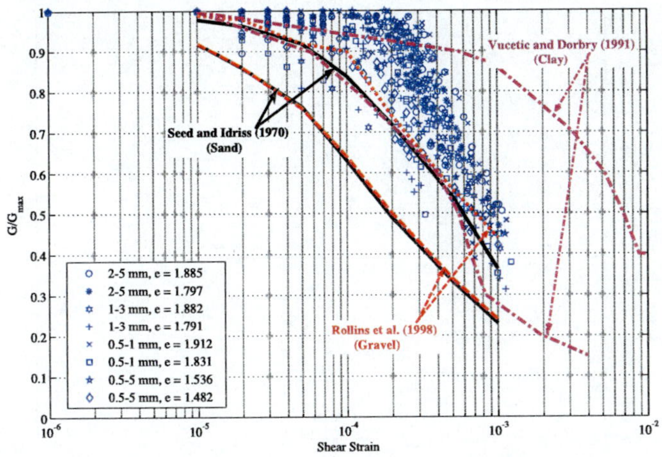

Fig. 2.12 Comparison of normalized shear modulus of silica gel versus natural soils

from 4×10^{-5} to 2.5×10^{-4}. For the sand, the elastic threshold shear strain level is in the range of 2×10^{-5} to 5×10^{-5}. For the clay, the elastic threshold shear strain level is in the range of 5×10^{-5} to 6×10^{-4} as shown from Fig. 2.12. Thus, the elastic threshold shear strain level for the silica gel is basically in the range of that for the natural soils. The behavior of soil at shear strains below elastic shear strain threshold is commonly treated linear or "elastic", even though the soils exhibit material damping at such low strain levels [16].

2.2.4 Damping Ratio of Silica Gel

Damping ratio as a typical index to characterize the energy dissipation properties of a material was also studied for the silica gel. The damping ratios were determined by the amplitude decay method according to ASTM D 4015. The damping ratio versus shear strain for all tested silica gel samples is shown through Figs. 2.13, 2.14, 2.15 and 2.16. In general, with the increasing of the shear strain, the damping ratio first displayed a slow linear increase and then followed by a rapid nonlinear increasing behavior. The results from Fig. 2.13 also showed that the loose samples (SG1-L) exhibit larger damp-ing ratio than the dense samples under the same confining stress. With the increasing shear strain, the damping data for the loose and dense samples tend to fall together. It also shows that the elastic threshold shear strain for the damping ratio of the dense silica gel is smaller than that for the loose silica gel, which means that dense silica gel enters into plastic state earlier than the loose silica gel. This is consistent with the conclusion obtained from the normalized shear modulus versus shear strain

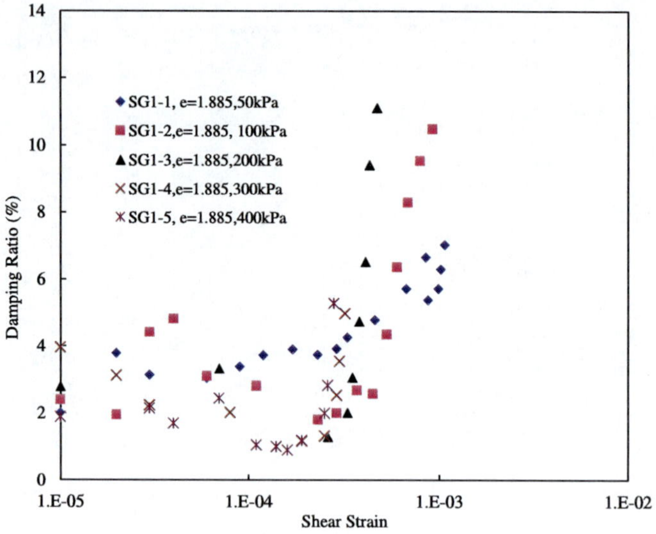

Fig. 2.13 Damping ratio versus shear strain for SG1

results. A similar trend was observed for the other tested silica gel samples. The elastic threshold shear strain amplitude for damping ratio is found to be in the same range (4×10^{-5} to 2×10^{-4}) as that for the shear modulus. Figure 2.13 displayed the damping ratio versus shear strain for the loose and the dense state of the silica gel with a grain size of 1–3 mm, respectively. It clearly shows that the elastic threshold shear strain for the damping ratio is confining stress dependent. The lower confining stress, the smaller is the elastic threshold shear strain. The similar trend can be observed from the testing data for the sample of other grain size [Figs. 2.14 (2–5 mm), 2.15 (0.5–1 mm), and 2.16 (0.5–5 mm)].

This conclusion is consistent with that observed from the normalized shear modulus. However, there is not a clear trend for the damping ratio to decrease with increasing confining pressures as that observed for Monterey sand by Chung et al. [11].

In order to make a comparison of the damping ratios of silica gel and the natural soils, the damping data for the sand from [20] for gravel from [17], and for clay from Vucetic and Dobry [23] were also plotted in Fig. 2.17. This plot shows that the damping ratio of clay, sand and gravel have overlapped zones. The damping ratio for the clay is covering a wider range than sand and gravel when shear strain getting large. The lower range of damping ratio for clay, sand and gravel are almost the same when the shear strain is less than 10^{-4}. The upper bound of damping ratio for sand is higher than that for clay and gravel. In Fig. 2.17, the damping ratios of silica gel fall into the range of natural soils when the shear strain is less than 10^{-4}. For the shear strain in the range of 1×10^{-4} to 4×10^{-4}, the damping ratio of silica gel is falling in the lower range of the three types of natural soils. When the shear strain is greater than about 4×10^{-4}, the damping ratio of silica gel is falling in the range

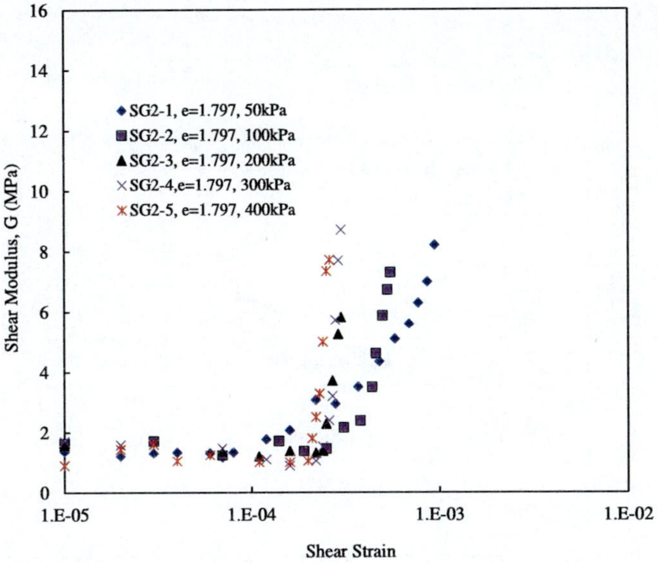

Fig. 2.14 Damping ratio versus shear strain for SG2

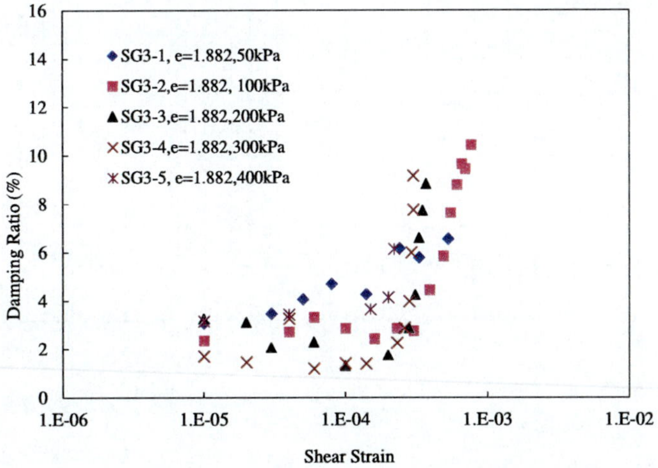

Fig. 2.15 Damping ratio versus shear strain for SG3

for that of gravel, which is in the lower range of sand and the upper range of clay. In general, the damping ratio of the silica gel is within the range for that of natural soils given by Rollins et al. [17], Seed et al. [20], and Vucetic and Dorbry [23] as concluded from the above.

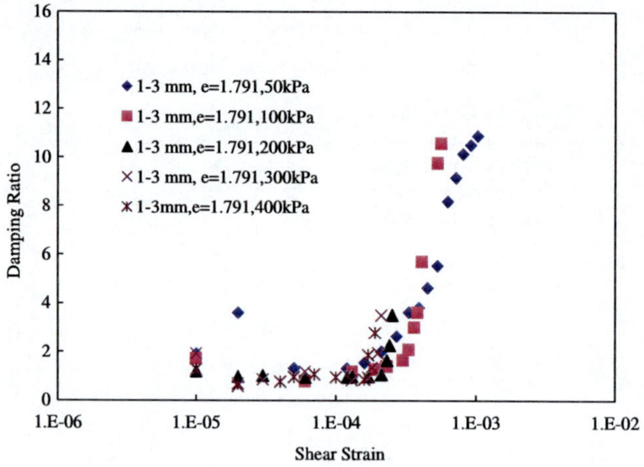

Fig. 2.16 Damping ratio versus shear strain for SG4

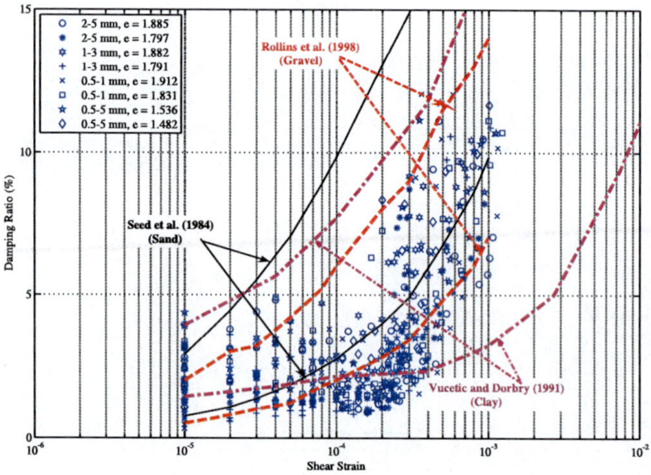

Fig. 2.17 Comparison of damping ratios of silica gel versus natural soils

Ishibashi and Zhang [15] proposed a unified formula for the damping ratio of sand after summarized the previous researcher's data. The formula expressed the damping ratio for sand as a function of G/G_{max} as follows:

$$\xi = 0.333\left[0.586\left(\frac{G}{G_{max}}\right)^2 - 1.547\left(\frac{G}{G_{max}}\right) + 1\right] \quad (2.5)$$

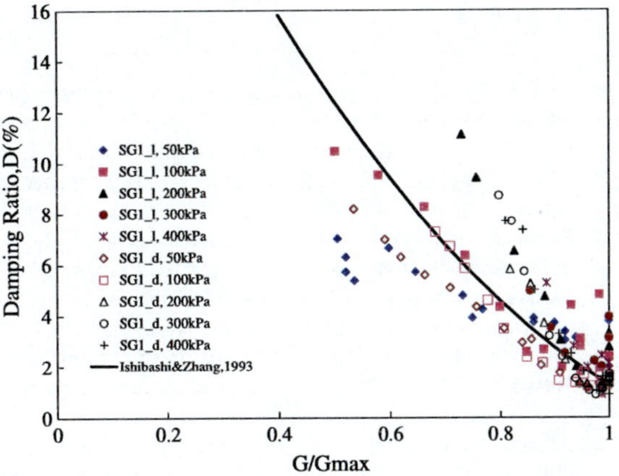

Fig. 2.18 Comparison of damping ratios of silica gel versus natural soils

The damping ratio of silica gel of different gradations versus G/G_{max} was plotted in Fig. 2.18. The fitting curve of Eq. (2.5) was plotted on each plot. It was clear that the fitting line by Ishibashi and Zhang (1993) can be proposed as an average line for the silica gel tested. This comparison also showed the similarity of the dynamic behavior of sand and silica gel.

2.3 Summary and Conclusions

In summary, silica gel is an appropriate transparent solid to make transparent soil with the similar behavior as the natural sand, especially the calcareous sand. The internal friction angle values are within the range of the natural sand. The shear and compression behavior are similar with the natural sand, though the silica gel is more compressible than natural sand.

The dynamic properties were investigated in detail. The relationship between the maximum shear modulus and confining pressures of silica gel follows the equations by Hardin [14] with different *S* value. The shear modulus of silica gel increases with confining pressures. The shear modulus of dense silica gel is greater than that of loose silica gel with the same grain size distribution. The relationship between the normalized shear modulus and shear strain and the relationship between the damping ratio and shear strain of silica gel were found to follow the general trend as those of natural soils. The linear elastic shear strain threshold of silica gel is confining stress dependent. Test findings show that silica gel could at a certain degree simulate the natural behavior of the natural soils though some differences exist between them.

References

1. Hunt R (1984) Geotechnical engineering investigation manual. McGraw-Hill, New York
2. Iskander M, Lai J, Oswald C, Mainnheimer R (1994) Development of a transparent material to model the geotechnical properties of soils. Geotech Test J 17(4):525–433
3. Iskander M, Sadek S, Liu J (2002) Optical measurement of deformation using transparent silica gel to model sand. Int J Phys Model Geotech 4(2):13–26
4. Sadek S, Iskander M, Liu J (2003) Accuracy of digital image correlation for measuring deformations in transparent media. ASCE J Comput Civ Eng 17(2):88–96
5. Welker A, Bowders J, Gilbert R (2000) Using a reduced equivalent diameter for a prefabricated vertical drain to account for smear. Geosynth Int 7(1):47–57
6. Liu J, Iskander MG (2010) Modelling capacity of transparent soil. Can Geotech J 47(4):451–460
7. Iskander M, Liu J (2010) Spatial deformation measurement using transparent soil. Geotech Test J 33(4):1–7
8. Ahmed M, Iskander M (2011) Analysis of tunneling induced ground movements using transparent soil models. J Geotech Geoenviron Eng 137(5):525–535
9. Ahmed M, Iskander M (2011) Evaluation of tunnel face stability by transparent soil models. Tunnel Undergr Space Technol 27(1):101–110
10. Borden RH, Shao L, Gupta A (1996) Dynamic properties of piedmont residual soils. J Geotech Eng 122(10):813–821
11. Chung RM, Yokel FY, Drnevich VP (1984) Evaluation of dynamic properties of sands by resonant column testing. Geotech Test J 7(2):60–69
12. Darendeli MB (2001) Development of a new family of normalized modulus reduction and material damping curves. PhD dissertation, Univ. of Texas at Austin, Austin, TX
13. Hardin BO, Drnevich VP (1972) Shear modulus and damping in soils: measurement and parameter effects. J Soil Mech Found Div ASCE 98(6):603–624
14. Hardin BO (1978) The nature of stress–strain behavior for soils. In: Proceedings of the ASCE geotechnical engineering division specialty conference earthquake engineering and soil dynamics. 1:3–90
15. Ishibashi I, Zhang X-J (1993) Unified dynamic shear moduli and damping ratios of sand and clay. Soils Found 33(1):182–191
16. Ishihara K (1996) Soil behavior in earthquake geotechnichs. Oxford University Press, Oxford
17. Rollins KM, Evans MD, Diehl NB, Daily WD (1998) Shear modulus and damping relationships for gravel. J Geotech Geoenviron Eng 124(5):396–405
18. Rowe PW (1971) Theoretical meaning and observed values of deformation parameters for soil. In: Parry RHG (ed) Stress–strain behavior of soils, proceedings of roscoe memorial symposium, pp 143–194. Cambridge University, Cambridge
19. Seed HB, Idriss IM (1970) Soil moduli and damping factors for dynamic response analysis. Report No. EERC 70–10, Earthquake Engineering Research Centre, University of California, Berkeley, CA.
20. Seed HB, Wong RT, Idriss IM, Tokimatsu K (1984) Moduli and damping factors for dynamic analysis of cohesionless soils. Report No. EERC 84–14, Earthquake Engineering Research Centre, University of California, Berkeley, CA.
21. Stokoe KH II, Hwang SK, Darendeli MB, Lee NJ (1995) Correlation study of nonlinear dynamic soils properties. Final Rep to Westinghouse Savannah River Company, Aiken, S.C
22. Stokoe KH II, Darendeli MB, Andrus RD, Brown LT (1999) Dynamic soil properties: laboratory, field and correlation studies. In: Proceedings of 2nd international conference on earthquake geotechnical engineering, vol 3, pp 811–845. Lisbon, Portugal
23. Vucetic M, Dorbry R (1991) Effect of soil plasticity on cyclic response. J Geotech Eng 117(1):89–107

24. Zhang J, Andrus DR, Juang CH (2005) Normalized shear modulus and material damping ratio relationships. J Geotech Geoenviron Eng 131(4):453–464
25. Zhao H, Ge L (2007) Dynamic properties of transparent soil. In: Proceedings of GeoDenver 2007, dynamic response of soil properties (GSP 160), Denver, Colorado, Feb 18–21, USA, pp 1–9

Chapter 3
Transparent Sand of Fused Quartz

Abstract Fused quartz was manufactured by firstly melting natural quartz crystals present in quartzite sand at approximately 2000 °C, and then cooling them. Its purity is high. In this chapter, the static and dynamic properties of transparent soil made of fused quartz are presented. Duncan-Chang model parameters were determined for some fused quartz. These provides valuable information for conducting research using transparent soil of fused quartz in both static and dynamic geomechanics.

3.1 Introduction

The mixture of transparent granular materials and pore fluid with the same refractive index has been used as a synthetic transparent soil to mimic the behavior of natural soil. Two main innovative transparent granular materials have been widely used in previous studies. Silica gel is used to model natural sand and its properties were reported in Sect. 2.2. However, silica gel particles are hydroscopic and thus affected by high humidity and water, which can cause the particles to break and become colored; its particles also deform plastically even under low confining pressure. Ezzein & Bathurst [28] conducted a series of investigations on fused quartz for modeling natural soil. The static and dynamic properties of fused quartz and the transparent soil manufactured by fused quartz and pore fluid have been studied by Kong et al. [19–21]. In this chapter, the static and dynamic properties of transparent soil made of fused quartz are presented. These provides valuable information for conducting research using transparent soil of fused quartz in both static and dynamic geomechanics.

3.2 Static Properties of Fused Quartz

Fused quartz was manufactured by firstly melting natural quartz crystals present in quartzite sand at approximately 2000 °C, and then cooling them. After that it is

H. Zhao et al., *Transparent Soil Modelling Technique and Its Application*, Springer Tracts in Civil Engineering, https://doi.org/10.1007/978-981-19-6825-9_3

broken into the required sizes. In this section, the investigations on static properties of fused quartz with matching pore fluid were presented.

3.2.1 Materials

The fused quartz used were manufactured by Xuzhou Xinyi Wanhe Minerals Limited, China. Most particles were angular shape. Two different sized fused quartz and one standard sand (Fujian standard sand) were tested in this study as shown in Fig. 3.1. One type of fused quartz is of standard gradation in size, which has the same gradation as Fujian sand. The second one is angular in shape with grain sizes between 0.5 mm and 1.0 mm. The specific gravity of fused quartz particles equals 2.186, which was approximately 80% of that of natural sand particles. The fused quartz of standard gradation was poorly graded, with 0.91 coefficient of curvature (C_c) and 2.59 coefficient of uniformity (C_u). Some major physical properties of the two transparent soils and Fujian standard sand are shown as follow: The minimum dry density of standard gradation, uniform size and Fuijan standard sand equal 1.119 g/cm^3, 0.97 g/cm^3, and 1.65 g/cm^3, respectively; The maximum dry density of standard gradation, uniform size and Fuijan standard sand equal 1.52 g/cm^3, 1.274 g/cm^3, and 1.864 g/cm^3, respectively; The specific gravity of fused quartz and Fujian standard sand equal 2.186 and 2.65, respectively; The refractive index of fused quartz equals 1.4584. The particle shape of graded fused quartz, uniform fused quartz and Fujian standard sand are angular, angular and round, respectively; The mineral composition of fused quartz and Fujian standard sand is SiO_2.

The matching liquid should be safe, colorless, stable and odorless. It also should have low volatility, low viscosity and preferably neutral pH values. In this study, the matching pore fluid is mixed by twelve alkane and white mineral oil, which the refractive index was 1.4585 at 24 °C.

Certain gradation fused quartz was selected, washed and dried. Then impurities and moisture were removed. The pore fluid was mixed based on twelve alkane: white mineral oil of 1:4 by mass at 24 °C. The mixture solution was stirred constantly by using a glass rod; and the solution was mixed homogeneously. Abbe refractometer

(a) Graded fused quartz (b) Uniform fused quartz (c) Fujian standard asnd

Fig. 3.1 Pictures of the particles of tested materials

was used to measure the refractive index of the mixture solution, and the proportion of the two liquids was fine-tuning until the value of refractive index was 1.4585, i.e. the refractive index matched glass sand particles. The fused quartz was slowly poured into the mixture solution with continuous stir. In the process, the liquids level was slightly higher than the surface of fused quartz particles. In order to remove the air in mixture, the mixture was placed in a vacuum chamber until the suspension liquid was transparent.

Consolidation undrained (CU) triaxial compression tests were performed on saturated transparent soil and Fujian standard sand in loose condition (the relative density equals 30%). A strain controlled triaxial apparatus was used in this study. All samples were 39 mm in diameter and 80 mm in height with saturation degree more than 95%. For each sample, the confining pressure varied from 50 to 400 kPa. All tests were performed at a shearing rate of 0.5 mm/min. Samples of Fujian standard sand with the same grain size distribution were also tested under the same conditions.

3.2.2 *Stress–Strain Curves of Transparent Soil of Fused Quartz*

(1) **Stress–strain curves at different relative density (D_r)**
For the fused quartz of 0.5 ~ 1.0 mm, stress–strain curves under 100, 200 kPa confining pressure with D_r of 30% and 70% were plotted in Fig. 3.2. From Fig. 3.2, it is observed that with the increase of relative density, the stress–strain relationship changes from strain hardening to strain softening. At loose state ($D_r = 30\%$), the soil basically displays a weak strain hardening type stress–strain relations, soil strength have the peak at axial strain about 20%. At dense state ($D_r = 70\%$), the soil displays a strain softening stress–strain relations, but the softening is weak, and the soil strength has the peak at axial strain about 15%. These are mainly because the fused quartz of 0.5 ~ 1.0 mm is has a uniform size and the pore size is large. At loose state, the compression caused the reduction of pore volume, the micro-structure of the transparent soil does not change, therefore, the shear strength is low. At dense state, the confining pressure not only caused the compression of the pore volume but also changed the micro-structure of the transparent soil, the soil particles have a more compact arrangement, the soil skeleton becomes stiff, which makes the soil resistance to deformation increase and the shear strength also increases, when the axial stress reaches its maximum bearing capacity, the stress decreases and the stress–strain curve displays a strain softening characteristics.

(2) **Stress–strain curves at different confining pressure**
Take fused quartz with particle size of 0.5 ~ 1.0 mm as an example, stress–strain curves for $D_r = 30\%$ and 70% are plotted in Fig. 3.3 with four confining pressures (50, 100, 200, 400 kPa). From Fig. 3.3, it can be seen that the initial elastic modulus has a significant relation with confining pressure, the larger the

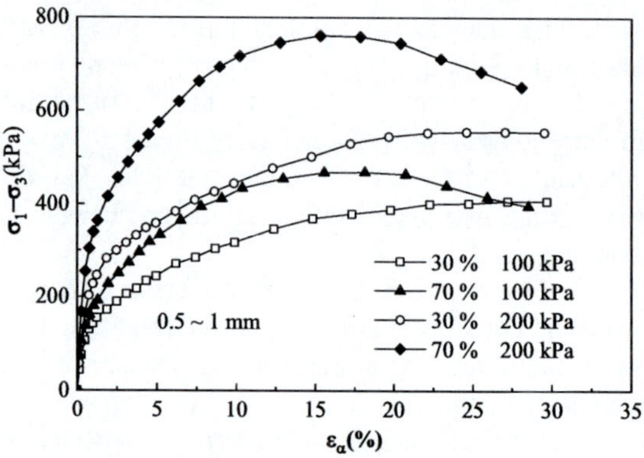

Fig. 3.2 Curves of stress–strain under different relative densities

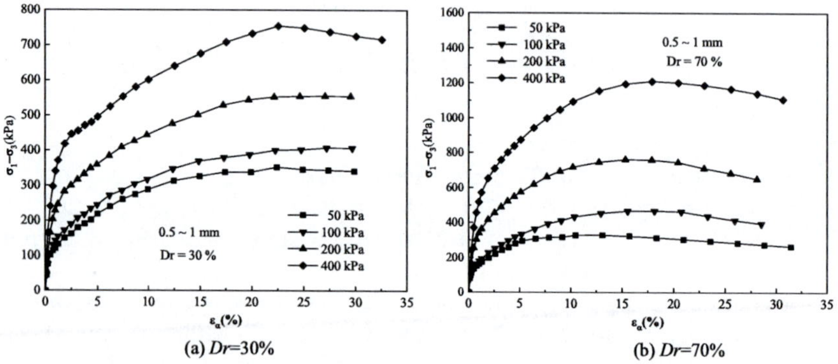

Fig. 3.3 Curves of stress–strain under different confining pressures

confining pressure, the larger the initial elastic modulus, the deviatoric stress increases with the confining pressure. These because that before shearing, the larger the confining pressure, the more compression; the higher the relative density, the stronger the bite ability and shear deformation resistance between particles, the greater the deformation modulus.

(3) **Stress–strain relations with different particle sizes**
Take the confining pressure = 100 kPa, 200 kPa as an example, the stress–strain curves for the samples of D_r = 30%, 70% with different particle sizes are plotted in Fig. 3.4. From Fig. 3.4, it is observed that transparent soil of standard gradation displays a strain hardening type and the peaks appears at axial strain about 12% at low relative density (D_r = 30%). When the relative density is large (D_r = 70%), the transparent soil displays a significant strain softening behavior, the soil strength has a peak at an axial strain about 7%. Compared with the transparent soil of 0.5 ~ 1.0 mm, the strength peak appears earlier, and the deviatoric stress is larger. These are because standard gradation

Fig. 3.4 Curves of stress–strain under different grain sizes

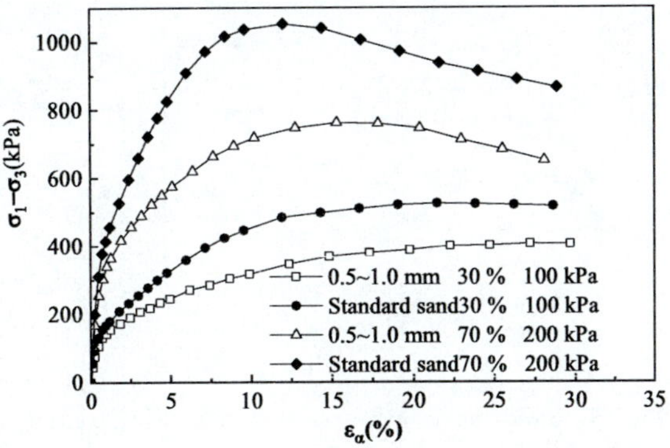

has more uniform particle size and the pore size is smaller which leads to less compression; therefore, the strength peak appears earlier; the arrangement of particles is more compact and denser, which makes the deformation resistance increases significantly and the shear strength also increases. The comparisons between loose ($D_r = 30\%$) and dense Fujian sand ($D_r = 70\%$) show that, due to different pressure development patterns, the dense sand fails at a much higher deviatoric stress in comparison to the loose sand. The failure lines for both dense and loose sands are the same.

(4) **Comparison with Fujian standard sand**
The stress–strain curves at the loose conditions ($D_r = 30\%$) under a constant confining pressure of 200 kPa are shown in Fig. 3.5a for transparent soil of standard gradation and 0.5 ~ 1.0 mm, as well as Fujian standard sand. It shows that for the Fujian standard sand the stress with strain develops more rapidly in the initial and peaks at smaller strain than those of transparent soil. The shape of fused quartz particles is angular, but the particles of Fujian standard sand are smoother than others. The shear failure pattern of Fujian standard sand and fused quartz are consistent, mostly as hardening type. In loose conditions, the strength of Fujian standard sand increases gradually with axial strain and the peak strength reaches the maximum value at 12% axial strain. The peak strength of standard gradation transparent soil, however, reaches earlier and at higher strains than the diameter of 0.5 ~ 1.0 mm. The higher strength of standard gradation transparent soil could be ascribed to the better particle size distribution. The effective stress path of transparent soil with a diameter of 0.5 ~ 1.0 mm under different relative density is shown in Fig. 3.5b.

3.2.3 *Shear Strength*

Shear strengths for triaxial tests carried out on transparent soil and standard sand are summarized. The effective internal friction angle of transparent soil of standard gradation, 0.1 ~ 0.5 mm, and Fujian standard sand equal to 35°, 31°, and 33°, respectively. For the same material, with increasing relative density, the shear strength

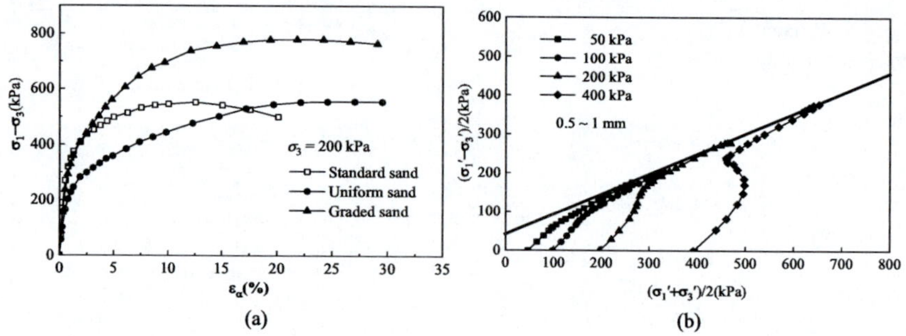

Fig. 3.5 Stress–strain curves: **a** comparative curves, **b** effective stress path

increases. Shear strength indexes ratio of transparent soil with standard gradation are larger than that of 0.5 ~ 1.0 mm size, especially in loose state ($D_r = 30\%$). Relative density is one of the major factors influencing shear strength. By comparing the test results from transparent soil of standard gradation to Fujian standard sand, it is found that shear strength of transparent soil is similar to that of Fujian standard sand at the same grading. It suggests that transparent soil can be used to simulate natural sand.

3.2.4 *Pore Pressure*

Typical pore pressure curves of transparent soil from consolidation undrained triaxial tests are shown in Fig. 3.6a. In the initial period the pore water pressure is risen at a fast rate until reaching the peak, and then slowly decreases until the end (the case of other conditions have similar rule). The above phenomenon indicates that the transparent soil of fused quartz is more dilative at higher confining pressure and more contractive at lower confining pressure.

In consolidated undrained triaxial test, the pore pressure increment changes with the deviatoric stress increment, so a pore pressure coefficient A value is not a constant, this study uses the shear broken of pore pressure coefficient A_f value to study the pore pressure characteristic of transparent soil. The change of pore pressure coefficient A_f value with confining pressure variation on different material and size distribution are shown in Fig. 3.6b. With the increase in the relative density, the skeletal structure of the soil is strengthened, the soil skeleton can sustain higher stress, which reduces the pore pressure coefficient A_f value. At loose state, transparent soil with single particle size has the highest A_f and Fujian standard sand has the smallest A_f Compared with Fujian standard sand, fused quartz is irregular subangular, particle porosity is large, the A_f value is increased; and the uniform sized fused quartz has higher porosity, so the A_f value is maximum. But under dense conditions, the A_f value of three samples has small difference.

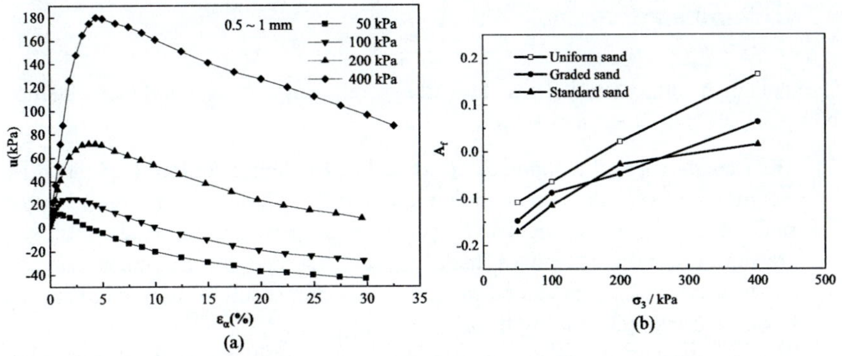

Fig. 3.6 Curves on pore water pressure and pore pressure coefficient versus axial strain

3.2.5 Deviatoric Stress and Stress Ratio

With lower density, the increasing trend in the deviatoric stress with confining pressure is relatively moderate as illustrated in Fig. 3.7a. For different particle size with the same relative density, transparent soil with standard gradation has the highest deviatoric stress compared with the others. This is due to the well gradation and angular shape of the fused quartz particles.

The change of the effective stress ratio with strain variation with 0.5 ~ 1.0 mm diameter transparent soil is shown in Fig. 3.7b. Effective stress ratio decreases with the increase in confining pressure. In loose state ($D_r = 30\%$), as the axial strain develops, the trend of effective stress ratio is rapidly growing until it is reached a stable value. While in dense state ($D_r = 70\%$), the effective stress rapidly increases to a stable value, and then slowly declines.

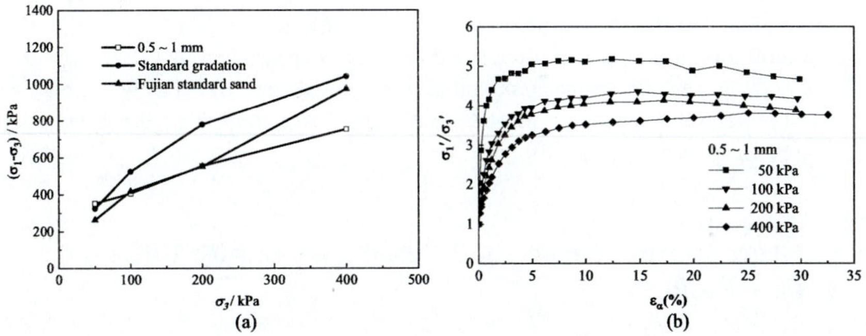

Fig. 3.7 Curves on stress: **a** the maximum deviatoric stress versus confining pressure, **b** effective stress ratio

3.2.6 Summary

The static geotechnical properties of the transparent soil of fused quartz are obtained through a series of laboratory tests.

(1) The transparent soil composed with fused quartz and mixed oil (1:4 blend by weight of twelve alkane and white mineral oil) has good transparency. At the same gradation, the physical and mechanical properties of this transparent soil are similar to Fujian standard sand. It is demonstrated that the transparent soil can be used to simulate natural sand. The transparent soil of fused quartz tested has a friction angle ranging from 31° to 35°.
(2) The relative density is one of the main factors influencing the deformation characteristics of transparent soil. The stress–strain relationship of transparent soil of fused quartz is a transition from the strain hardening to the strain softening with the increase of relative density. The pore pressure coefficient A_f value is reduced, the deformation modulus is increased, and the increasing trend in the deviatoric stress with confining pressure is faster with the increase of relative density.
(3) Relative density is also one of the major factors influencing the shear strength of transparent soil of fused quartz. The shear strength is increased with the increasing of relative density. The friction angle of transparent soil with dense state is about 2° larger than that of loose state. With the same relative density, the internal friction angle of transparent soil with water fluid specimens is about 2° smaller than that of dry specimens; the internal friction angle with oil specimens is about 1° larger than that of dry specimens.
(4) Peak strength of transparent soil with standard gradation reaches earlier and at higher strains than the diameter of 0.5 ~ 1.0 mm. The stress of standard sand with strain develops more rapidly in the initial and peaks at smaller strain than those of transparent glass soil. For different sands with the same relative density, standard gradation has the highest deviatoric stress compared with the other two sands. At the same relative density and confining pressure conditions, the deformation modulus of Fujian standard sand is larger than that of transparent soil. At loose state, transparent soil with single particle size has the highest A_{f} and Fujian sand has the smallest A_{f}. But under dense conditions, the A_{f} value of three samples has small difference.

3.3 Geotechnical Properties of Fused Quartz with Different Pore Fluid

Several pore fluids can be used to make transparent soil with fused quartz. What effect these pore fluids have on the geotechnical properties of transparent soil is a major concern for researchers. This section presents studies of the effects of three different pore fluids (mixed oil, $CaBr_2$ solutions, sucrose solution) on the static properties of transparent soil of fused quartz. These results are going to provide the support information for the transparent soil modelling technique.

3.3.1 *Fused Quartz and Pore Fluid*

The fused quartz tested has a grain size of 0.5 ~ 1.0 mm (the commonly used grain size). Its specific gravity is 2.186. The minimum dry density is 0.970 g/cm^3 tested by the funnel method following the Specification of Soil Test (1999), the maximum dry density is 1.274 g/cm^3 with the vibration method. The tested sample has a D_r = 60% (Studies show that at D_r = 60%, the transparent soil has the best transparency).

The selected pore fluid should be colorless, transparent, safe and stable, no chemical interaction with the solid particles, and it has the same refractive index with the solid particles. The mixture of 15 white oil and n-12 alkane, $CaBr_2$ solution, and sucrose solution are selected as three typical pore fluids. Their refractive index was measured by Abbe refractometer, making it to be the same with the fused quartz as 1.4585. At 25 °C, the relationships between the refractive indexes of three typical pore fluids and the mixing ratios are shown in Fig. 3.8.

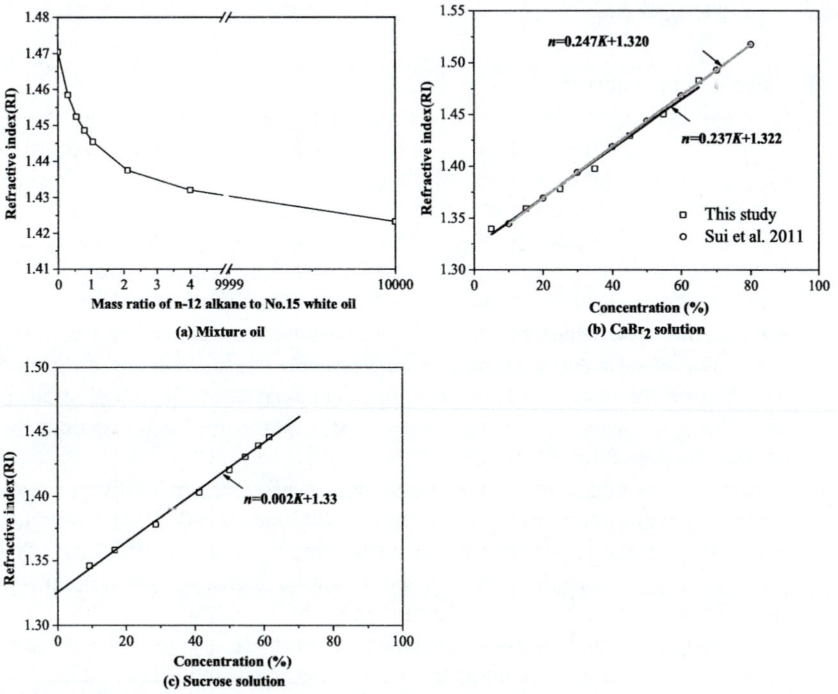

Fig. 3.8 Curves of refractive index versus concentration

Table 3.1 Conditions of laboratory tests

Test No.	Pore fluid	Confining pressure σ_3/kPa	Relative density D_r/%	Test type
I	Mixture oil	50 ~ 400	60	CD
II	$CaBr_2$ solution	50 ~ 400	60	CD
III	Sucrose solution	50 ~ 400	60	CD
IV	Pure water	50 ~ 400	60	CD
V	Dry	100 ~ 200	60	UU

3.3.2 Experimental Program

The testing apparatus is TSY10 type strain controlled triaxial testing machine. For three typical pore fluids, transparent soil of $D_r = 60\%$ with particle diameter of 0.5 ~ 1.0 mm was conducted consolidation drained triaxial test with different confining pressures (50, 100, 200, 400 kPa). For comparison, the dry fused quartz and water saturated fused quartz were also tested. The testing program is listed in Table 3.1.

3.3.3 Testing Results

(1) Stress–strain relationships

Under confining pressure of 100 kPa, the stress–strain curves for transparent soil made of fused quartz and different pore fluids are shown in Fig. 3.9. From Fig. 3.9, at the same axial strain, the deviatoric stress is maximum for the transparent soil of sucrose pore fluid, the least for the transparent soil of mixture oil, the deviatoric stresses for the transparent soil of $CaBr_2$ solution and for the transparent soil of water are close. The increase of deviatoric stress for the transparent soil of sucrose solution is possibly because the viscosity of sucrose solution is larger, which constrained the rearrangement of the solid particles and increases the shearing resistance. However, the mixture soil has lubrication on the solid particles which reduces the shear resistance. The axial strain is about the same where the deviatoric stress peak occurs for the transparent soil of different pore fluids.

Stress–strain curves for the transparent soil of different pore fluids are shown in Fig. 3.10 with confining pressure of 50, 100, 200 and 400 kPa. It is known from Fig. 3.10 that the stress–strain curves have the similar variation trend. Increasing the confining pressure, the peak deviatoric stress increases. In addition, the stress-strain curve displays a hardening type at high confining pressure, which is mainly because that the number of fine particles increases due to breakage of solid particles. Under low confining pressure, the stress-strain curves all display a strain softening behavior after the peak stress, but the softening is weak. This is because when the dense fused quartz is subjected to shearing, some particles

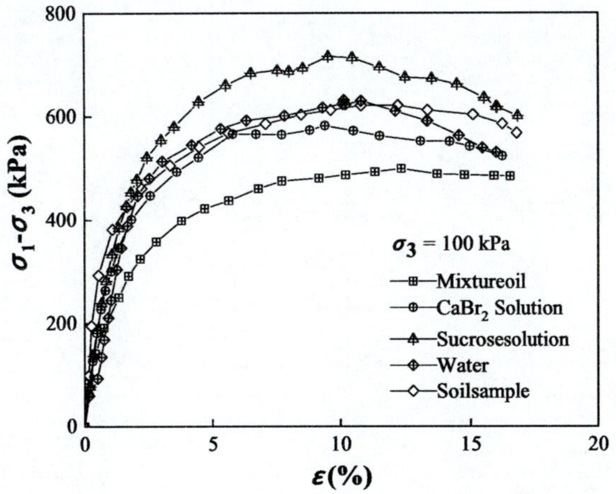

Fig. 3.9 Curves of stress–strain with different pore fluids

are rotating over the other, which need to overcome the biting force between particles, therefore it has a higher shear strength. Once the particles passed, the structure becomes loose, the shear strength decreased, hence it shows a softening behavior. This phenomenon is similar to the nature of natural sand.

(2) Volume change behavior

The volume variation of the transparent soil made of different pore fluids and fused quartz are plotted in Fig. 3.11 under different confining pressures. The volume variation is similar generally for all the tested specimens. The testing specimen has a dilation in volume when shearing under low confining pressure, with the increase of the confining pressure and the breakage of solid particles, the shear dilation is restricted and it displays shear shrinkage. The shear band in the specimen is not obvious during the triaxial test. From Fig. 3.11, the dilation during the test continues to a large strain value and the dilation rate does not decrease.

At confining pressure of 100 kPa, the volume change curve was plotted in Fig. 3.12 for transparent soils of different pore fluids. From Fig. 3.12, the transparent soil prepared with mixed oil and sucrose solution shows a greater shear dilation trend, in contrast, the dilation trend of transparent soil prepared by calcium bromide solution is not very obvious. This is largely related to the properties of pore fluid and the effect of rubber membrane embedding on volume deformation. During the testing, it was found that mixture soil and sucrose solution has obvious reaction with the rubber membrane and the rubber membrane has a large deformation, but the $CaBr_2$ solution almost has no reaction with the rubber membrane. Because the rubber material can be easily eroded by the organic solvent, mixture oil and sucrose are organic solvent, therefore, the rubber membrane is swollen and deforms greatly. While $CaBr_2$ solution is inorganic solvent, it cannot make the rubber membrane eroded. In addition, the rubber membrane is embedded into the side surface of the testing specimen and expelled the pore water out, so as to produce additional drainage of the sample.

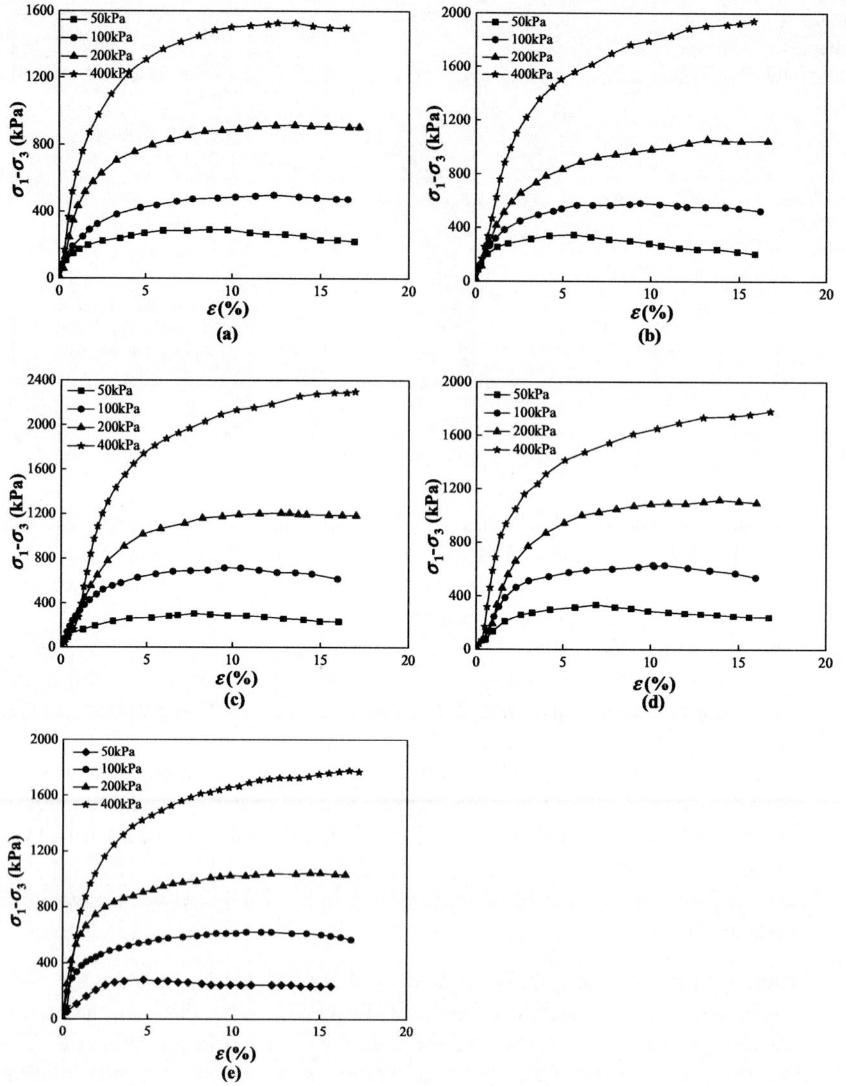

Fig. 3.10 Curves of stress–strain under different confining pressures: **a** mixture oil, **b** $CaBr_2$ solution, **c** sucrose solution, **d** water, and **e** dry sample

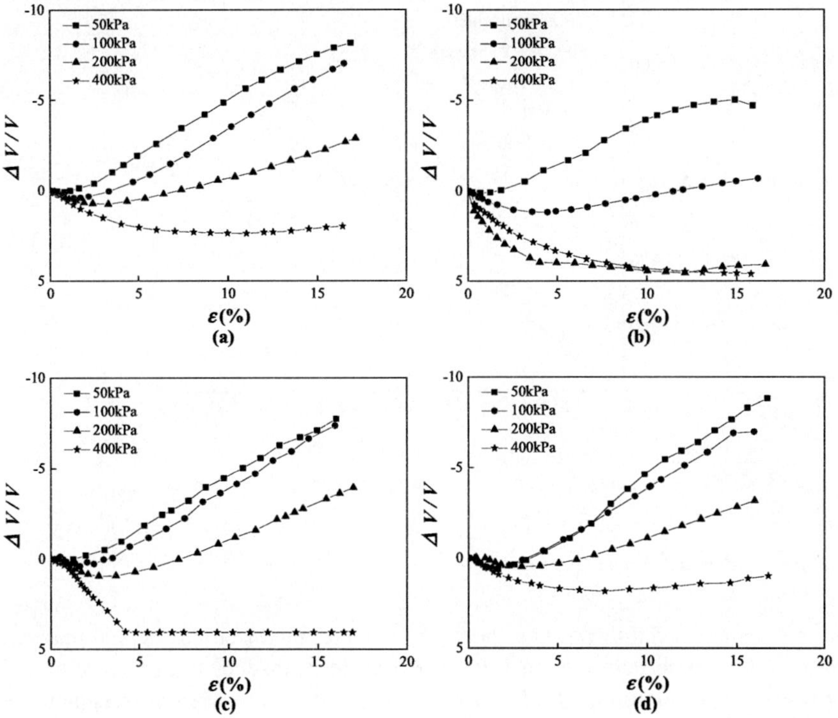

Fig. 3.11 Volume change under different confining pressures: **a** mixture oil, **b** $CaBr_2$ solution, **c** sucrose solution, and **d** water

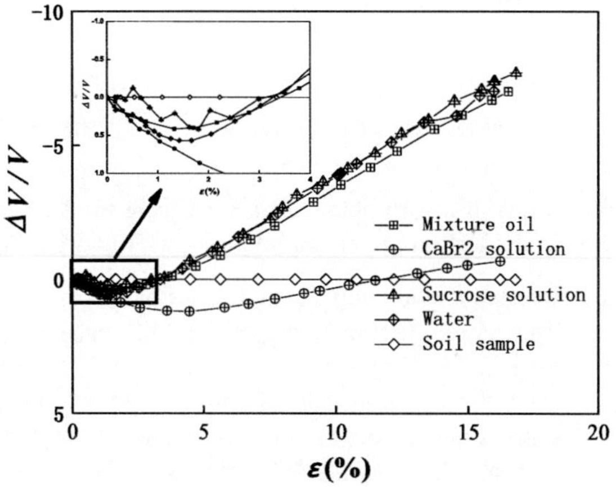

Fig. 3.12 Comparative curves of volume change influenced by pore fluid

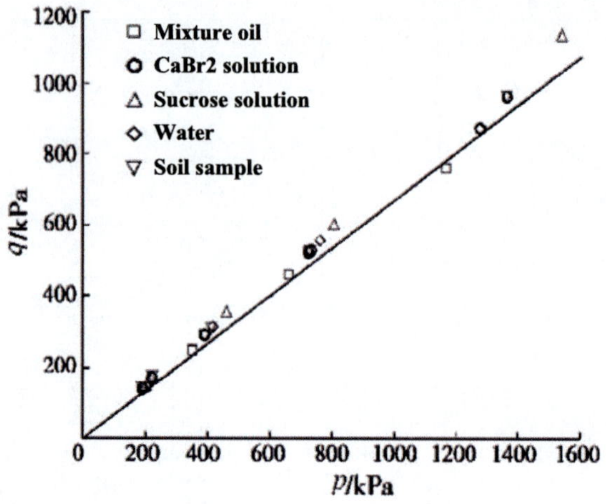

Fig. 3.13 Critical state strength line on p versus q space from end points of tests

3.3.4 Critical State Line

In traditional triaxial test, it is difficult to get the critical state of dilation material because too much deformation leads to nonuniform loading before reaching the critical state. In addition, dilation also has a significant influence on stress path. As shown in Fig. 3.13. From Fig. 3.13, the critical state line is simulated by samples made of different pore liquids under different confining pressure. From Fig. 3.13, the failure lines for transparent soils made with fused quartz and different pore fluids are basically coincident.

3.3.5 Duncan-Chang Model for Transparent Soils

From triaxial test to obtain Duncan-Chang model parameters, they can be used in the triaxial test numerical simulations and other applications.

(1) Internal friction angle φ and cohesion c
The internal friction angle values φ of transparent soil with different pore fluids are listed in Table 3.2, the cohesion $c = 0$. The obtained internal friction angle for the transparent soil of different pore fluids varies from 48° to 55°. The transparent soil of mixture oil is the smallest. The fused quartz particles are very angular. The internal friction angle obtained falling within the range of 46° ~ 57° for the angular sand.

Table 3.2 Internal friction angles of transparent soil samples obtained by CD tests

Test No.	Pore fluid	φ/(°)
I	Mixture oil	48
II	$CaBr_2$ solution	52
III	Sucrose solution	50
IV	Pure water	55

(2) K, n, R_f

In Duncan-Chang model, the deviatoric stress and the axial strain $(\sigma_1 - \sigma_3)$–ε_a from triaxial testing is fitted with the hyperbola function as the following:

$$\sigma_1 - \sigma_3 = \frac{\varepsilon_a}{a + b\varepsilon_a} \tag{3.1}$$

Take the transparent soil made of mixture oil as an example, as shown in Fig. 3.10, the hyperbola curve can be transformed into a straight line plotted in Fig. 3.14, where the slope b and intercept are obtained through $a = \frac{1}{E_i}$, $b = \frac{1}{(\sigma_1-\sigma_3)_u}$, E_i is the initial tangent modulus, $(\sigma_1 - \sigma_3)_u$ is the asymptotic value of the principal stress difference. Testing results indicate that E_i is changing with σ_3. $\lg(E_i/P_a)$ has approximately a linear relationship with $\lg(\sigma_3/P_a)$ as shown in Fig. 3.15 with intercept lg*K and slope n*, therefore $\lg(E_i/P_a) = \lg K + n \lg(\sigma_3/P_a)$.

$(\sigma_1 - \sigma_3)_u$ is the value at $\varepsilon_a \to \infty$, in fact, ε_a cannot be infinite, the specimen would fail at a certain value of ε_a, the failure stress difference is denoted as $(\sigma_1 - \sigma_3)_f$, let R_f called failure ratio, which takes different value for different σ_3, using the average value. K, n, and R_f for the transparent soil of different pore fluids are shown in Table 3.3.

$$R_f = \frac{(\sigma_1 - \sigma_3)_f}{(\sigma_1 - \sigma_3)_u} \tag{3.2}$$

(3) G, F, D

The load is applied vertically in triaxial test, so the lateral strain ε_r is negative. The ε_a–$(-\varepsilon_r)$ curve is fitted with the hyperbola function as the following:

$$\varepsilon_a = \frac{-\varepsilon_r}{f + D(-\varepsilon_r)} \tag{3.3}$$

Take the example of the transparent soil made of mixture oil, as shown in Fig. 3.16, ε_a—$-\varepsilon_r$ is plotted in Fig. 16a, while $\frac{-\varepsilon_r}{\varepsilon_a}$–$(-\varepsilon_r)$ is plotted in Fig. 16b which is a straight line, its slope is D and its interception is f.

From Eq. (3.3), when $-\varepsilon_r \to 0$, the following expression is obtained,

$$f = \left(\frac{-\varepsilon_r}{\varepsilon_a}\right)_{\varepsilon_a \to 0} = \nu_i \tag{3.4}$$

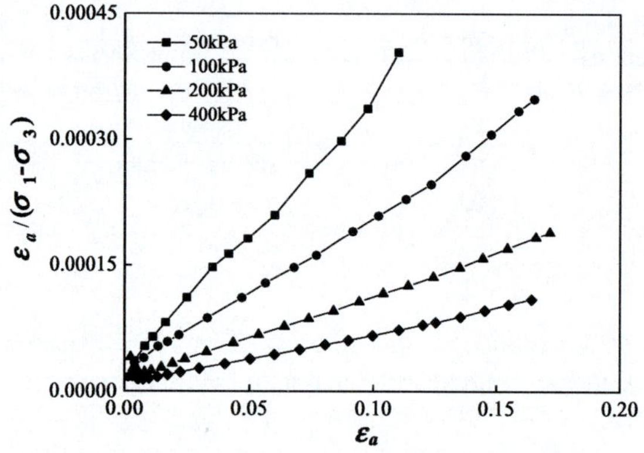

Fig. 3.14 Curves of stress versus strain of transparent soil samples (mixed oil)

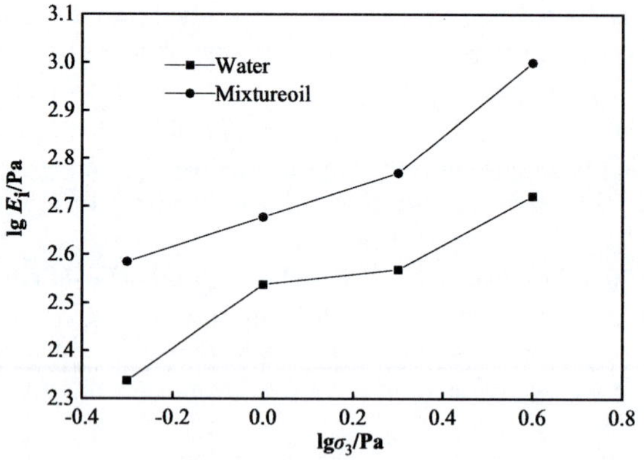

Fig. 3.15 Curves of stress versus initial modulus

Table 3.3 Values of K, n, R_f of transparent soil

Testing No.	Pore fluid	K	n	R_f
I	Mixture oil	491.25	0.45	0.92
II	$CaBr_2$ solution	374.97	0.42	0.89
III	Sucrose solution	410.96	0.52	0.88
IV	Pure water	303.39	0.39	0.78

where ν_i is the initial Poisson' ratio, for different σ_3, there are different ν_i. ν_i has an approximately linear relation with $\lg(\sigma_3/P_a)$, as shown in Fig. 3.17. Its intercept is G and its slope is F, hence $\nu_i = G\text{-}F\ \lg(\sigma_3/P_a)$ F, G, D values for transparent soil of different pore fluids are summarized in Table 3.4.

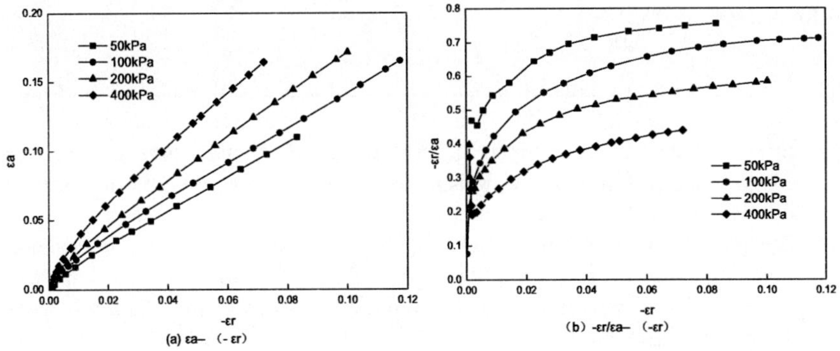

Fig. 3.16 Lateral strain versus axial strain of transparent soil made of mixed oil

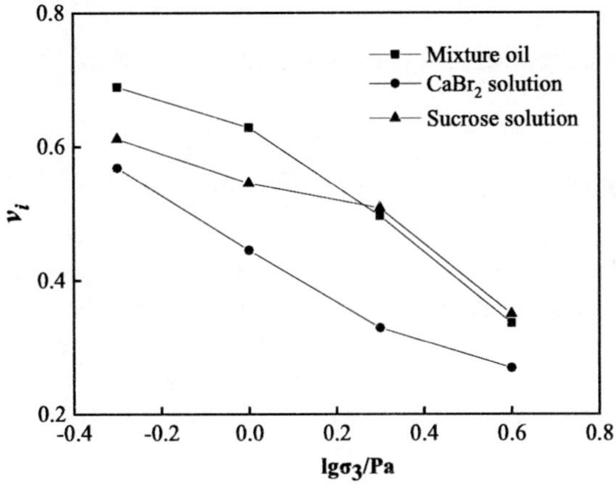

Fig. 3.17 Curves of Poisson's ratio versus stress

Table 3.4 Values of *F*, *G*, *D* of transparent soil

Testing No.	Pore fluid	*F*	*G*	*D*
I	Mixture oil	0.39	0.59	3.48
II	CaBr2 solution	0.34	0.45	8.42
III	Sucrose solution	0.27	0.54	4.33
IV	Pure water	0.37	0.59	4.67

3.3.6 *Summary*

(1) The stress–strain relationships for transparent soil of different pore fluids are similar, but there are certain differences. At the same strain, the stress of transparent soil of sucrose solution is the largest, while the transparent soil of mixture oil has the smallest, and the transparent soil made of $CaBr_2$ solution has the similar value as that of water saturated.
(2) The obtained internal friction angle in this study is from 48° ~ 52° for the dense angular transparent soil with different pore fluids. The transparent soil made of mixture oil and sucrose solution has more dilation, while the dilation is weak for that of $CaBr_2$ solution.
(3) From the triaxial testing results, Duncan-Chang model parameters were determined for the transparent soil made of different pore fluids and fused quartz.

3.4 Dynamic Properties for Transparent Soil of Fused Quartz

3.4.1 *Experiment*

The dynamic properties of transparent soil made of fused quartz are important for the application in earthquake related design and research. A comprehensive experimental program was conducted to investigate the dynamic behavior of fused quartz. A torsional resonant column test apparatus was used for the tests according to ASTM D 4015. The working principle for this device has been described in Chapter 2. The soil samples were 30 mm at the inner diameter, 70 mm at the outside diameter and 100 mm in height.

The transparent soil was manufactured by fused quartz and pore fluid with the same refractive index (RI = 1.4585). The fused quartz used in this study is angular in shape. The particle size used ranged from 0.5 mm and 1.0 mm, and the relative density is about 60%. Pore fluid was blended (75:25 by weight) with Norpar® 12 (RI = 1.4235) and Drakeol® 15 white mineral oil (RI = 1.4688), the most widely used in model test application till now (Iskander et al. 2015). The transparent soil samples were manufactured as follows: Firstly, the latex membrane was tied on one end of the platform through the soil sample preparation mold, the bottom of latex membrane was kept sealed. Secondly, pore fluid was poured into the mold until reached 1/3 of the maximum height. Thirdly, measured the weight of fused quartz material, then rained them into the pore fluid slowly. Finally, the top latex membrane was tied on the other end of the platform, the mold was removed and the transparent soil sample was manufactured. An Abbe refractometer was used to measure the refractive index of the mixture's solution.

3.4.2 Shear Modulus and Damping Ratio of Fused Quartz

Figure 3.18a shows the shear modulus, G, as a function of shear strain amplitude determined from resonant column tests and dynamic torsional shear tests with a confining pressure of 100 kPa for dry and oil saturation samples. The graphs indicate that the dry specimen has a higher shear modulus than that of the oil saturation specimen. The maximum shear modulus, G_{max}, at a strain level of 10^{-6} under 100 kPa confining pressure is about 60 MPa for both the dry and oil saturation specimen.

The normalized shear modulus versus shear strain of tested fused quartz and transparent soil is shown in Fig. 3.18b. In order to produce a comparative analysis, the normalized shear modulus versus shear strain of silica gels, 0.5–1.0 mm in particle size and with a 100 kPa confining pressure [1], and gravels under undrained and drained conditions [3] are also shown in Fig. 3.18b. The results indicate that the normalized shear modulus for fused quartz has a similar distribution with silica gel materials and gravels.

The damping ratios, ζ, were determined by the amplitude decay method according to ASTM D 4015. The damping ratio as a function of shear strain amplitude determined from resonant column tests with a confining pressure of 100 kPa is given in Fig. 3.19 for dry and oil saturation samples. The damping ratio versus shear strain for silica gel [1] and gravels [3] is also shown in Fig. 3.19 for comparative analysis. The damping ratio firstly displayed a slow linear increase, and then showed a rapid nonlinear increasing behavior with the increasing of the shear strain; the dry specimen of fused quartz has a lower damping ratio than that of the oil saturation specimen. The results from Fig. 3.19 also showed that the damping ratios of both fused quartz and silica gel were similar to those of the natural soil reported by Rollins et al. [3].

Unified formulas for the damping ratio of sand after summarized the previous researcher's data were proposed by Ishibashi & Zhang [18], and Zhang et al. [16]. The formulas expressed the damping ratio for sand or saturated soils (from South Carolina, North Carolina, and Alabama) as a function of G/G_{max} as follows:

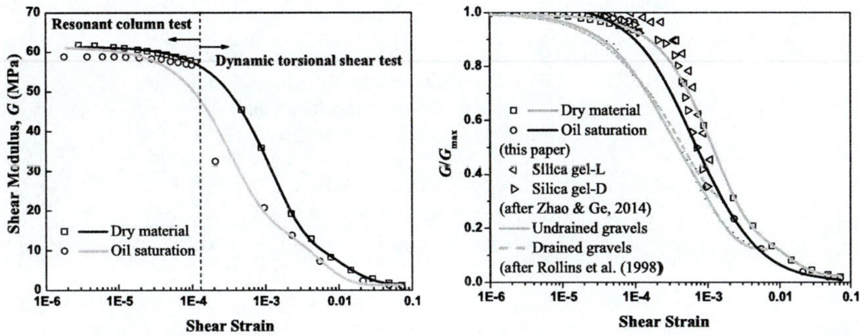

Fig. 3.18 Shear modulus versus shear strain: **a** dry and oil saturation condition; and **b** normalized comparison

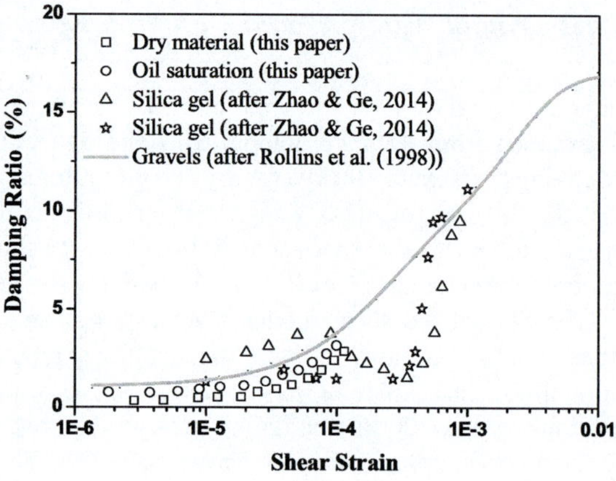

Fig. 3.19 Comparison of damping ratios

$$\zeta = 0.333\left[0.586\left(\frac{G}{G_{\max}}\right)^2 - 1.547\left(\frac{G}{G_{\max}}\right) + 1\right] \quad (3.5)$$

$$\zeta - \zeta_{\min} = 10.6(G/G_{\max})^2 - 31.6(G/G_{\max}) + 21.0 \quad (3.6)$$

The damping ratios (both fused quartz of dry and oil saturation specimens, and silica gel) versus G/G_{max} are shown in Fig. 3.20, on which the fitting curve of Eqs. (3.5) and (3.6) are plotted. Although a short of range of $G/G_{\max}$ data has been obtained, existing laboratory test data for transparent soil are distributed closer to the fitting curve (based on the data of sand) by Ishibashi & Zhang [18].

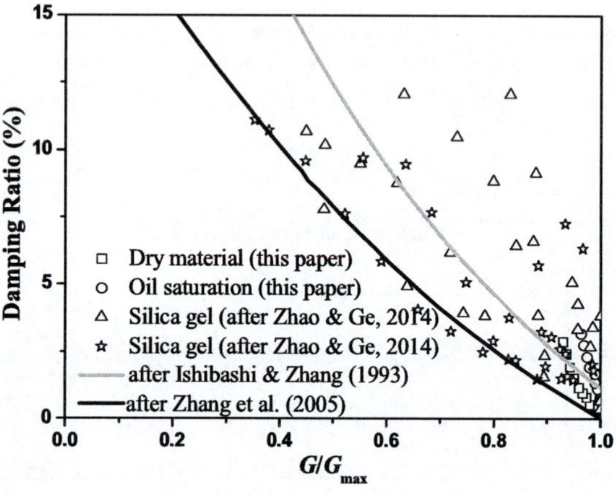

Fig. 3.20 Damping ratios versus normalized shear modulus

3.5 Shear Modulus and Damping Ratio of Transparent Soils with Different Pore Fluids

3.5.1 Pore Fluids

Previous application researches show that amorphous silica, silica gel, hydrogel (aquabeads), fused quartz, and laponite [6] are five types of transparent soil families; and in which silica gel and fused quartz are the two main innovative transparent granular particles materials; mixed oil, calcium bromide ($CaBr_2$) solution and sucrose solution are the three main pore fluids which are appropriate for manufacturing transparent soils [2].

For further application of fused quartz to substitute sand and to track particle movement inside soil mass during dynamic events (such as, liquefaction resulting from earthquake), it is necessary to know the dynamic shear modulus and damping ratios of transparent soils manufactured by fused quartz and three different pore fluids (mixed oil, CaBr2 solution, and sucrose solution). The results of the dynamic shear modulus and damping ratios of transparent soils conducted through a series of RCT and dynamic torsional shear tests (TST) are presented to examine their dynamic properties.

Fused quartz used in this study is angular in shape. The particle sizes distribution of fused quartz used in this study and previous researches are compared and shown in Fig. 3.21 according to ASTM D2488.

In general laboratory conditions, the RI of fused quartz was constant at 1.4585. The chosen pore fluids should not only have the fitting RI with fused quartz, but also have similar viscosity with water. Through screening, three different matching pore fluids were chosen in this study: (1) mixed oil (which is mixed by alkane 12 (RI = 1.4235) and white mineral oil $15^{\#}$ (RI = 1.4688)); (2) $CaBr_2$ solution; and (3)

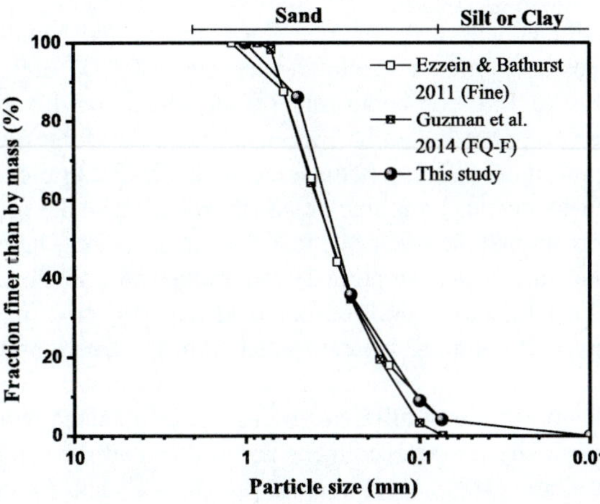

Fig. 3.21 Particle size distributions of transparent granular materials

sucrose solution. Abbe refractometer (BM-2WAJ Digital Handheld Refractometer), Cannon–Fenske viscometer tubes were used to measure the RI and viscosity of the three pore fluids at atmosphere temperature (approximately 20 °C), respectively. The RI versus solution concentration (or mass ratio for mixed oil) for the three pore fluids, the viscosity versus solution concentration for $CaBr_2$ solution and sucrose solution at 20 °C temperature are shown in Fig. 3.22a, b and c, respectively. At 20 °C temperature, the RI of $CaBr_2$ solution and sucrose solution show nearly a linear increase with the increasing of the solution concentration; the viscosity of $CaBr_2$ solution and sucrose solution show a non-linear increase with the increasing solution concentration.

3.5.2 Testing Methods

Medium dense transparent soil specimens (with 60% relative density, D_r) were prepared with three pore fluids and one dry condition, and the summaries of the tests conducted are provided in Table 3.5.

The RCT could amplify the rotational deformation with a shear strain resolution ranging from 10^{-6} to 10^{-4} as suggested in the tests following ASTM D 4015. The schematic illustration is shown in Fig. 3.23a. The apparatus of RCT used in this study is similar as that used in Chap. 2 for testing the dynamic properties of silica gel. To minimize the degree of inherent anisotropy of the specimens, the sample preparation was carried out carefully by pouring the weighed mass of pore fluids and air-dried fused quartz particles into a mold successively. The falling height was kept constant throughout the pluviation process in order to obtain highly uniform density specimens. Back pressure method was used to saturate specimens, and Skempton's *B*-values of more than 0.93 were observed in the specimens used in the tests. The torsional force could be applied by a torsional drive at first-mode resonance with frequencies ranging from 20.0 to 150.0 Hz. The prepared soil specimen is a solid cylinder with a diameter of 39.1 mm in and height of 80.0 mm. The confining pressures for each specimen vary with 100, 200, 300, 400, and 500 kPa.

The TST non-contacting sensor which could provide a shear strain resolution ranging from 10^{-4} to 10^{-1} was used in the tests following ASTM STP 1213. A general view of the TST apparatus is shown in Fig. 3.23 (after Georgiannou et al. [8]). These devices create harmonic torsional excitation on the specimen (solid or hollow) at the top through the electronic load system or motor. The frequency curve was measured once a constant amplitude load applied to a torsional harmonic frequency range. Shear modulus could be obtained from the first mode resonance frequency. The material damping is determined from the free vibration attenuation curve, which can be achieved by removing the forced vibration. The range of shear modulus and damping ratio of different strains can be obtained with different harmonic torsional amplitude. These specimens are hollow and the symbols of axial load W_n, torque M_T, and external and internal pressures P_0 and P_i are adopted. The same internal and external pressure of the torsional shear test is applied in this study. During

Fig. 3.22 Curves on refractive index and viscosity versus concentration of pore fluids used for transparent soils

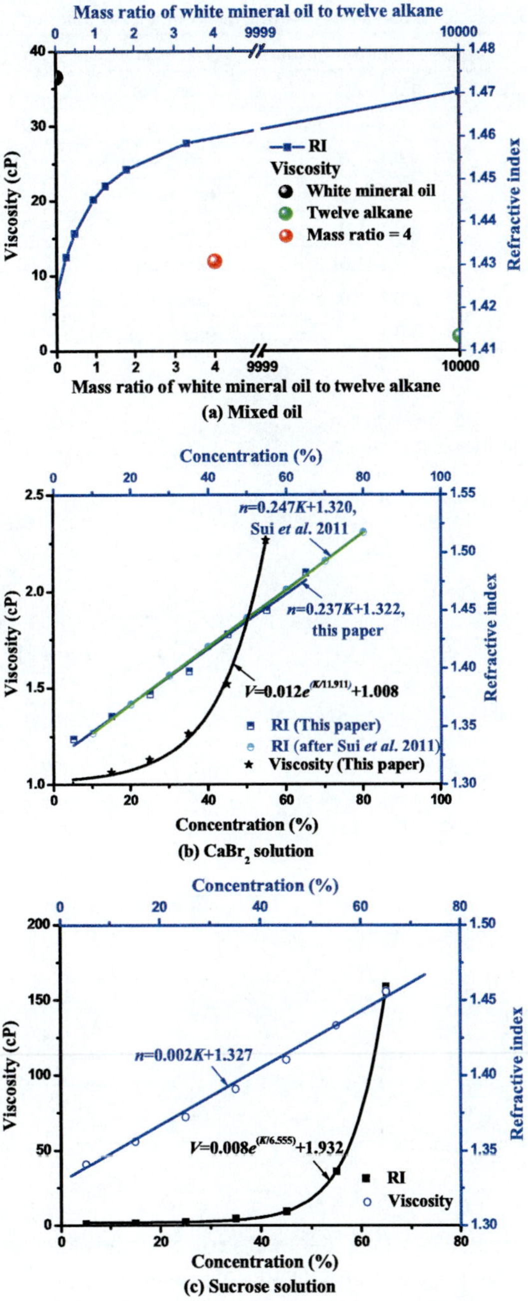

Table 3.5 Summary of experimental testing program

Test type	Grain size (mm)	Pore fluid	D_r (%)	Confining pressure (kPa)
RCT	0.1–1.0	None	60	100, 200, 300, 400, 500
	0.1–1.0	Mixed oil	60	100, 200, 300, 400, 500
	0.1–1.0	$CaBr_2$ solution	60	100, 200, 300, 400, 500
	0.1–1.0	Sucrose solution	60	100, 200, 300, 400, 500
TST	0.1–1.0	None	60	100, 200, 300
	0.1–1.0	Mixed oil	60	100
	0.1–1.0	$CaBr_2$ solution	60	100, 200, 300
	0.1–1.0	Sucrose solution	60	100

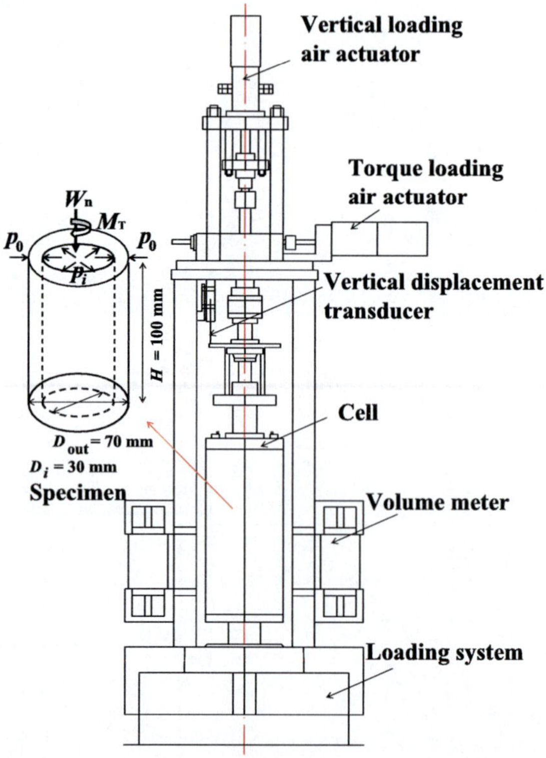

Fig. 3.23 The schematic illustration of Torsional shear test (TST, after Georgiannou et al. [8])

the shearing stage, the total vertical stress increments were kept close to zero. To minimize the degree of inherent anisotropy in the radial direction of the hollow cylindrical specimens, the sample preparation was carried out carefully by pouring the air-dried fused quartz particles into a mold, while moving the nozzle of the pluviator radially and circumferentially at the same time in alternative directions.

The falling height was kept constant throughout the pluviation process in order to obtain highly uniform density specimens. The sand specimens having the dimensions of 30 mm inner diameter, 70 mm outer diameter, and 100 mm height, were prepared and then saturated through combined methods (CO_2 pluviation, de-aired water and back pressure methods). For water-soluble pore fluids (such as, $CaBr_2$ solution, and sucrose solution), the weighed mass solutes were uniformly mixed in air-dried fused quartz particles, and pouring into mold together; CO_2 pluviation, de-aired water and back pressure methods were used to saturating the specimens. Skempton's *B*-values of more than 0.95 were observed in the specimens used in the tests. For non-water-soluble pore fluid (such as, mixed oil), the weighed mass mixed oil was pouring into the mold first, then pouring into air-dried fused quartz particles. The saturation methods were the same with normal methods. Unfortunately, a little of water will seepage into specimens during back pressure method. Skempton's *B*-values of about 0.91 were observed in mixed oil saturated specimens used in the tests. In order to study the behavior of the "soil" specimens under seismic conditions, cyclic torsional loading with a constant amplitude of shear stress was applied at a constant shear strain rate of about 2.5%/min. The torsional force at frequencies is ranging from 0.01 to 10.0 Hz, and 0.1 Hz was used here. The confining pressure for each specimen varies with 100, 200, and 300 kPa.

3.5.3 Shear Modulus Influenced by Pore Fluids

The maximum shear modulus (G_{max}) is the shear modulus value at a strain level of 10^{-6}. If the lowest strain level of the tests was larger than 10^{-6}, the values could be determined through regression analysis. The functions of the normalized shear modulus value versus shear strain had been described by Stokoe et al. [9] and Borden et al. [10], respectively. In this study it is found that both approaches lead to the similar values of G_{max} for fused quartz, which are almost identical to the shear modulus at a shear strain level of 10^{-5}.

The G_{max} of fused quartz with dry specimens and three different pore fluids saturated specimens under five different confining pressures (100, 200, 300, 400, and 500 kPa) are shown in Fig. 3.24. It shows that G_{max} tends to increase with the increasing of confining pressures for each specimen. The dry specimens have a little of higher G_{max} values than those of the pore fluids saturated specimens under the same confining pressures; the $CaBr_2$ solution saturated specimens have a little of higher G_{max} values than those of the mixed oil saturated specimens. The trend shows that the G_{max} values of the sucrose solution saturated specimens increase with the increasing of confining pressures which is a little faster than those of the other specimens.

One empirical equation to fit the test data between G_{max} and confining pressure (σ_0) had been described by Rowe [11], and the equation is shown as follow.

$$G_{max} = \frac{S \cdot \mathrm{OCR}^k}{(0.3 + 0.7e^2)} \sigma_0^n P_a^{(1-n)} \tag{3.7}$$

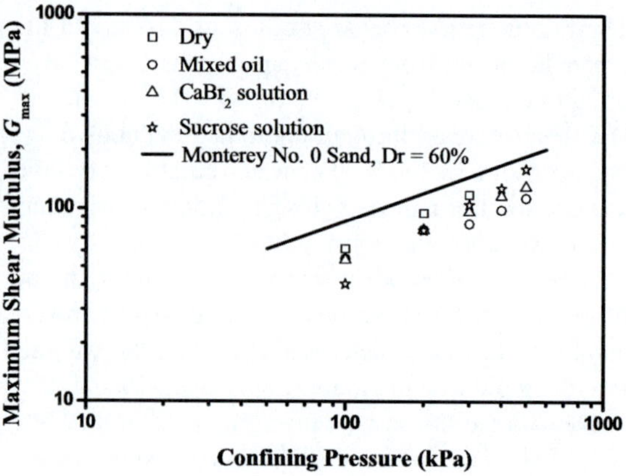

Fig. 3.24 The maximum shear modulus of fused quartz versus confining pressure with different pore fluids

where S is a non-dimensional stiffness coefficient, n is the power of the stress, OCR is the over consolidation ratio (taken as 1 for cohesionless soils here), k is a parameter depending on the plasticity of the soil (taken as 0 for cohesionless soils here y), P_a is the atmosphere pressure, e is the void ratio (taken as 0.8 here).

Table 3.6 summarized parameters of S and n values for all the tests. It shows that the parameter S ranges between 314 and 464, while the parameter n ranges between 0.43 and 0.82. Monterey No. 0 sand of $D_r = 60\%$ with parameter S value of 523 and n value of 0.48 is also shown in Fig. 3.24 for comparative analysis [12]. It shows that the G_{max} values of transparent soils are slightly lower than those of natural sand. When the viscosity of specimens is high, k value in Eq. (3.7) can't be chosen as 0 which is only suitable for cohesionless soil.

The shear modulus as a function of shear strain amplitude determined from RCT and TST with confining pressures of 100, 200, and 300 kPa for dry specimens and $CaBr_2$ solution saturated specimens are given in Fig. 3.25a, and b, respectively. The shear strains resolution ranging from 10^{-6} to 10^{-4} were achieved by RCT, and the shear strains resolution ranging from 10^{-4} to 10^{-1} were obtained by TST. It is evident that the shear modulus is confining stress dependent. The higher the confining stress, the higher is the shear modulus measured (e.g. [3, 9]). The dry specimens have a relatively higher shear modulus than that of the $CaBr_2$ solution saturated specimens,

Table 3.6 Parameters S and n

Specimens	S value	n value
Dry	464	0.58
Mixed oil	408	0.43
$CaBr_2$ solution	396	0.55
Sucrose solution	314	0.82
Monterey No. 0 sand ($D_r = 60\%$) (after Chung et al. [35])	532	0.48

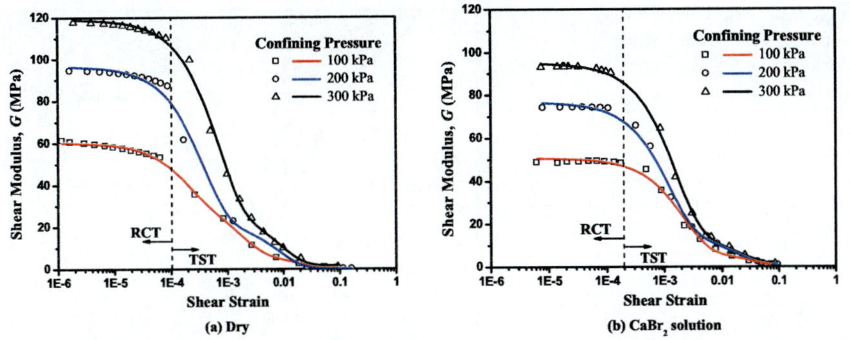

Fig. 3.25 The influence of confining pressure on the shear modulus of fused quartz: **a** dry specimens; and **b** $CaBr_2$ solution saturated specimens

which was also observed for the natural soil specimens [14]. When the mixed oil saturated specimen with 100 kPa confining pressure was tested, some mixed oil had been leaking into confining pressure control system. This phenomenon will somehow influence the confining pressure control system. When saturating specimens, a little of water might also seepage into specimen during back pressure stage. Hence, the other two confining pressures (200, and 300 kPa) for mixed oil saturated specimens have not been carried out.

The normalized shear modulus versus shear strain of tested specimens under 100 kPa confining pressures are shown in Fig. 3.26. The normalized shear modulus versus shear strain (reported by Cao et al. [7], $D_r = 70\%$, practical sizes ranging from 0.1 to 2.0 mm) of fused quartz with water saturated specimens under 100 kPa confining pressures are also shown in Fig. 3.26. Many investigators had studied the relationship between normalized shear modulus and shear strain amplitude of natural soils. Approximate upper and lower bounds relationship between normalized shear modulus and shear strain have been reported by Rollins et al. [3] for gravels (with D_r from 40 to 100%), and by Seed and Idriss [14] for sand (with D_r from 30 to 90%, effective vertical stress $\sigma_v' = 144$ kPa, and coefficient of lateral earth pressure at rest $k_0 = 0.5$), and reported by Vucetic and Dorbry [15] for clay (with OCR from 1 to 15) are also shown in Fig. 3.26 for comparison. The normalized shear modulus versus shear strain for three different aging groups (Quaternary soil, Tertiary and older soil, and Residual/saprolite soil, with plasticity index equals 0, and confining pressure equals 100 kPa) reported by Zhang et al. [5] are also included in Fig. 3.26. It would appear from Fig. 3.26 that the attenuation trends of normalized shear modulus for gravel and sand are very similar. While most of the normalized shear modulus of clay is located above that of sand and gravel, but when the shear strain is larger than 1×10^{-3}, the region of sand and gravel was located in clay. The normalized shear modulus of the dry and mixed oil saturated specimens falls within the region for sand and gravel in the elastic region, while the sucrose solution saturated specimens lie within the region for clay and the $CaBr_2$ solution saturated specimen locates above

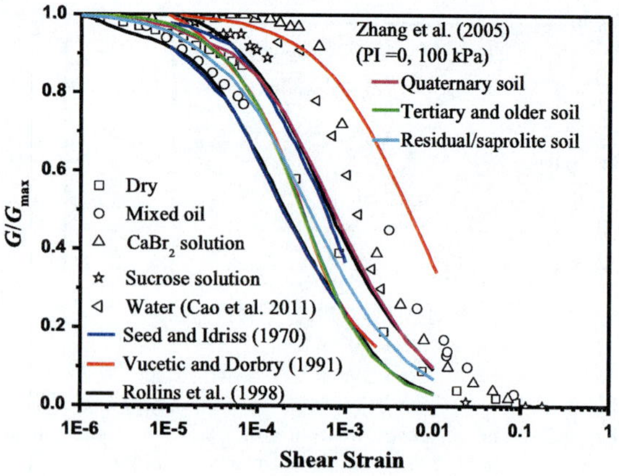

Fig. 3.26 Normalized shear modulus versus shear strain for fused quartz and natural soils

that of clay. In the non-elastic region, all the normalized shear modulus falls within the region for clay. Thus, the normalized shear modulus response of the dry and the mixed oil saturated specimens behave more like sand in the elastic region, while the others behavior are more like clay. It is also shown in Fig. 3.26 that the elastic threshold shear strain level ranges from 2×10^{-5} to 5×10^{-5} for sand, and 5×10^{-5} to 6×10^{-4} for clay. Since the elastic threshold shear strain for the fused quartz ranges from 2×10^{-5} to 3×10^{-4}, the elastic threshold shear strain level for the fused quartz is basically in the similar range of that for the natural soils.

3.5.4 Damping Ratios Influenced by Pore Fluids

Damping ratios were also studied for the dry specimens and three pore fluids saturated specimens. The amplitude decay method was used to determine the damping ratios of specimens according to ASTM D 4015. Damping ratios for all tested specimens at different confining pressures are plotted in Fig. 3.10. In general, the damping ratio values increase slowly at first and appear to have exponential growth when the shear strain reaches a certain value. For the same specimens, the damping ratio values of specimens under low confining pressure are a little of higher than those of specimens under high confining pressure. It is clear that the damping ratio values are influenced by pore fluids obviously. The damping ratio values of the mixed oil saturated specimens are the largest; the values of sucrose solution saturated specimens are a little larger than those of the dry fused quartz specimens; the $CaBr_2$ solution saturated specimens have the lowest damping ratio values.

The damping ratios data for natural soils from literature (as reported by Rollins et al. [3] for gravel; Vucetic and Dorbry [15] for clay; Seed et al. [17] for sand; and Zhang et al. [5] are also plotted in Fig. 3.27. It can be observed that all the plots of the mixed oil saturated specimens and sucrose solution saturated specimens fall

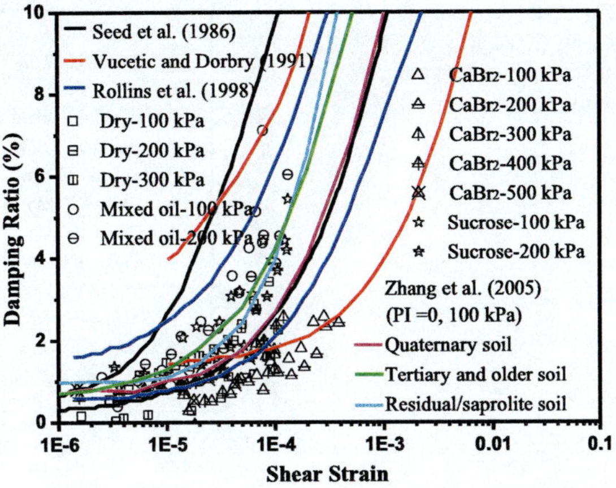

Fig. 3.27 Damping ratio versus shear strain for fused quartz and natural soils

into the range of natural soils when the shear strain is about larger than 3×10^{-5}; and the damping ratios of the two specimens fall within the range of sand and gravel when the shear strain is smaller than 3×10^{-5}. Majority of the damping ratios of the dry specimens and the $CaBr_2$ solution saturated specimens fall in the lower range of the three types of natural soils, especially when the shear strain is smaller than 10^{-4}. In general, the damping ratios of the fused quartz are within the range for those of natural soils in Rollins et al. [13], Vucetic and Dorbry [15], and Seed et al. [14] as concluded from the above discussion.

Unified formulas for the damping ratio of sand after summary of the previous researcher's data was proposed by Ishibashi and Zhang [18], and Zhang et al. [5]. The formulas expressed the damping ratio for sand or soils (from South Carolina, North Carolina, and Alabama) as a function of $G/G_{\max}$ as follow:

$$\zeta = 0.333[0.586(G/G_{\max})^2 - 1.547(G/G_{\max}) + 1] \tag{3.8}$$

$$\zeta - \zeta_{\min} = 10.6(G/G_{\max})^2 - 31.6(G/G_{\max}) + 21.0 \tag{3.9}$$

The damping ratios versus G/G_{max} for all the specimens are shown in Fig. 3.28. The fitting curves of Eqs. (3.8) and (3.9) were plotted on each plot. Although a short of range of $G/G_{\max}$ data obtained in this study, existing laboratory test data for transparent soil are distributed closer to the fitting curve by Ishibashi and Zhang [18], which is suitable for sand.

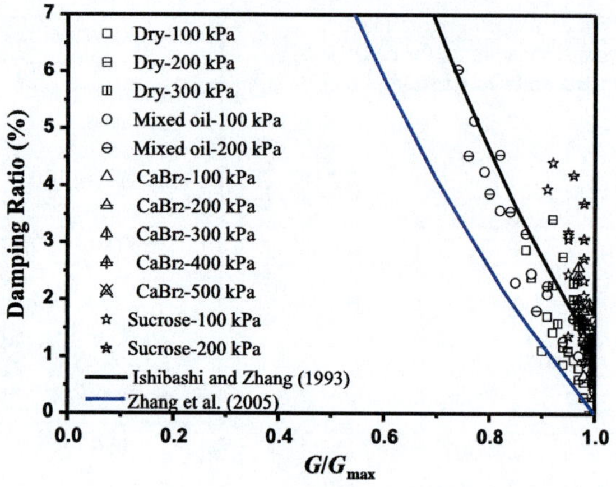

Fig. 3.28 Damping ratio versus G/G_{max} for fused quartz with different pore fluids

3.6 Cyclic Undrained Behavior and Liquefaction Resistance of Transparent Sand Made of Fused Quartz

Fused quartz, manufactured by the Jiangsu Kaida Silica Co. LTD in China, is chosen as transparent granular particles to mimic natural sand particles. The particle size distribution of fused quartz used for this study and in previous research are compared in Fig. 3.29 according to ASTM D 2488. The particle size distribution of definition liquefaction-prone sand by Japan Code is also shown in Fig. 3.29. The grain-size characteristics and classification of transparent granular particle materials and Ottawa sand are shown in Table 3.1. Fused quartz specimens are saturated by pure water.

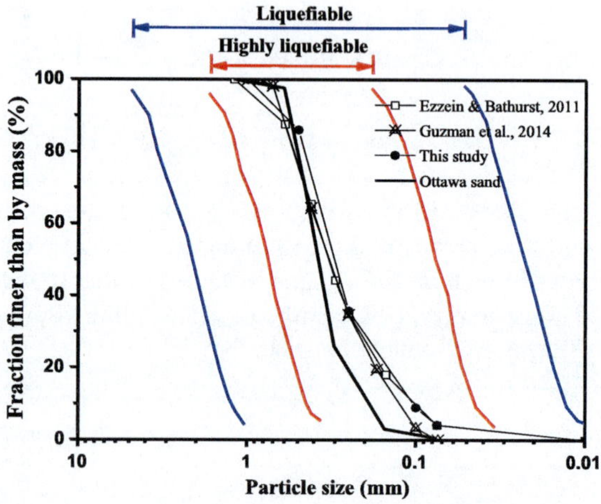

Fig. 3.29 Particle size distributions of fused quartz, and the liquefaction-prone sand (Design standard for port and harbour facilities and commentaries, Japan Port and Harbour Association; 1999 [in Japanese])

3.6.1 Testing Methods

The TST with a shear strain resolution ranging from 10^{-4} to 10^{-1} was conducted following ASTM STP 1213. The TST devices perform harmonic torsional excitation on the hollow specimen at the top through the electronic load system. The frequency curve was measured once a constant amplitude load was applied to a torsional harmonic frequency range. The symbols of axial load W_n, torque M_T, and external and internal pressures P_0 and P_i were adopted. The internal and external pressure of the torsional shear test applied in this study was the same. During the shearing stage, the total vertical stress increments were maintained close to zero. The frequency of torsional force was 0.1 Hz in these tests. The transparent sand specimen was a hollow cylindrical specimen with 30 mm in inside diameter, 70 mm in outside diameter, and 100 mm in height. The saturation of the samples was ensured by combined methods (CO_2 pluviation, de-aired water and back pressure method). The confining pressure was increased step-by-step from 100 to 200 and 300 kPa. The relative density of specimens equals 63.8%.

3.6.2 Results and Analysis

Undrained cyclic behaviour

The typical time histories of a shear stress–strain relationship and the effective stress paths obtained from the tests under the confining pressure of 100, 200 and 300 kPa are shown in Fig. 3.30. The cyclic shear stress amplitude equals to 40 kPa, 60 kPa and 75 kPa, respectively. Figure 3.30a shows that the initial liquefaction occurred once the shear stress is larger than the threshold value. Cyclic mobility [24], which followed the initial liquefaction, is observed in the effective stress path of all the specimens, where the shear stress recovered repeatedly after reaching the minimum effective stress Fig. 3.30a. The pore water pressure stops to build up when it reaches about 87 ~ 93% of the initial confining pressure which is similar to silty sands. Figure 3.30b shows that the shear strain remains constant at the initial stage, then has an important increase after several cyclic loadings. The amplitude shear stress has a reduction with the increase of the shear strain and the reduction is 9.3 ~ 13.4% at the failure of specimens. It is interesting to note that the hysteresis loop depicted in Fig. 3.30b is asymmetric, which may be caused by the dilation of samples subjected to cyclic loading, but with the confining pressure increasing this asymmetry is reduced.

The relationship between monotonic and cyclic undrained behaviors of sand under torsional loading has been implemented by Lombardi et al. [26]. The effective stress ratio mobilized at the stage of initial liquefaction is the same as that mobilized along the failure envelope under monotonic loading. The mobilized angle of shearing resistance in the effective stress path of transparent sand with specimen has been conjectured and the result, which is around 42.6°, is shown in Fig. 3.30a. Some

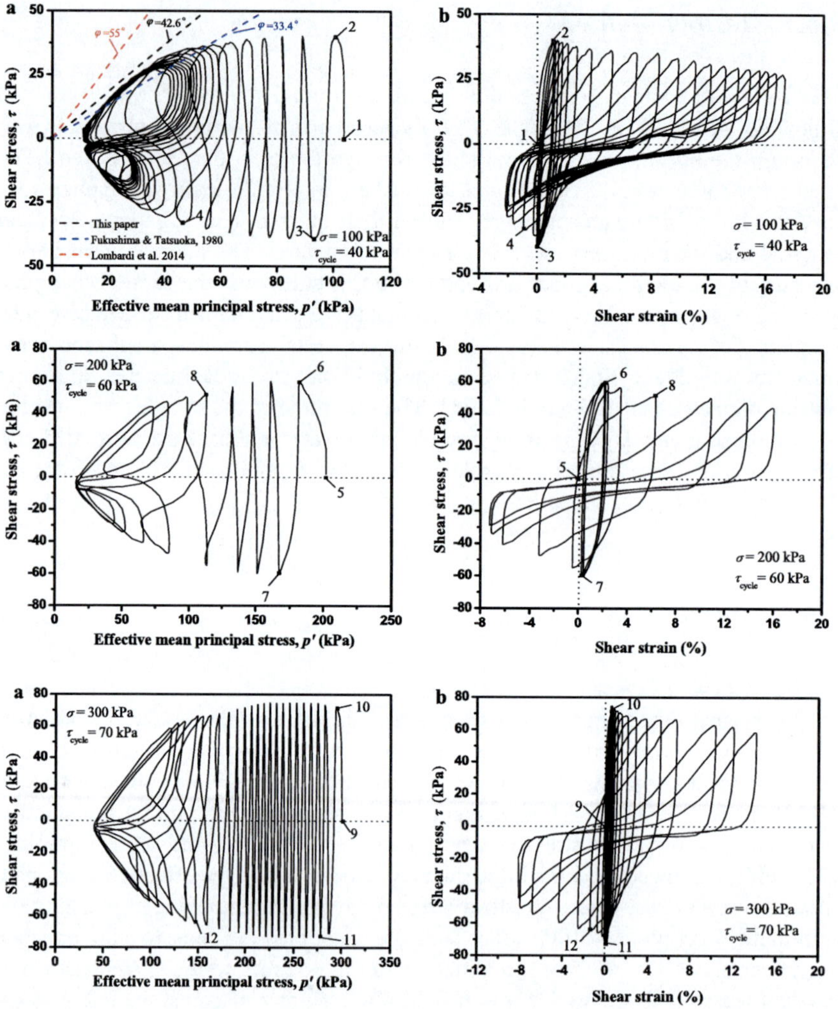

Fig. 3.30 **a** Shear stress versus effective mean principal stress and **b** shear stress versus shear strain of specimens

results reported by previous researchers for natural angular sand are also shown in Fig. 3.30a. Based on static undrained torsional shear test method, Fukushima and Tatsuoka [25] noted the mobilized angle of shearing resistance in the effective stress path of Toyoura sand (with D_r equal to 40 and 84% under the initial mean effective stress of 98 kPa) range from 33.4 to 37.4°. Lombardi et al. [26] investigate the undrained cyclic behavior of Redhill 100 sand ($D_r = 67\%$ and the initial mean effective stress is 100 kPa) by means of multi-stage cyclic triaxial tests. Redhill 100 is a poorly graded fine-grained silica sand that is similar to the fused quartz.

The mobilized angle of shearing resistance about the Redhill 100 is around 55°. Figure 3.30a shows that the ratio of shearing failure line of transparent sand is a little larger than that of the Toyoura sand, but smaller than that of Redhill 100. It can speculate that the mobilized angle of shearing resistance of the transparent sand manufactured by fused quartz is the same as the natural sand which has angular shape.

Excess pore pressure

Figure 3.31 illustrates the function of excess pore pressure against the number of cycles of the specimens. Several selected points have been marked along the effective stress path and shear strain, and the corresponding excess pore pressure against the number of cycles in Figs. 31a and c. On unloading from maximum shear stress of the first (or second) cycle (point 2, 6, 10), a significant increase in the accumulated excess pore pressure (associated mainly with reloading in the opposite direction) is observed and marked by point 3 (or point 7 and 11) on the excess pore pressure against the number of cycles in Fig. 3.31. During the following cycles loading, the excess pore pressure builds up at a constant rate, lower than that corresponding to the first cycle, until the shear strain near the 5% double amplitude strain (point 4, 8 and 12) and then undergoes a second significant increase. However, this phenomenon is not obvious for specimen with the confining pressure of 100 kPa. Some degree of softening takes place for specimens when it has a high dynamic shear stress and confining pressure. This phenomenon was also noted by Lombardi et al. [26] for Toyoura sand, but it occurred near the end of the test and the specimens have a higher density.

Liquefaction resistance

In order to consider deformation, liquefaction and cyclic mobility phenomena, two criterion of liquefaction are chosen according to Ishihara [4], Alarcon-Guzman et al. [22] and Tatsuoka et al. [27] (double amplitude strain required at 5% and 10%, respectively). The ultimate liquefaction resistance normalized by the initial mean effective stress which is defined as shear stress ratio ($\tau_{z\theta}/P'$) has been plotted against the number of cycles (N_f) to each specimen in Fig. 3.32. The failure depends on the level of cyclic stress applied to the specimen and the confining pressure. The liquefaction resistance of the specimens has the same trend with the variation of number of cycles with two criterion. The liquefaction resistance of specimen is reduced by the increasing of the number of cycles or the confining pressure. The reduction of liquefaction resistance versus the number of cycles decreases with the increasing of confining pressure. The shear stress ratios versus the number of cycles for natural sands are also shown in Fig. 3.32 for comparison. It shows that the specimens have a close trend with natural sands [23, 27].

3.7 Summary

The fused quartz particles have the angular shape. The static properties of fused quartz are similar to that of natural sand. The internal friction angle for the transparent soil of fused quartz is varying from 31° to 55°, which depends on the shape, grain size,

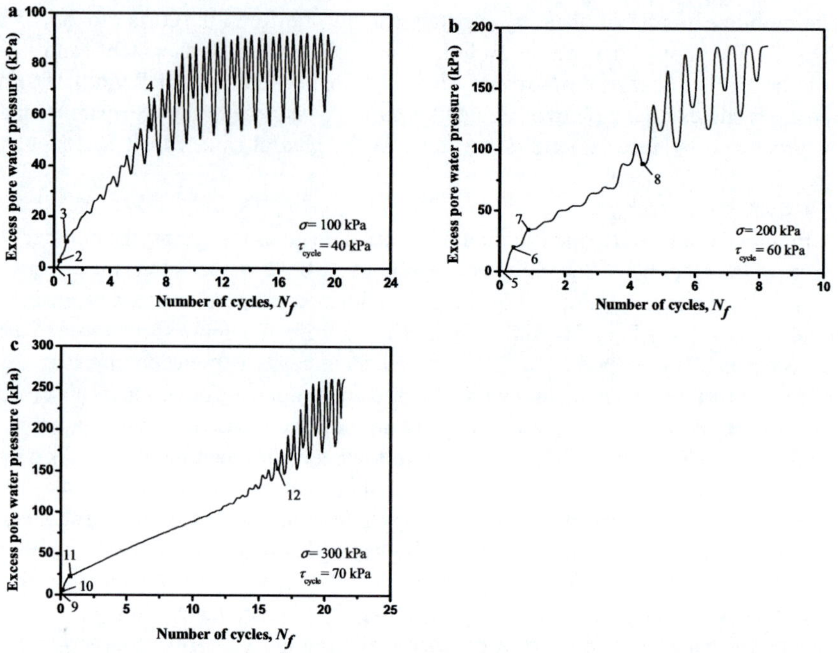

Fig. 3.31 Excess pore pressure versus number of cycles for saturated specimens under difference confining pressure **a** 100 kPa, **b** 200 kPa and **c** 300 kPa

Fig. 3.32 Shear stress ratio versus the number of cycles for both fused quartz and natural sand at the double amplitude (DA) shear strain equals to 5 and 10%

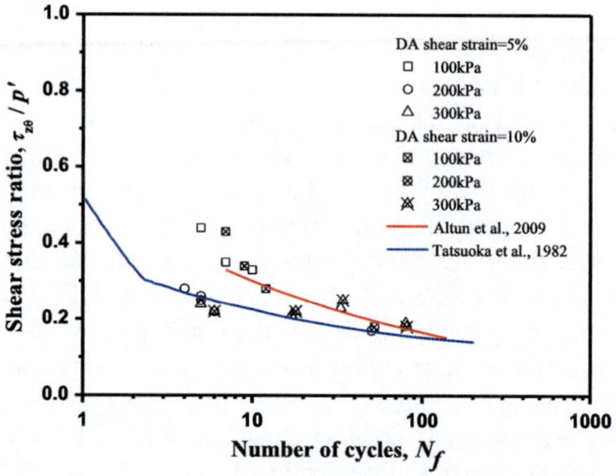

relative density and pore fluid. The stress–strain curve displays a strain hardening behavior at loose state and a strain softening behavior at dense state, which also depends on the grain size distribution. The compressibility of the fused quartz is similar as those of natural sand.

The shear modulus and damping ratio of fused quartz with dry and oil saturation are comprehensively studied. The dry specimen has a higher shear modulus than that of the oil saturation specimen. The maximum shear modulus, G_{max}, at a strain level of 10^{-6} under a 100 kPa confining pressure is about 60 MPa for both dry and oil saturation specimens. The relationships between the normalized shear modulus (or damping ratio) and shear strain of fused quartz are found to follow the same general trend as those of natural soil.

For the fused quartz, the relationships between the normalized shear modulus and shear strain, damping ratio and shear strain are found to be consistent with those general trends of natural soils. The dry specimens have a higher shear modulus than those of the $CaBr_2$ solution saturated specimens, which is similar to that of natural soils. The damping ratio values of the mixed oil saturated specimens are the largest; the sucrose solution saturated specimens are a little larger than those of dry specimens; and the $CaBr_2$ solution saturated specimens have the lowest damping ratio values. The results show that the dynamic properties of transparent soil manufactured by fused quartz and mixed oil are most similar to those of natural sand, and hence could have a wider application in dynamic problems.

The normalized shear modulus response of the dry and the mixed oil saturated specimens behave more like sand in the elastic region, while with the other pore fluid, its behavior are more like clay. The elastic threshold shear strain level for the fused quartz with different pore fluids is basically in the similar range of that for the natural soils. The damping ratio values of transparent soil are influenced by pore fluids obviously. The damping ratio values of the mixed oil saturated specimens are the largest; the values of sucrose solution saturated specimens are a little larger than those of the dry fused quartz specimens; the $CaBr_2$ solution saturated specimens have the lowest damping ratio values. In general, the damping ratios of the fused quartz with different pore fluids are within the range for those of natural soils. Although a short range of G/G_{max} data obtained in this study, existing laboratory test data for transparent soil are distributed closer to the fitting curve by Ishibashi and Zhang [18], which is suitable for sand.

Through cyclic torsional shear tests (TST), some common features of cyclic undrained behaviour have been observed for saturated transparent sand under cyclic torsional loading with different confining pressure, which is similar to natural sands. The angle of shearing failure line of transparent sand is around 42.6°, which is a normal value in the angular sand. The excess pore pressure shows a relatively high rate of accumulation at the initial stage; increases at a lower rate until the specimen nears failure, where the excess pore pressure undergoes a second significant increase. Some degree of softening takes place for the transparent sand and natural sands at a high confining pressure. The liquefaction resistance of transparent sand is reduced by the increasing of the number of cycles or the confining pressure and the reduction decrease with the increasing confining pressure, which is similar to the natural sands.

References

1. Zhao H, Ge L (2014) Investigation on the shear moduli and damping ratios of silica gel. Granul Matt 16:449–456
2. Iskander M, Bathurst RJ, Omidvar M (2015) Past, present, and future of transparent soils. Geotech Test J 38:557–573
3. Rollins KM, Evans MD, Diehl NB, Daily WD (1998) Shear modulus and damping relationships for gravel. J Geotech Geoenviron Eng 124:396–405
4. Ishihara K, Zhang XJ (1993) Unified dynamic shear moduli and damping ratios of sand and clay. Soils Found 33:182–191
5. Zhang JF, Andrus RD, Juang CH (2005) Normalized shear modulus and material damping ratio relationships. J Geotech Geoenviron Eng 131:453–464
6. Wallace JF, Rutherford CJ (2015) Geotechnical properties of LAPONITE RD. Geotech Test J 38(5):574–587
7. Cao ZH, Liu JY, Liu HL (2011) Transparent fused silica to model natural sand. Pan-Am CGS Geotechnical Conference
8. Georgiannou VN, Tsomokos A, Stavrou K (2008) Monotonic and cyclic behavior of sand under torsional loading. Geotechnique 58(2):113–124
9. Stokoe KH II, Darendeli MB, Andrus RD, Brown LT (1999) Dynamic soil properties: laboratory, field and correlation studies. In: Proceedings of 2nd international conference on earthquake geotechnical engineering, vol 3, pp 811–845, Lisbon, Portugal
10. Borden RH, Shao L, Gupta A (1996) Dynamic properties of piedmont residual soils. J Geotech Eng 122(10):813–821
11. Rowe PW (1971) Theoretical meaning and observed values of deformation parameters for soil. In: Parry RHG (ed) Stress-strain behavior of soils, proceedings of roscoe memorial symposium, pp 143–194, Cambridge University, CA
12. Chung RM, Yokel FY, Drnevich VP (1984) Evaluation of dynamic properties of sands by resonant column testing. Geotech Test J 7(2):60–69
13. Rollins KM, Evans MD, Diehl NB, Daily WD (1998) Shear modulus and damping relationships for gravel. J Geotech Geoenviron Eng 124(5):396–405
14. Seed HB, Idriss IM (1970) Soil moduli and damping factors for dynamic response analysis. Report No. EERC 70–10, Earthquake Engineering Research Centre, University of California, Berkeley, CA
15. Vucetic M, Dorbry R (1991) Effect of soil plasticity on cyclic response. J Geotech Eng 117(1):89–107
16. Zhang JF, Andrus RD, Juang CH (2005) Normalized shear modulus and material damping ratio relationships. J Geotech Geoenviron Eng 131(4):453–464
17. Seed HB, Wong RT, Idriss IM, Tokimatsu K (1986) Moduli and damping factors for dynamic analysis of cohesionless soils. J Geotech Eng 112(11):1016–1032
18. Ishibashi I, Zhang XJ (1993) Unified dynamic shear moduli and damping ratios of sand and clay. Soils Found 33(1):182–191
19. Kong G, Liu L, Liu H-L, Zhou H (2013) Triaxial tests on deformation characteristics of transparent glass sand. Chinese J Geotech Eng 35(6):1140–1146
20. Kong G, Sun X, Li H, Cao Z (2016) Effect of pore fluid on strength properties of transparent soil. Chinese J Geotech Eng 38(2):377–384
21. Kong G, Sun X, Xiao Y, Zhao H (2016) Comparative experiments on compressive deformation properties of transparent soil and standard sand. Chinese J Geotech Eng 38(10):1895–1903
22. Alarcon-Guzman A, Leonards GA, Chameau JL (1988) Undrained monotonic and cyclic strength of sands. J Geotech Eng 114(10):1089–1109
23. Altun S, Goktepe AB, Ansal AM, Akguner C (2009) Simulation of torsional shear test results with neuro-fuzzy control system. Soil Dyn Earthq Eng 29(2):253–260
24. Castro G, Poulos SJ (1977) Factors affecting liquefaction and cyclic mobility. J Geotech Eng Div 103:501–516

25. Fukushima S, Tatsuoka F (1980) Specimen size effects in undrained torsional simple shear test on sand. In: Proceedings of 35th annual symposium on civil engineering: Sapporo (Japan), pp 295–296
26. Lombardi D, Bhattacharya S, Hyodo M, Kaneko T (2014) Undrained behaviour of two silica sands and practical implications for modeling SSI in liquefiable soils. Soil Dyn Earthq Eng 66:293–304
27. Tatsuoka F, Muramatsu M, Sasaki T (1982) Cyclic undrained stress-strain behaviour of dense sand by torsional simple shear test. Soils Found 22(2):55–70
28. Ezzein FM, Bathurst RJ (2011) A transparent sand for geotechnical laboratory modeling. ASTM Geotech Testing J 34(6):590–601

Chapter 4
Transparent Clay of Carbopol U10

Abstract Carbopol® Ultrez™ 10 (referred to as U10) was the newly found material as a transparent clay surrogate for the possible use in physical geotechnical model tests. This synthetic clay was studied and used in the field of geotechnical engineering in a landslide by (Bates et al. in Environ Fluid Mech 14:501–518, 2014). However, the geotechnical properties of the synthetic clay as a standalone transparent clay surrogate at low effective stress have not yet been investigated. The transparency, speckle pattern and the geotechnical properties were studied and presented in this chapter.

4.1 Introduction

Silica gel and fused quartz are used to manufacture transparent soils to substitute sand in model testing or other geotechnical investigations. To model clay, Iskander et al. [1] used amorphous silica powder mixed with refractive index matching pore fluid. Laponite RD® was a material proposed by Wallace and Rutherford [2] to make transparent soil for investigating the deformation problem in clay. Wallace and Rutherford [2] have measured the geotechnical properties of Laponite RD®. The comparison of the geotechnical properties of amorphous silica with natural clay were carried out by Iskander et al. [1, 3, 4], Sadek et al. [5], and Liu et al. [6]. The deficiency of this materials is that physical modeling is limited by the transparent depth. The transparent depth of clay manufactured amorphous silica (<0.074 mm) is limited to 5 cm according to Iskander et al. [3]. Synthetic clay made by Laponite® RD, with properties similar to hydrogels is characterized for geotechnical properties [2]. Laponite RD® has been used to investigate torpedo anchors [7], drag embankment anchors [8] and failure surfaces of shear measurement devices [9]. However, the speckle pattern within the soil is a result of reflectivity index mismatches between the pore fluid and either air pockets or unsaturated Laponite® RD. Besides these two main types of clay, aquabead was used to model the marine deposits to investigate the multiphase flow. However, the bearing capacity of transparent soil manufactured by aquabead is nearly about zero, which is inconsistencies with natural soil. The accuracy and reliability of physical geotechnical models were limited by

H. Zhao et al., *Transparent Soil Modelling Technique and Its Application*, Springer Tracts in Civil Engineering, https://doi.org/10.1007/978-981-19-6825-9_4

the characteristics of synthetic transparent clay and the differences between synthetic transparent clay and natural clay. Hence, it is still necessary to find a new type of synthetic transparent clay which holds better optical transparency and geotechnical properties more similar to the natural one.

Carbopol® Ultrez™ 10 (referred to as U10, manufactured by Lubrizol Corporation, USA) was the newly found material as a transparent clay surrogate for the possible use in physical geotechnical model tests. This synthetic clay was studied and used in the field of geotechnical engineering in a landslide by Bates et al. [10]. However, the geotechnical properties of the synthetic clay as a standalone transparent clay surrogate at low effective stress have not yet been investigated. The transparency, speckle pattern and the geotechnical properties were studied and presented in this chapter.

4.2 Materials and Manufacture Process

4.2.1 Raw Materials

Carbopol® Ultrez™ 10 is one of insoluble acrylic acid polymers, which become stiff gels upon neutralisation in an aqueous medium [11]. By employing organic amines as a neutralising agent, it is possible to gel many semipolar liquids or mixtures of these liquids with water. Compatibility of the polymer with non-aqueous liquids depends on the formation of ion pairs with the amine. U10 is a cross-linked polyacrylic acid polymer that provides efficient rheology modification with encased self-wetting for ease of use [12]. U10 has many typical features and benefits such as, short flow properties, high viscosity, and high suspending ability. The particle size, specific gravity, and specific surface area of U10 used are 50 ~ 100 nm, 1.4543, and 0.21 m^2/g, respectively. The photos of U10 powder materials used to manufacture synthetic transparent clay is shown in Fig. 1a. The 1×10^4, 5×10^4 and 1×10^5 SEM images of the U10 powder materials is shown in Fig. 1b, c and d, respectively.

U10 polymer powders shrink in contact with water. When the powder disperses in water, the cross-linked acrylic resin chain begins to spread. After neutralizing with an alkaline substance, the polymer becomes expanding structure. From dry polymers to gelled polymers, the volume may increase to 1000 times bigger. The viscosity of gelled polymers versus mixtures of U10 and NaCl at the room temperature is shown in Fig. 4.2 (after instructions by the producer). The viscosity of synthetic clay increased from 1×10^4 to 8×10^4 mPa·s with increasing mixture of U10 (0.1 ~ 1.0% Mix). An opposite trend for NaCl mixture was also shown in Fig. 4.2. The results show that U10 powder is not conducive to the fluxility of the synthetic clay.

NaOH powder of fused white particles was chosen as the neutralized material, purity of which is greater than 98% for analytical reagent (AR). The solution of the NaOH is strongly alkaline.

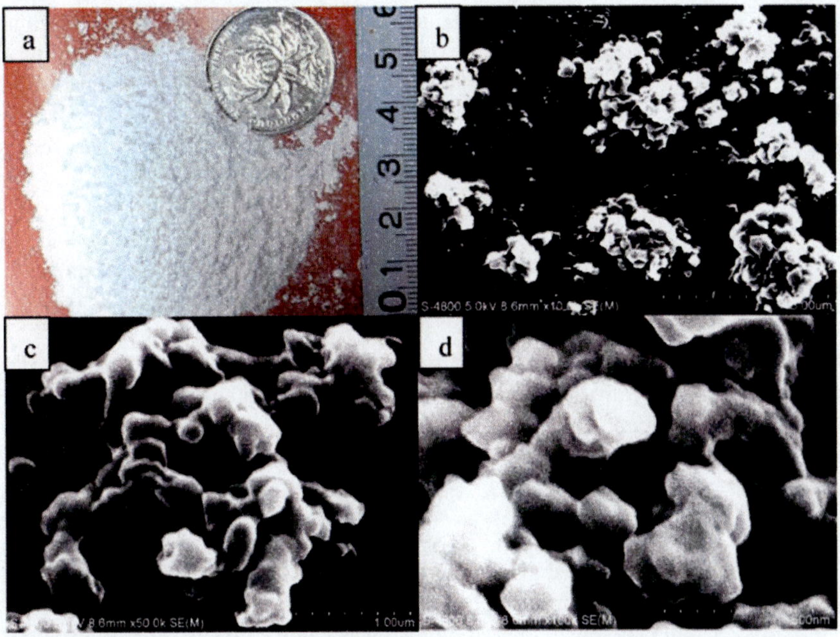

Fig. 4.1 Photos of Carbopol® Ultrez™ 10 polymer material: a powder; b SEM of 1×10^4; c 5×10^4; and d 1×10^5 magnification

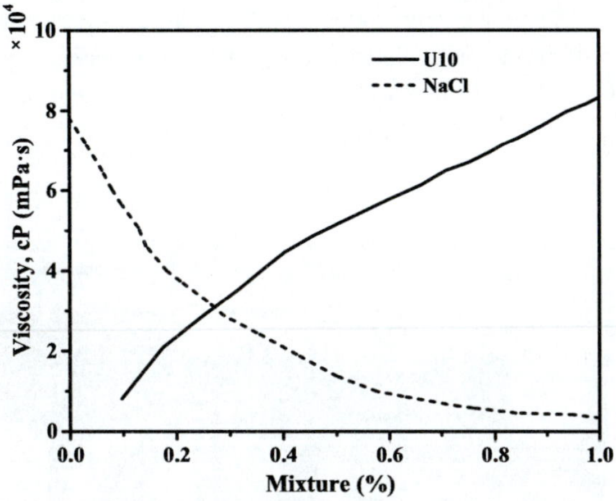

Fig. 4.2 Viscosity of synthetic clay versus mixture of U10 or NaCl (after instructions by producer)

Speckle field can be used in synthetic clay to investigate the internal deformation problems. Nano materials (as additional materials *i.e.* nano SiC, carbon nanotubes, nano silica fume, nano alumina and nano calcium oxide) can be incorporated into the synthetic transparent clay as tracking particles if necessary. In this study carbon nanotube was chosen as speckle material in transparent clay, the amount of which is about 0.01 ~ 0.1% by mass, the amount of nano materials can be negligible in the mix design. Nanotubes are multi walled (outside diameter and inside diameter are equal to 8 ~ 15 and 3 ~ 5 mm, respectively) materials with appearance of black powder. The specific surface area and density of the material are 233 m^2/g and 0.15 g/cm^3, respectively.

4.2.2 *Manufacture Processes*

The synthetic transparent clay was manufactured by U10 powder, NaOH powder and distilled water. The synthetic transparent clay was prepared at initial mixtures ranging with 0.25, 0.5, 0.75, and 1.0% of U10 powder by weight calculated as follows:

$$\text{Mixture } (\%) = \frac{m_{\text{U10}}}{m_{\text{U10}} + m_{\text{H}_2\text{O}} + m_{\text{NaOH}}} \times 100\% \tag{4.1}$$

Photo of the manufactured synthetic transparent clay with 0.75% initial mixture after 7 days self-consolidation through 300 mm thickness was shown in Fig. 4.3. In the figure, transparent clay manufactured by U10 was compared with other transparent soils. The manufacture processes of synthetic transparent clay used in this study contain the following steps (0.75% Mixture was taken as an example):

(1) Safe preparation: Before the manufactured operation, equipment including anti-fog mask, rubber gloves and goggles were prepared in order to prevent the effects of fine particulate materials on human health.
(2) Materials preparation: weighing the amount of U10 powder, NaOH powder, and distilled water according to the Eq. (4.1), respectively. For example, 0.35 kg U10

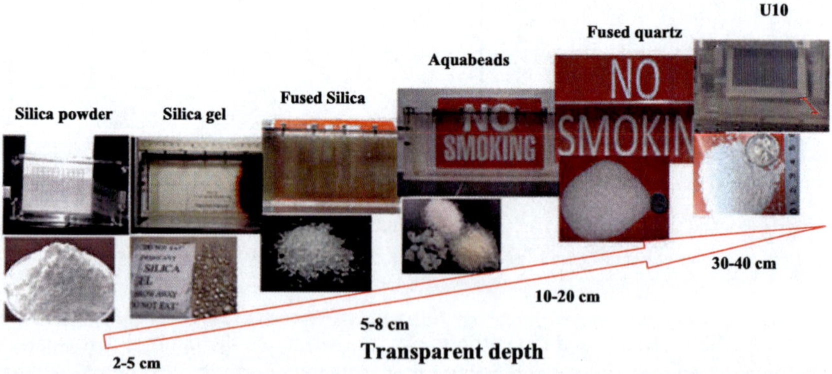

Fig. 4.3 Photo of the transparent clay (0.75% mixture) after 7 days self-consolidation through 300 mm thickness, in comparison with other transparent materials

powder, 0.15 kg NaOH powder, and 19 L distilled water were chosen in this paper.

(3) Mixture of U10 powder: Heating the distilled water to 60 ~ 70 °C for 3 ~ 4 h firstly. Then pouring the U10 powders into the heated water, mixing 30 ~ 40 min by low power agitator, and stand for 3 ~ 4 h in airproof condition at room temperature.
(4) Mixture of the NaOH-water solution: dissolving NaOH powder into 0.5 L distilled water firstly. Then pouring the NaOH-water solution into the U10-water mixed liquor, and mixing 30 ~ 40 min by high power agitator. Stop mixing when the mixture becomes a smooth paste.
(5) Bubbles clear: (If the speckle field is necessary, Carbon nanotubes were sprinkled into the transparent clay served as tracer particle.) The air bubbles mixed in the sample were removed through vacuuming the sample. The synthetic transparent clay sample was manufactured at last.

4.3 Optical Properties of Synthetic Clay

4.3.1 *Transparency Analysis*

Figure 4.3 indicated that transparent clay manufactured by U10 has superior optical transparency compared with other transparent materials. The transparent depth of this material is about 30 ~ 40 cm. To further confirm the transparency characters of synthetic transparent clay, a series of comparative tests are conducted through using the Modulation Transfer Function (MTF) method [13]. MTF is a measure of the modulation (or contrast) that is transferred from the object to an image. The resulting image will be degraded due to the partial transparency when objects are observed through transparent materials. Therefore, bright areas in images will not appear as bright as those in the objects, and dark areas will not be as black as those in the objects. MTF is based on a calibration target consisting of black and white line pairs. The target is defined by spatial frequency (number of black and white areas) referred to as line pairs per millimeter (lp/mm). As described detailed in Black and Take [13], individual line pairs turn from visible in the image to a uniform color due to the limit of resolution when the spatial frequency increases. Schematic diagram of calculation of MTF is shown in Fig. 4.4. The MTF defined as follow:

$$\mathrm{MTF} = \frac{(I_{\max(\mathrm{I})} - I_{\min(\mathrm{I})})/(I_{\max(\mathrm{I})} - I_{\min(\mathrm{I})})}{(I_{\max(\mathrm{o})} - I_{\min(\mathrm{o})})/(I_{\max(\mathrm{o})} - I_{\min(\mathrm{o})})} \tag{4.2}$$

where $I_{\max(\mathrm{I})}$ and $I_{\min(\mathrm{I})}$ are maximum and minimum intensity of an actual object with repeating structure having maximum and minimum intensity $I_{\max(\mathrm{o})}$ and $I_{\min(\mathrm{o})}$, as shown in Fig. 4.4. The value of MTF represents the similarity of calibration target in transparent clay to the actual one (not in). The higher MTF value means better transparency for transparent clay.

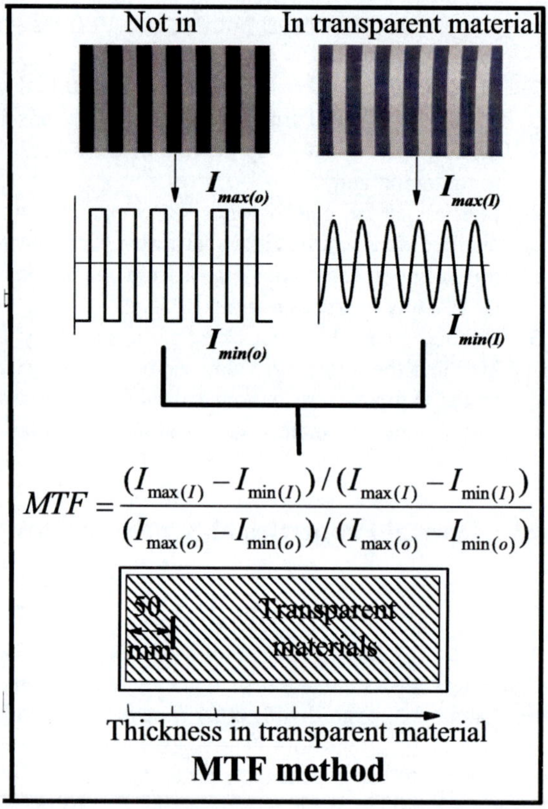

Fig. 4.4 Schematic diagram of calculation of MTF (after [13])

At low spatial frequencies, the modulation is close to 1 and generally falls as the spatial frequency increases until it reaches zero. Both high and low frequency are necessary in the measurement for the rational range of influence. Targets of 0.02, 0.04, 0.25 and 0.50 lp/mm (referred to Black and Take [13]) were used in the transparency calibration. The thickness of transparent clay has a significant effect on the transparency. Therefore, thicknesses of 50, 100, 150, and 200 mm for 0.75 % mixture of U10 powder were used to calibrate the transparency, respectively. Figure 5a shows the MTF values versus sample thickness and spatial frequency in synthetic transparent clay. The results indicate that a linear decrease of transparency with the increasing of the sample thickness. The decrease slop is -0.17 per 50 mm given an increase of 0.75 % in the thickness. Furthermore, the transparency is reduced by increasing the usage of U10 powder mixture, shown in Fig. 5b. The MTF values of synthetic transparent clay are decreased with the increasing of U10 powder usage and the thickness of samples. Take 150 mm thickness sample with 0.5 % and 0.75 % U10 powder usage for example, the MTF values of synthetic transparent clay decrease from 0.583 to 0.498, respectively. Take 0.75 % U10 powder usage for example,

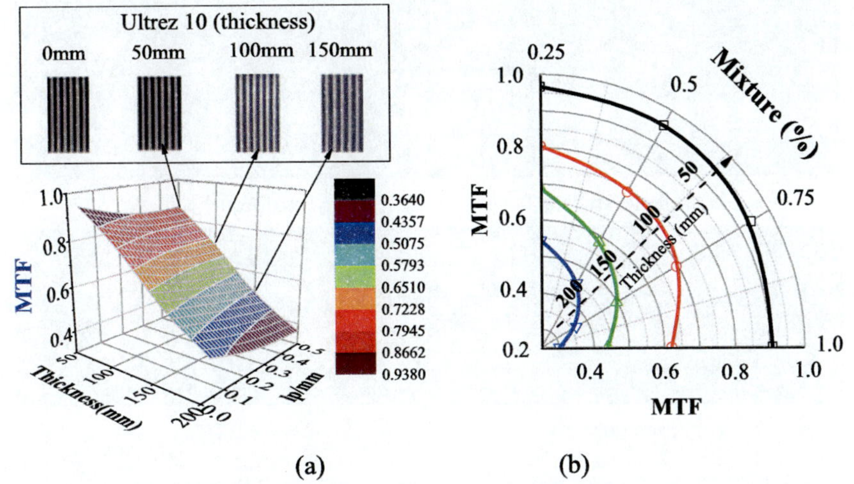

Fig. 4.5 Transparency of synthetic clay versus soil thickness: **a** spatial frequency through MTF method for 0.75% initial mixture; **b** varying mixtures of different thickness

under the thickness of 50 and 200 mm, the MTF value of synthetic transparent clay decreases from 0.938 to 0.365, respectively.

4.3.2 *Speckle Pattern*

For different investigation aims, speckle pattern could be chosen. Carbon nanotubes can be sprinkled into the transparent clay served as tracer particle if necessary *i.e.* deformation problem. Distinctive speckle pattern under laser sheets can be obtained by optical system. Optical system is composed by transparent soil, plexiglas mold, laser, computer control system, CCD camera. Light sheet laser with 2-W maximum power was chosen in this study. The resolution of CCD camera equals 1280 × 960pixels, and its frame exposure time equals 100 μs ~ 30 s; the photos can be automatic continuous obtained by computer control system. The model is sliced optically using a laser light sheet. The interaction of laser light and transparent soils produces a distinctive speckle image. A sequence of speckle images is captured during loading using a digital camera. Digital image correlation is used to generate the two-dimensional deformation field within each slice from consecutive images. Speckle pattern of four different transparent soils (amorphous silica gels, aquabead, fused quartz and U10) under 2-W laser irradiation are shown in Fig. 4.6. The results show that the speckle is smaller than others due to the finer particle of speckle materials.

Testing about the mechanical characteristics have performed with and without speckle materials (0.01 ~ 0.1% by mass). Results show that few nano materials have

(a) Amorphous silica gels (b) Aquabead

(c) Fused quartz (d) U10

Fig. 4.6 Comparison of speckle pattern for transparent soils under 2-W laser irradiation

almost no impact on the character. Take shear strength and thermal conductivity for example, the detailed description is described in following sections. This indicates that the new synthetic transparent clay provides a new possibility for modeling clay in consideration of the deformation problem in the physical model test.

4.4 Geotechnical Properties of Synthetic Clay

4.4.1 Shear Strength

A series of laboratory vane shear tests were performed to investigate the following properties of the transparent clay: the increase in strength with time (thixotropy), the change in shear strength with the mixture concentration and with depth, the sensitivity, and the measured strength with consolidation after vane insertion. Vane shear test is a relatively suitable measurement to measure the strength of very soft clay compared with other traditional measurement like direct shear. As used in Wallace and Rutherford [2], vane shear test was used to estimate the shear strength (0.15 ~ 0.5 kPa). The laboratory vane shear device used here is a portable type (PS-VST-P). The vane is a four-blade, 50.8 mm by 101.6 mm square vane with a rod diameter of 10 mm. The measurement range of the vane shear apparatus is approximately 0 ~ 8.125 kPa. Typically soil is pre-conditioned through consolidation, thus the shear envelope over a wider range of stress regimes can expand the applicability. Clay under different loading stress can represent the clay at different depth for actual soil. However, the new material is used to provide a new thought to model clay in view of basic strength parameters.

Synthetic transparent clay with 0.5, 0.75, and 1.0% initial U10 powder mixture was manufactured in the 100 mm × 100 mm × 150 mm cube, and allowed to set or self-weight consolidate for 6 ~ 336 h. Laboratory vane tests were conducted where the top of the vane was at the depth of 25 mm below the surface (minimum of one vane diameter below the surface as recommended by ASTM D4648/D4648M-13(2013)). The shear rate is 1° per 10 s in these testing. The on-set of static holding is defined as the moment where the specimen is taken out from the vacuum environment. The sample with 0.75% initial U10 powder mixture exhibits an increase in the peak undrained shear strength which was measured by laboratory vane, $S_{u,peak}$, with the increasing time as shown in Fig. 7a. Moreover, transparent clay mixed with nano materials was also tested in Fig. 7a. The results indicated that tiny amount of nano materials don't affect the strength property of transparent clay. The thixotropy strength ratio of synthetic clay (0.75% mixture), $S_{u,peak}/S_{uo}$, in comparison with other natural clays is shown in Fig. 7b. The temporal evolution of the thixotropy strength ratio of synthetic clay is higher than the natural one. Table 4.1 indicates that the average peak shear strength of synthetic clay that is manufactured by both U10 and Laponite® RD is within the scope of those of Harbor mud at liquid consistency [14]. The average peak shear strength of samples undergoing self-weight consolidation for 7 days is 0.22 kPa, 0.28 kPa, and 0.43 kPa for 0.5%, 0.75%, and 1.0% mixtures, respectively, as shown in Fig. 4.8. This observation indicates that with self-consolidation alone, the transparent clay remains very soft for an approximate 120 mm high sample size. The average peak strength of samples undergoing self-weight consolidation for 14 days is 0.3 kPa. Therefore, a 7-days setting time was used for testing due to the relatively small increase in undrained shear strength during a second 7-days.

The transparency of samples decreases with the increase of mixture concentration, and the undrained shear strength increases with the increase of mixture concentration. Figure 4.8 shows mixture optimization interval selection through S_u and MTF

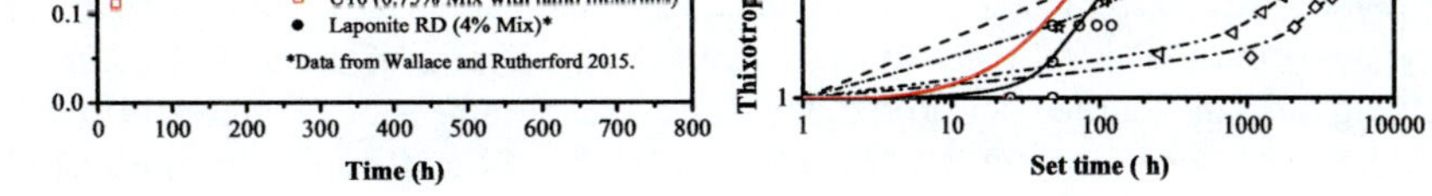

(a) (b)

Fig. 4.7 Thixotropy measured by laboratory vane shear test for 0.75% initial mixture: **a** peak undrained shear strength; **b** thixotropy strength ratio

Table 4.1 Average peak shear strength of synthetic transparent clay and natural clay

Materials	Mixture (%)	S_u (kPa)
Synthetic clay manufactured by Carbopol® Ultrez™ 10 (this paper)	0.50	0.23 ~ 0.28[a]
	0.75	0.26 ~ 0.30[a]
	1.00	0.42 ~ 0.45[a]
Synthetic clay manufactured by Laponite RD® (after [2])	4.00	0.30[a]
	4.25	0.34[a]
	4.50	0.38[a]
Harbor mud (liquid consistency)		0.25 ~ 0.43[b]

Notes [a]Peak strength at 7 days by laboratory vane shear test; [b]Shear strength at liquid consistency from Schlue et al. [14]

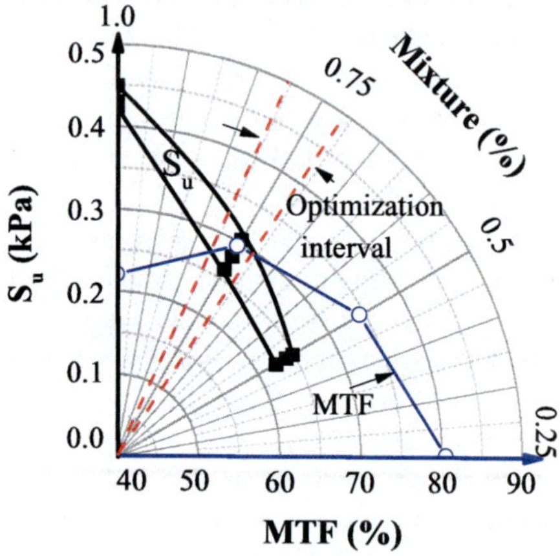

Fig. 4.8 Mixture optimization interval selection through S_u and MTF at 100 mm versus U10 powder mixture

at 100 mm versus U10 mixture. Optimization interval selection for mixture proportioning was determined through undrained shearing strength and MTF versus U10 powder mixture, through which the U10 powder mixture of 0.75% was chosen in the following tests. Although this study focuses on the geotechnical properties of 0.75% U10 powder mixture to describe the rationality, the mixture used in physical model tests can still be determined by the model scale and strength requirement.

The sensitivity of soil is an indication of the reduction in shear strength of soil when disturbed, *e.g.* when it is remoulded or when it is subjected to monotonic or cyclic loading. The soil sensitivity is defined as the ratio of the undrained shear strength of undisturbed soil to the undrained shear strength of the remoulded soil at the same water content, *e.g.* $S_t = S_u/S_{ur}$ (where S_u and S_{ur} is the undisturbed undrained shear strength, and remoulded undrained shear strength, respectively). The ratio of the peak undisturbed strength to the remoulded strength, as determined by the unconfined compression test, was used initially by Terzaghi [15] as a quantitative assessment of sensitivity. However, the strength of some clays is so low that

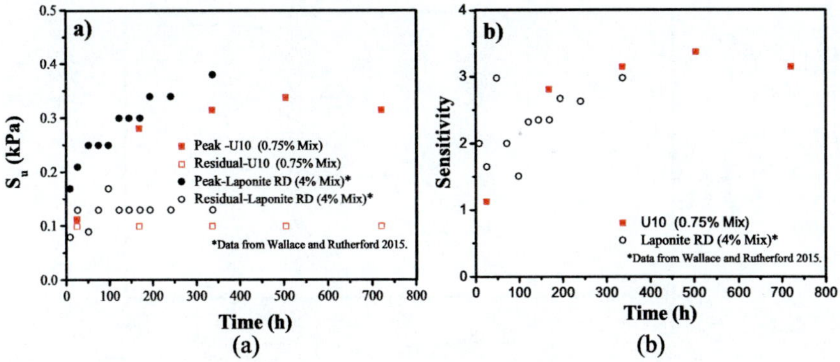

Fig. 4.9 The measured values by laboratory vane shear test: a Peak and residual strength versus time, b sensitivity versus time

unconfined compression test specimens cannot be used. Therefore, the laboratory vane shear test was used in this study. The remolded shear strength was measured by manually rotating after obtaining the peak shear strength and performing an additional laboratory test. Figure 9a shows that the peak strength and residual strength change versus time, while the corresponding change of sensitivity with time developing is shown in Fig. 9b. According to classification of Skemption and Northey [16] or Rosenquist [17], the sensitivity of synthetic clay ranges from low to medium sensitivity. Nevertheless, the synthetic clay belongs to low sensitive clay for other classification standards (*e.g.* Swedish (2004) or USA standard).

Previous studies on Norwegian marine clays [18], highly plastic sensitive clays [19] from Sweden, and two silty clays [20] from Saint Alban and Saint Louis, found that an increase in measured shear strength occurs with the increase in the delay between vane insertion and vane rotation, which is caused by the consolidation occurring in the soil surrounding the vane. The delay was introduced into the testing procedure to investigate the consolidation behavior of the synthetic clay post vane insertion. Samples with the 0.75% initial mixture were allowed to consolidate for 7 days before a laboratory vane was inserted, and the rotation onset was then delayed to 1080 min. Figure 4.10 shows an increase in S_u with the increase in delay time, indicating that excess pore water pressure generates in the synthetic clay during vane insertion, and is dissipated during the delay time that is comparable to that of natural soils. The ratio of $S_{u,measured}/S_{uo}$ is slightly higher than that of natural soils due to the strength increase under self-weight consolidation, but smaller than that of synthetic clay manufactured by Laponite RD®.

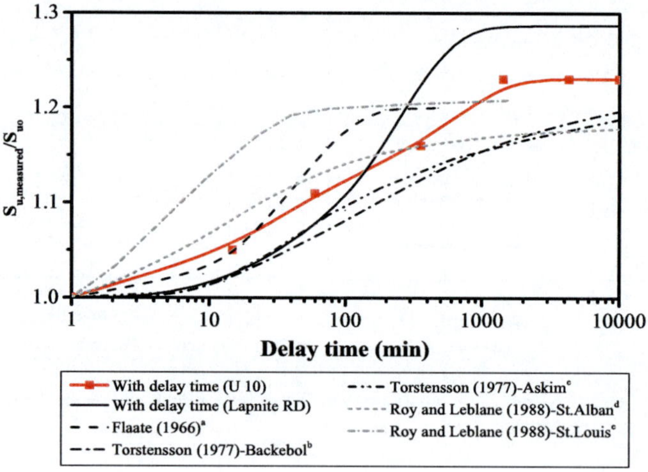

Fig. 4.10 Effect of delay time on measured S_u of 0.75% initial mixture samples where the vane was inserted after 7 days of self-weight consolidation

4.4.2 Consolidation

The compressibility parameters for transparent clay were determined for the 0.75% initial mixture by consolidation tests. The fully hydrated slurries were poured into consolidometer and allowed to self-weight consolidate for 7 days prior to specimen preparation. The samples size in the tests is 30 cm × 30 cm × 2 cm. The step load was set as 6.25 kPa by dead weight in the test. Dial gauge with an accuracy of 0.01 mm was used to monitor quantitatively the deformations of the samples. The top of consolidometer was covered by preservative films to prevent water evaporation. Duration of each load increment is sufficiently long enough (8 h in these tests) so that end of primary void ratio versus log effective vertical stress (e-logσ'_v) curves can be constructed. The pressure history (OCR = 1, 2, 3, and 4), loading steps, and time intervals for each specimen with 0.75% U10 powder mixture are shown in Table 4.2.

Time-settlement data from incremental load tests on both mixtures shows that consolidation behavior of the transparent clay is similar to that of natural clay, as shown in Fig. 4.11. The influence of OCRs to the synthetic clay samples was also shown in Fig. 4.12. Figure 4.12 shows that the synthetic clay of higher pre-consolidation pressure mixture has a smaller compressibility, detailed in Table 4.3.

The e-logσ'_v curves for synthetic transparent clays with 4 different OCR values are shown in Fig. 4.13. Through the e-logσ'_v curves, the coefficient of compression a_v obtained from secant slope of curves, and the modulus of compression E_s calculated by $(1 + e_0)/a_v$ is shown in Table 4.2. The time-deformation curve by Taloy (square root of time) method at 25 kPa surcharge was used as consolidation parameters. The coefficient of consolidation, c_v, in the consolidation test is 3.82×10^{-4} cm^2·s^{-1} for

Table 4.2 The compression testing conditions of synthetic transparent clay with 0.75% U10 powder mixture

Number	OCR	Loading steps (kPa)	Time intervals (h)
CT-1-1	1	6.25 → 12.50 → 18.75 → 25.00 → 31.25 → 37.50 → 43.75 → 50.00 → 56.25	8
CT-1-2	2	12.50 → 6.25 → 12.50 → 18.75 → 25.00 → 31.25 → 37.50 → 43.75 → 50.00 → 56.25	8
CT-1-3	3	18.75 → 6.25 → 12.50 → 18.75 → 25.00 → 31.25 → 37.50 → 43.75 → 50.00 → 56.25	8
CT-1-4	4	25.00 → 6.25 → 12.50 → 18.75 → 25.00 → 31.25 → 37.50 → 43.75 → 50.00 → 56.25	8

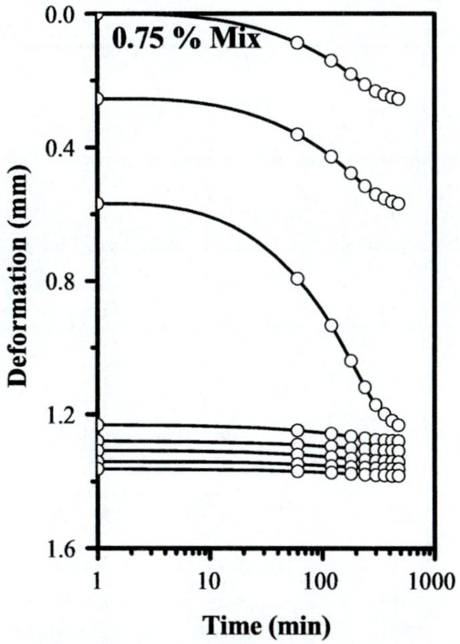

Fig. 4.11 Curves on deformation versus time for 0.75% initial mixture samples

0.75% initial mixture. The coefficient of consolidation of the natural mud ranges from 10^{-3} to 10^{-4} $cm^2 \cdot s^{-1}$ [21]. This result shows that the synthetic clay has a similar c_v to natural clay (Table 4.3).

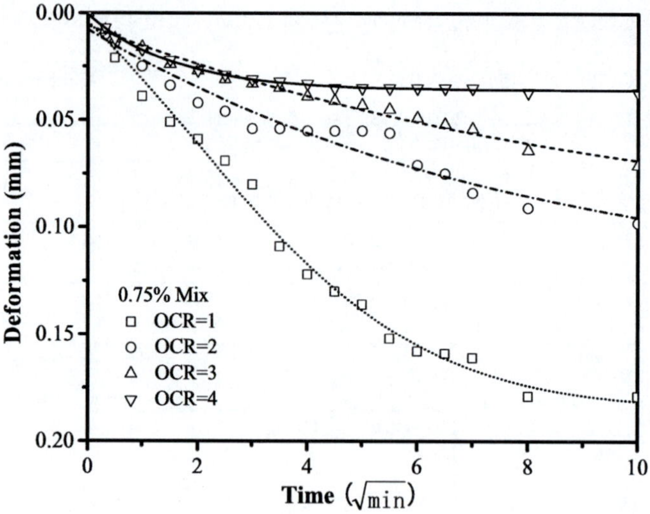

Fig. 4.12 Curves on consolidation versus time of synthetic transparent clay with different OCRs

Table 4.3 Compressibility parameters of synthetic transparent clay with 0.75% U10 powder mixture

OCR	1	2	3	4
a_v	0.19	–	–	–
E_s (kPa)	716	–	–	–
C_v ($\times 10^{-4}$cm^2·s^{-1})	3.82	2.62	2.98	1.03
C_c	6.40	18.40	12.32	31.81
C_r	–	5.36	23.41	14.86
C_s	–	1.30	6.20	7.99
C_a	–	0.17	0.30	1.40
C_r/C_c	–	0.29	1.90	0.46
C_a/C_c	–	0.009	0.024	0.044

Notes a_v is the coefficient of consolidation; E_s is the one-dimensional compression module; C_v is the coefficient of compression; C_c is the compression index; C_r is the recompression index; C_r is the rebound index; C_a is the secondary compression index

4.4.3 Hydraulic Conductivity

The hydraulic conductivity of the synthetic clay was determined by the falling head permeability test. Results obtained from the tests are shown in Figs. 4.14 and 4.15. Different self-weight consolidation time for synthetic clays was investigated in the test. The testing results with delaying time are shown in Fig. 4.15. It demonstrates that the consolidation process after 7 days has limited impact on hydraulic conductivity.

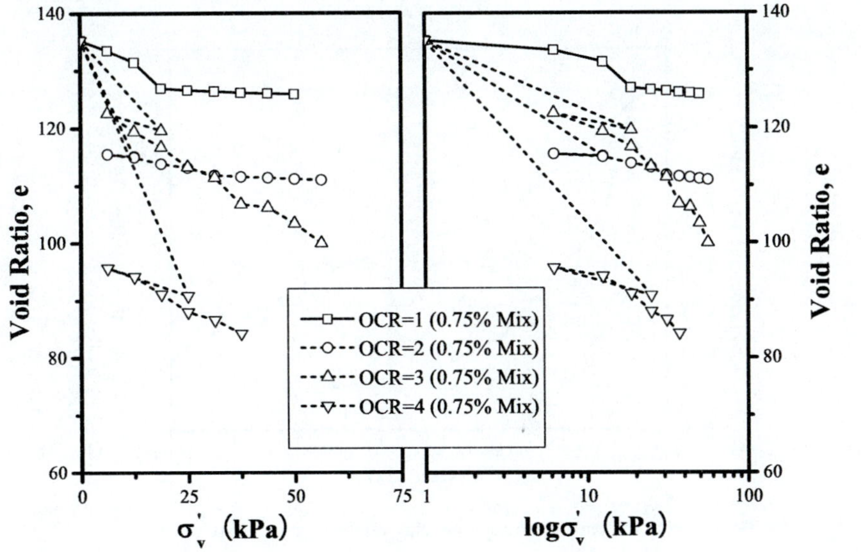

Fig. 4.13 Curves on e-σ'_v, e-logσ'_v of synthetic transparent clay with different OCRs

Therefore, a self-consolidation time of 7 days was taken as time in the permeameter test.

Permeameter tests of synthetic clay with different initial mixtures were also performed. The results of permeability tests show a linear decrease of permeability with the increases in U10 powder concentration. For the falling head tests, the synthetic clay was poured directly into the permeameter and allowed to self-weight consolidate for 7 days. Figure 4.15 presents the permeability data ranged from 2.1×10^{-7} to 1.2×10^{-6} cm/s, by way of indication, the permeability of natural clays is near

Fig. 4.14 *e-k* graph for the permeability test

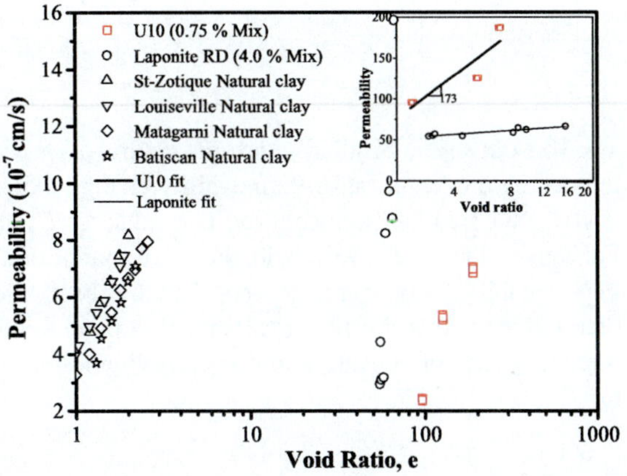

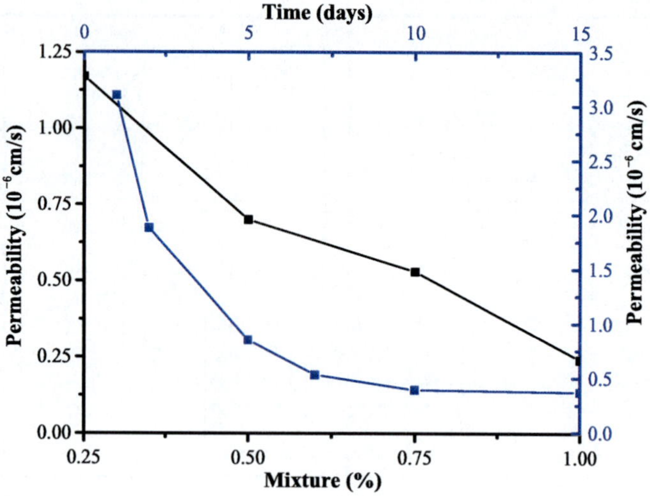

Fig. 4.15 Permeability of synthetic transparent clay versus initial mixture

Fig. 4.16 Thermal conductivity of synthetic transparent clay versus temperature

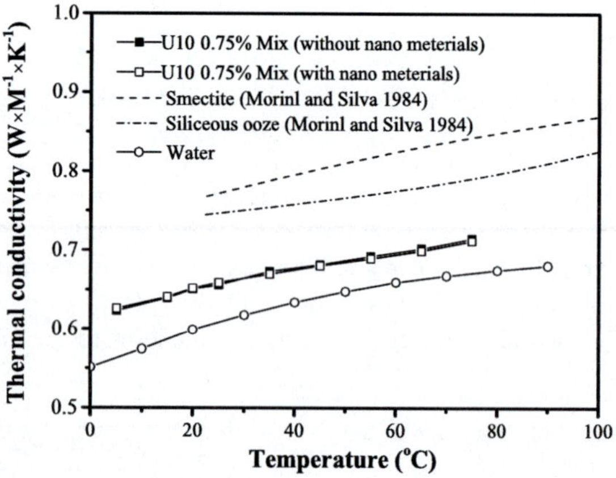

1×10^{-7} cm/s [26]. Similar with some natural clays, the permeability decreases with the decrease of void ratio. Permeability change index, $C_k = \Delta e/\Delta \log k$, is used to describe the trend of permeability. The values of C_k are between $0.93e_0$ and $1.73e_0$. For a given decrease in void ratio, the small change of C_k will result in large decreases in permeability, which is consistent with the observation by Terzaghi et al. [19]. The data indicates that the new synthetic transparent clay provide a new possible for modeling clay in consideration of consolidation characters in the physical model test.

4.4.4 Thermal Conductivity

The test system of thermal-physical properties includes KD2 pro (from Decagon Company, USA), synthetic clay samples and HD-5015 electric bath incubator. The measurement method uses an instantaneous linear heat source. The KD2 pro contains two probes (diameter is 1.288 mm, length is 3 mm, and the distance of two is 6 mm) with an accuracy of 0.001 °C. The electric bath incubator takes advantage of the loop water to maintain the temperature in the testing and realizes a control accuracy of 0.1 °C. The clay samples are set as a cylinder with 50 mm in diameter and 70 mm in height. After 6 days of hydration, the samples were covered by the preservative film and self-consolidate in an incubator for 24 h.

The synthetic clay samples under temperature ranges of 5 to 75 °C were tested by KD2 probe, and the results are shown in Fig. 4.16. The thermal conductivity ranges from 0.62 to 0.72 $W{\cdot}M^{-1}{\cdot}K^{-1}$. The change of thermal conductivity of Smectite and Siliceous ooze [22] versus temperature are also presented in Fig. 4.16. What's more, transparent clay mixed with nano materials was also shown in Fig. 4.16. The results indicated that the thermal property of transparent clay was not affected by tiny amount of nano materials. Similar with other materials, the thermal conductivity of synthetic clay nearly linear increases with the increasing of temperature. The values of thermal conductivity of synthetic clay are a little of smaller than those of Smectite and Siliceous ooze. Table 4.4 shows the comparison of the synthetic clay with the natural clays: Fairbanks bank [23], Clay loam soil [24], Kimmeridge clay and Oxford clay [25]. The thermal conductivity value of synthetic clay (0.62 ~ 0.71 $W{\cdot}M^{-1}{\cdot}K^{-1}$) is approach to that of the peat (0.49 $W{\cdot}M^{-1}{\cdot}K^{-1}$, after [23]), and Clay loam soil (0.16 ~ 0.71 $W{\cdot}M^{-1}{\cdot}K^{-1}$, after [24]). The thermal property datum in the test indicate that the synthetic clay can be considered as an alternative of natural soft clays.

Table 4.4 Thermal conductivity of synthetic transparent clay and natural clay

Materials	Thermal conductivity ($W \times M^{-1} \times K^{-1}$)	References
Water	0.55 ~ 0.68[a]	–
Synthetic clay (this paper)	0.62 ~ 0.71	–
Healy clay	1.30	[23]
Fairbanks peat	0.49[b]	[23]
Clay loam soil	0.16 ~ 0.71[b]	[24]
Kimmeridge clay	1.42 ~ 1.60	[25]
Oxford clay	1.54 ~ 1.60	[25]

Notes [a] Thermal conductivity with temperature of 0 ~ 90 °C; [b] Under saturated and undrained condition ($K_{sat(u)}$)

4.5 Discussions and Conclusions

Considering the above testing and analysis, the advantages of the new transparent clay manufactured by U10 (0.75% mixture as an example) could be drawn as: (1) geotechnical properties (*e.g.* thixotropy, consolidation characters, hydraulic conductivity and thermal conductivity) similar to natural clay or mud; (2) high optical transparency and (3) finer speckle effect if nano materials were mixed as speckle materials. However, there are also defects of the new transparent clay: the high void ratio (greater than 50) and low strength (0.26 to 0.30 kPa). As described in Wallace and Rutherford [2], while permeability data of clay (made by Laponite RD®) are higher than 10^{-7} cm/s at which soils are usually considered practically impervious or homogenous clays, the void ratios are also greater than 50. Moreover, the relationship between permeability and void ratio was analyzed. As a result of high void ratio, transparent clay present relative low shear strength. This strength characteristic limits its similarity to natural clay except mud or other very soft soil with extremely low strength. Therefore, the strength similarity problem should be considered for new transparent clay's possible application in the future.

The synthetic transparent clay manufactured by Carbopol® Ultrez™ 10 (referred to as U10) is capable of providing a new thought to model natural clays and investigate the internal problem in clay in view of basic parameters, although it does not match with a particular natural clay in all of geotechnical properties. The geotechnical properties of synthetic transparent clay can be modified to match required properties when used in a specific physical geotechnical model. The main conclusions about the synthetic clay are as follow:

(1) The transparency of the synthetic transparent clay measured by MTF method increases with the increasing of U10 powder mixture. The vane shear strength of the synthetic transparent clay decreases with the increasing of the usage of U10 powder mixtures. It is concluded that mixture of synthetic transparent clay of 0.75% U10 powder mixture is reasonable, given the change trend of strength and transparency.

(2) The peak strength of synthetic transparent clay (0.75% U10 powder mixture) by laboratory vane shear test during 7 days measures from 0.26 to 0.30 kPa, which is within the range of those of Harbor mud at liquid consistency (0.25 ~ 0.43 kPa, after Schlue et al. [14]). The obvious thixotropy occurrence is quite consistent with natural clay of low or middle sensitive with sensitivity of 2.8 ~ 3.4.

(3) The e-σ'_v and e-$\log\sigma'_v$ curves of the synthetic transparent clay are similar to those of the natural clay according to results of the compression tests. The permeability of transparent clay is comparable to that of natural clay with a permeability change index of 0.93 ~ 1.73 e_0. With 0.75% U10 powder mixture in this study, the permeability of synthetic transparent clay measures 5.3×10^{-7} cm/s; the coefficient of consolidation in the consolidation test is 3.82×10^{-4} $cm^2 \cdot s^{-1}$, which is similar to that of the natural mud (10^{-3} ~ 10^{-4} $cm^2 \cdot s^{-1}$, after Fang 2015).

(4) Thermal conductivity of synthetic transparent clay (0.75% U10 powder mixture) measures 0.62 ~ 0.72 $W \cdot M^{-1} \cdot K^{-1}$. Although thermal conductivity of synthetic transparent clay is relatively lower than that of majorities of natural clays, the value is close to that of the peat (0.49 $W \cdot M^{-1} \cdot K^{-1}$, after Côté [23]), and Clay loam soil (0.16 ~ 0.71 $W \cdot M^{-1} \cdot K^{-1}$, after [24]). Additionally, there is a similar temperature variation pattern for both synthetic transparent clay and natural clay.

References

1. Iskander M, Lai J, Oswald C, Mannheimer R (1994) Development of a transparent material to model the geotechnical properties of soil. Geotech Test J 17(4):425–433
2. Torstensson BA (1977) Time-dependent effects in the field vane test. In: Proc Int Symp on Soft Clay. Bangkok: Asian Institute of Technology, pp. 387–397
3. Iskander M, Liu J, Sadek S (2002) Transparent silica gels to model the geotechnical properties of sand. Int J Phys Modell Geotech 2(4):27–40
4. Iskander M, Liu J, Sadek S (2002) Transparent amorphous silica to model clay. J Geotech Geoenviron Eng 128(3):262–273
5. Sadek S, Iskander M, Liu J (2002) Geotechnical properties of transparent silica. Can Geotech J 39(1):111–124
6. Liu JY, Iskander M, Sadek S (2003) Consolidation and permeability of transparent amorphous silica. Geotech Test J 26(4):1–12
7. Shelton JT (2007) OMNI-max anchor development and technology. In: Proc of MTS/IEEE Oceans Conf., IEEE, Vancouver, Canada, September 29-October 4, Sponsor, Sponsor location
8. Beemer RD, Aubeny C (2012) Digital image processing of drag embedment anchors in translucent silicate gel", Proceeding of GeoManitoba (CD-ROM), Winnipeg, Canada, September 30-October 3, 2012. Canadian Geotechnical Society, Richmond, British Columbia, Canada
9. Chini CM, Wallace JF, Rutherford CJ, Peschel JM (2015) Shearing failure visualization via particle tracking in soft clay using a transparent soil. Geotech Test J 38(5):1–17
10. Bates B, Ancey C, Busson J (2014) Visualization of the internal flow properties and the material exchange interface in an entraining viscous Newtonian gravity current. Environ Fluid Mech 14(2):501–518
11. Islam MT, Rodríguez-Hornedo N, Ciotti S, Ackermann C (2004) Rheological characterization of topical carbomer gels neutralized to different pH. Pharm. Res-Dordr 21(7):1192-1199
12. Fresno MJC, Ramırez AD, Jiménez MM (2002) Systematic study of the flow behaviour and mechanical properties of Carbopol® Ultrez™ 10 hydroalcoholic gels. Eur J Pharm Biopharm 54(3):329–335
13. Black JA, Take WA (2015) Quantification of optical clarity of transparent soil using the modulation transfer function. Geotech Test J 38(5):588–602
14. Schlue BF, Moerz T, Kreiter S (2010) Influence of shear rate on undrained vane shear strength of organic Harbor mud. J Geotech Geoenviron Eng 136(10):1437–1447
15. Song Z, Hu Y, O'Loughlin C, Randolph MF (2009) Loss in anchor embedment during plate anchor keying in clay. J Geotech Geoenviron Eng 135(10):1475–1485
16. Skemption AW, Northey RD (1952) The sensitivity of clays. Géotechnique 3(1):30-53
17. Rosenqvist IT (1953) Considerations on the sensitivity of Norwegian quick-clays. Géotechnique 3(5):195–200
18. Flaate K (1966) Factors influencing the results of vane tests. Can Geotech J 3(1):18–31
19. Terzaghi K (1944) Ends and means in soil mechanics. Harvard University

20. Roy M, Leblanc A (1988) Factors affecting the measurements and interpretation of the vane strength in soft sensitive clays. Vane Shear Strength Testing in Soils: Field and Laboratory Studies, ASTM International
21. Fang HG (2005) The study on consolidation mechanism and model of structural soft soil in Shenzhen bay. Jilin University
22. Morin R, Silva AJ (1984) The Effects of high pressure and high temperature on some physical properties of ocean sediments. J Geophys Res: Solid Earth 89(B1):511–526
23. Côté J, Konrad JM (2005) A generalized thermal conductivity model for soils and construction materials. Can Geotech J 42(2):443–458
24. Ghuman BS, Lal R (1985) Thermal conductivity, thermal diffusivity, and thermal capacity of some Nigerian soils. Soil Sci 139(1):74–80
25. Bloomer JR (1981) Thermal conductivities of mudrocks in the United Kingdom. J Eng Geol Hydroge 14(4):357–362
26. Terzaghi K, Peck RB, Mesri G (1996) Soil mechanics in engineering practice, 3rd Edn. John Wiley & Sons Inc

Chapter 5
Transparent Rock

Abstract Fused silica blocks were manufactured by melting the highly pure silica dioxide above 1760°C and then quickly cooled down. By this process, it transforms the crystal form of silica dioxide into amorphous glass, which has more stable physical properties. This material has similar physical properties of brittle rocks, stable chemical properties and strong brittleness. In this chapter, the uniaxial compression tests and Brazilian tensile tests were conducted to characterize the deformation and strength behavior of the fused silica blocks, the size effect on the mechanical properties of this material was also investigated with the specimens being prepared with varying diameters and height-to-diameter (*H/D*) ratios. AE monitoring and camera monitoring were also performed during the loading of the specimens.

5.1 Introduction

The brittle failure behavior and crack growth mechanisms of rocks have been studied by many researchers through fracture mechanics or damage mechanics [14, 16, 24, 40]. However, most of these studies focus on the expansion of 2D cracks due to the opaque nature of the rocks. To study the 3D crack expansion in rocks, computed tomography (CT) or X-ray scanning methods were generally used in some research [20, 21, 33, 43]. These methods can accurately detect cracks inside rocks. However, the complexity and high cost of these technologies limit their use. Acoustic emission (AE) technology is another feasible and effective method to detect the initiation and propagation of internal cracks in rocks through the measurement of energy release during fracturing [22, 41]. However, the AE test cannot assist with the visualization of cracks induced in rock.

Alternatively, using substitute such as transparent materials of similar properties to replace natural rocks is also an effective method that can be used to study the mechanisms of propagation of 3D cracks. This method is usually simple, quick and inexpensive. Glass is a typical transparent material. It has been used to study the failure mechanisms in compression tests to infer what could occur in natural rocks [4, 5, 7, 34]. Frozen transparent Poly(methyl methacrylate) (PMMA) is another material suitable for representing natural brittle rocks. Crack extension from flaws, 3D crack

H. Zhao et al., *Transparent Soil Modelling Technique and Its Application*, Springer Tracts in Civil Engineering, https://doi.org/10.1007/978-981-19-6825-9_5

propagation and open hole instability mechanisms have been investigated by using PMMA [1, 12, 13, 39, 44, 47]. Nevertheless, there are limitations in the use of glass and PMMA. Glass has low transparency due to the high content of metallic mineral elements. PMMA should be frozen before the testing to increase its brittleness. In contrast, fused silica has high transparency, stable physical and chemical properties and strong brittleness. Thus, it is one of the most suitable materials for replacing brittle rocks or concrete to study fracture mechanisms.

Fused silica blocks have firstly been used to investigate the injection of chemicals into rock fractures [30] with success. Fused silica blocks were manufactured by melting the highly pure silica dioxide above 1760 °C and then quickly cooled down. By this process, it transforms the crystal form of silica dioxide into amorphous glass, which has more stable physical properties. This material has similar physical properties of brittle rocks, stable chemical properties and strong brittleness. However, the basic mechanical properties of this transparent material have not been studied, which limits its application to a certain extent.

The uniaxial compressive strength (UCS) and the deformation behavior and Brazilian tensile strength (BTS) are important design parameters in rock engineering. Therefore, uniaxial compression tests and Brazilian tensile tests were conducted to characterize the deformation and strength behavior of the transparent brittle material. Reviewing of the literatures shows that these mechanical parameters were significantly influenced by specimen size [6, 19, 28, 42, 44]. Thus, in the study, the size effect on the mechanical properties of this material was also investigated with the specimens being prepared with varying diameters and height-to-diameter (*H/D*) ratios. AE monitoring and camera monitoring were also performed during the loading of the specimens.

5.2 Testing Methodology

5.2.1 *Materials and Specimens*

Fused silica is the main material used to prepare the specimens. It comprises predominantly silicon dioxides (99.7%) and metal oxides such as ferric oxide, lithium oxide, aluminum oxide and potassium oxide (0.3%). The physical properties of the material were measured before the tests. Its density ranges between 2.203 and 2.215 g/cm^3, and the P-wave velocity ranges between 5892.06 and 6073.62 m/s.

To experimentally examine the strength and deformation behavior and the size effect, specimens were prepared with varying diameters and *H/D* ratios. In total, 9 specimens in 3 groups with different diameters were prepared according to the ISRM [18] standards for determining the UCS and 14 specimens in 5 groups with different diameters and *H/D* ratios were prepared according to the ISRM [18] standards for indirectly determining BTS. The details of the specimens are summarized in Tables 5.1 and 5.2, respectively. Generally, each group except group V for the

Table 5.1 Parameters and test results of specimens under uniaxial compression test

Types	No	*D*(mm)	*H*(mm)	*H/D* Ratio	$E_{t\,max}$ (GPa)	*v*	UCS (MPa)	Average strength (MPa)
I	U1-1	40.08	78.21	2.0	31.03	0.02	39.70	63.64
	U1-2	39.99	78.50		37.61	0.03	54.94	
	U1-3	39.91	78.59		37.61	0.03	96.28	
II	U2-1	45.01	89.55	2.0	47.96	0.01	54.71	68.77
	U2-2	44.65	89.33		59.77	0.02	72.49	
	U2-3	44.87	89.32		48.18	0.05	79.11	
III	U3-1	50.13	98.95	2.0	71.17	0.07	135.74	109.70
	U3-2	50.16	99.29		65.04	0.01	100.68	
	U3-3	49.71	99.37		64.6	0.03	92.67	

Notes: *D* is the diameter; *H* is the height; *H/D* ratio is the height-diameter ratio; $E_{t\,max}$ is the maximum tangent modulus; UCS is the uniaxial compressive strength

Brazilian tensile test has 3 specimens. Group V has only two specimens with the third one being accidentally damaged. The specimens in each group have the same diameters and *H/D* ratios. The cylindrical specimens of uniaxial compressive tests have the diameter of 40 mm, 45 mm and 50 mm, and *H/D* ratios of 2.0. The circular discs prepared for the Brazilian tensile test have diameters of 40 mm, 50 mm and 100 mm, and the *H/D* ratios of 0.3, 0.5 and 1.0, respectively.

5.2.2 Test Facilities and Processes

The uniaxial compressive and Brazilian tensile loadings were both achieved by utilizing an RMT-150C multifunction automatically rigid servo testing machine constructed at the Institute of Rock and Soil Mechanics, Chinese Academy of Sciences, as shown in Fig. 5.1. The schematic diagrams of the test facilities are illustrated in Fig. 5.2. The specimen was located between the two loading platens of equal size before the uniaxial compressive test. Vaseline was evenly applied to the contact surfaces between the ends of the specimen and the loading platens to reduce the end effect. The loading was applied by a hydraulic servo-system in a displacement control manner at 0.01 mm/s. During loading, 3 special extensometers were used to measure the axial and radial displacements and the loading stress was recorded by a linear variable differential transformer (LVDT). The extensometers were carefully calibrated before the tests. The axial extensometer contacted the upper loading platen, and the two lateral extensometers were placed radially at the central part of the specimen. The nonlinearity of the two lateral extensometers was controlled as small as possible during the installation to reduce measurement errors due to the asymmetry between the two lateral extensometers. The measurement ranges for the axial and

Table 5.2 Parameters and test results of specimens under Brazilian tensile test

Types	No	*D* (mm)	*H* (mm)	*H/D* Ratio	P_{max} (kN)	BTS (MPa)	Average strength (MPa)
I	B1-1	39.99	19.81	0.5	9.57	7.70	7.40
	B1-2	40.74	20.08		8.31	6.47	
	B1-3	40.55	20.13		10.28	8.04	
II	B2-1	49.97	15.32	0.3	11.29	9.40	6.92
	B2-2	50.62	15.14		8.74	7.29	
	B2-3	50.73	15.05		4.91	4.08	
III	B3-1	50.00	25.21	0.5	10.39	5.25	6.17
	B3-2	50.61	25.43		14.3	7.08	
	B3-3	50.60	25.23		5.58	2.79	
IV	B4-1	49.75	50.23	1.0	9.30	2.37	2.48
	B4-2	49.94	50.12		10.22	2.60	
	B4-3	50.75	50.13		9.89	2.47	
V	B5-1	98.47	48.15	0.5	39.03	5.23	5.93
	B5-2	99.28	51.04		52.72	6.63	

Notes D is the diameter; *H* is the height; *H/D* ratio is the height-diameter ratio; P_{max} is the load of failure; BTS is the Brazilian tensile strength

lateral extensometers are 0 ~ 5 mm and 0 ~ 2.5 mm, respectively. The accuracy of all the extensometers is 0.15%. Furthermore, three digital cameras were used to record the whole process of initiation and propagation of internal cracks from different directions as shown in Fig. 5.2a. The distance from the lens to the specimen center was maintained at 600 mm for all the cameras. This distance was chosen to ensure the accuracy and to reduce perspective distortions of the photos. The angles between the cameras were selected, according to the actual situation of the laboratory, to ensure that the cameras had a clear field of view during shooting. All the cameras were turned on before loading started. Since the cameras were not connected via a computer, the loading start time cannot be obtained directly from the videos. Thus, a special light was designed to hit the specimen when the loading started. The light, which indicates the beginning of loading, can be captured by the cameras at the same time.

For the Brazilian tensile tests, the specimens were placed in a loading box with a flexible upper platen and a fixed bottom platen. Two steel rods were placed into the upper and bottom platens, respectively, as shown in Fig. 5.2b. Then, the loading box was placed in the loading machine. The load and vertical displacement were measured during the test. The tests were stress-controlled at a rate of 0.01 N/s. The indirect tensile strength (σ_t) is then determined by Eq. (5.1) according to ASTM (2008) or ISRM (1978):

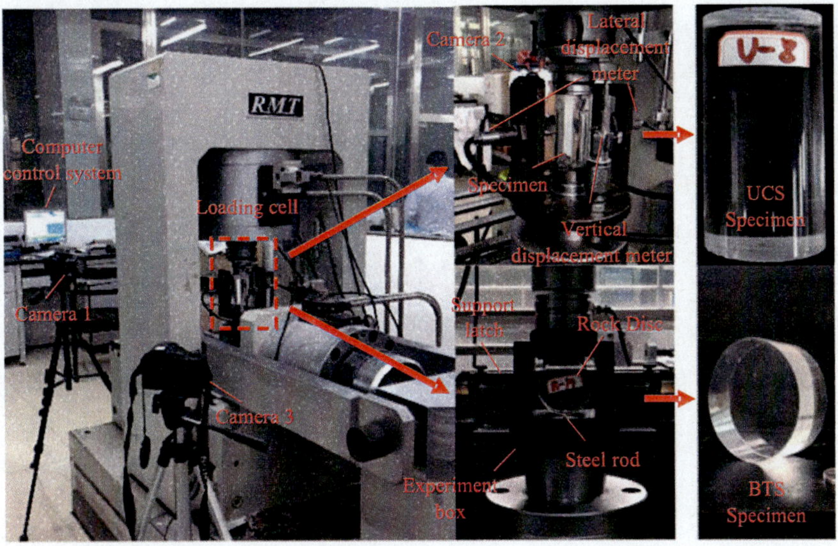

Fig. 5.1 The equipment and specimens of tests

$$\sigma_t = \frac{2P}{\pi DH} \tag{5.1}$$

where P is the load at failure, D is the diameter of the specimen, and H is the thickness of the specimen.

An AE sensor and a channel AE signal processing system were used to record the AE events that occurred in the specimens during the Brazilian tensile test. The AE data were used to supplement the measured data during the test. The AE sensor was in direct contact with the specimen. A small amount of Vaseline was applied to the interface between the specimen and the sensor to reduce the end effect of the platen. The location of the AE sensor is illustrated in Fig. 5.2b. The AE sensor has a resonant frequency of 125 kHz and its operating frequency ranges from 15 kHz to 1 MHz. The AE signals obtained from the AE sensors were amplified by a gain of 40 B to filter out the background noises.

5.3 Experimental Results and Discussions

5.3.1 Uniaxial Compression Test

The deformation behavior.

Figure 5.3 shows the measured axial stress-lateral strain curves for specimens with different diameters of the uniaxial compressive tests. It can be observed that

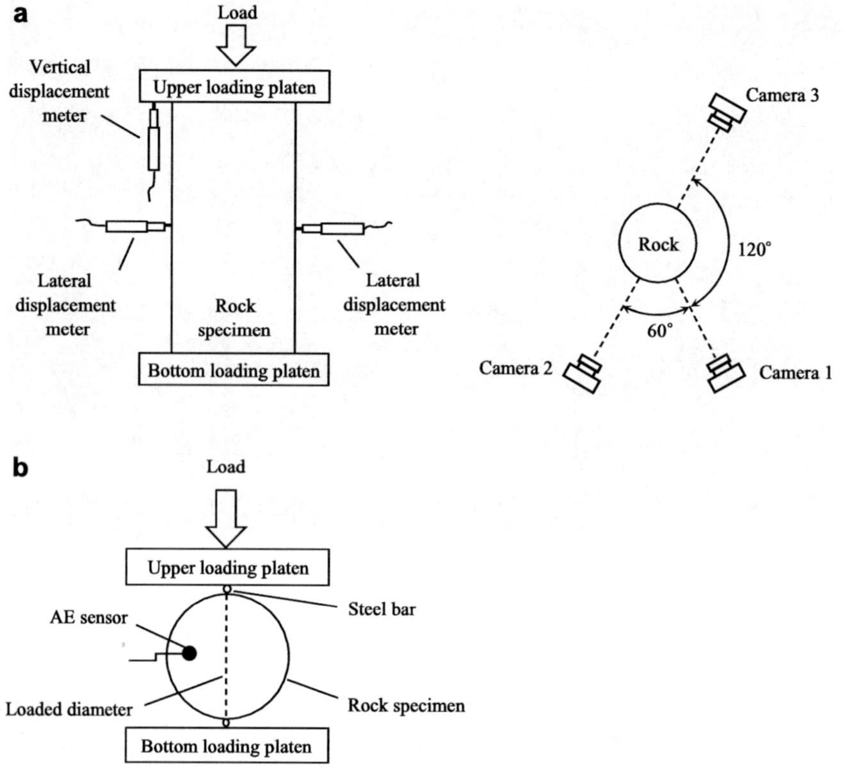

Fig. 5.2 The schematic of tests: **a** Uniaxial compression test; **b** Brazilian tensile test

the axial stress-lateral strain curves vary linearly with increasing axial stress, which is coincident for the different specimens. The average final lateral strains for each group are 0.3×10^{-3}, 0.2×10^{-3} and 0.3×10^{-3}, respectively. This indicates that the size effect on the lateral deformation properties of the specimens is negligible. The very small lateral deformation means that the transparent material has a much lower Poisson's ratio. The Poisson's ratio of the specimens is about 0.01 ~ 0.07, as summarized in Table 5.1. The axial stress-axial strain curves are also plotted in Fig. 5.3. Although these curves show the same trend, there is a certain diversity within each group of curves. The curves of the specimens in the same group do not completely coincide and the failure strength exhibits a certain variability. Two main reasons may have caused this diversity. One is invisible joint surfaces resulted from the sequence of quartz slurry pouring during the preparation of the specimens. The other is that small damages were produced in the specimens during polishing due to their fragility. Difference in the positions and degrees of damage lead to differences in the deformation behaviors and failure strength.

Specimen U2-3 is taken as an example to show the detailed experimental results, and its axial stress-axial strain curve is shown in Fig. 5.4. The failure process of intact

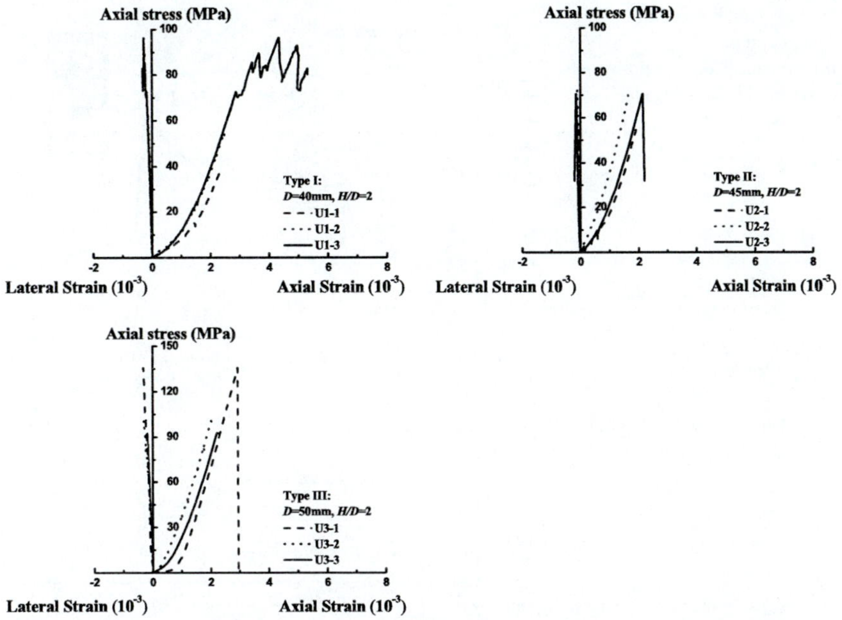

Fig. 5.3 Curves of axial stress versus axial and lateral strain under uniaxial compression tests

brittle rock material could be categorized into five stages, as proposed by a previous investigation [24]. Three stress threshold values, crack initiation stress σ_{ci}, dilatancy stress σ_{cd} and peak strength σ_f, are the dividing points of these stages. However, the axial stress-axial strain curve of U2-3 can only be divided into four regions: compaction, elastic deformation, stable crack growth and post-peak deformation. The crack pictures corresponding to those stress-dividing points are presented in Fig. 5.4. Initially, a nonlinear segment that represents the connection between the specimen and the instrument is obtained on the curve. Following this stage, a linear, homogeneous deformation occurs, which is defined as an elastic region. Cracks will not occur in these two stages, and the pictures corresponding to these stages in Fig. 5.4 also prove this inference. The elastic modulus (E) can be obtained from the elastic region. However, it is very difficult to define an unambiguous elastic modulus (E) of the material because the stress–strain curves of specimens are most nonlinear even when there is a short elastic region. Parameswaran and Jones [27] reported the same phenomenon for polycrystalline ice and found that the maximum tangent modulus (E_t) was close to the elastic modulus (E). Thus, the elastic modulus was also replaced with the maximum tangent modulus in this study.

The stage of stable crack growth begins until the stress value reached the initiation crack stress (σ_{ci}). The cracks can be easily observed in the specimen due to the transparency of the material. Thus, the initiation crack stress (σ_{ci}) was determined from the digital images recorded during the test. As a comparison, the stress was also

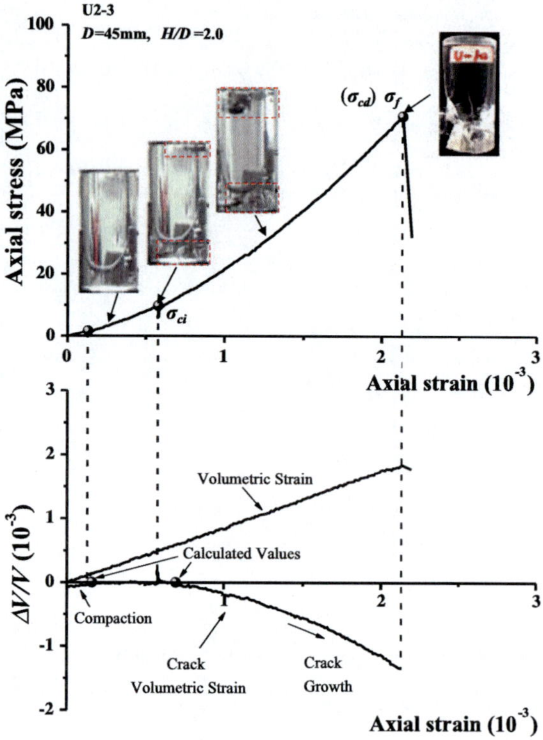

Fig. 5.4 Crack evolution process and the corresponding stress–strain curves of U2-3 for uniaxial compression test

calculated from the crack volumetric strain curve during the whole loading process according to the method proposed by Martin and Chandler [24]. The results are also shown in Fig. 5.4. It can be observed that the stress values obtained by the two methods are very close, which justified the feasibility of using a digital image to obtain the initiation crack stress (σ_{ci}). Brace et al. [8] found that the initiation crack stress (σ_{ci}) is approximately 30 ~ 50% of the peak strength. However, the initiation crack stress (σ_{ci}) of this material is only approximately 15% of the peak strength which may be attributed to its strong brittleness.

During the stage of stable crack growth, tensile and shear cracks appear sharply growing. The axial stress level (dilatancy stress σ_{cd}), at which the total volumetric strain reversal occurs, marks the beginning of the yield of the specimen. However, the reversal of the total volumetric strain is not observed from the volumetric strain-axial strain curve of the transparent specimen. The specimen directly failed after reaching the peak stress and did not yield. Thus, the dilatancy stress σ_{cd} is equal to the peak strength σ_f. Hallbauer et al. [15] and Martin and Chandler [24] noted that the dilatancy stress σ_{cd} indicates the most significant structural changes to the specimen, with the density of microcracks increasing. The dominant mechanism resulting in the reversal of the total volumetric strain is sliding along inclined surfaces, which causes large lateral deformation. However, due to the strong brittleness of this material, crushing damage occurred when its structure changed. Thus, the specimen lost bearing capacity beyond the dilatancy stress σ_{cd} and the axial stress decreased abruptly. Consequently, the failure process of this transparent material was categorized into four stages according to the stress threshold values. The same phenomenon was also proposed by Palchik [26] for carbonate rocks.

Hudson et al. [17] reported that the stress–strain curves of rocks are a function of the specimen size and shape. The typical stress–strain relationships of the specimens with different diameters are shown in Fig. 5a. The elastic moduli (*E*) for all the specimens are presented in Table 5.1 and Fig. 5b. It can be seen from Fig. 5.5 that the stress–strain relations of the specimens with different diameters have a similar trend. The elastic modulus increases linearly with increasing specimen diameter. To evaluate the similarity of the deformation behavior of fused silica with that of natural brittle rocks, the stress–strain relationships of charcoal gray granite, westerly granite and Beishan granite [36, 37, 45] are also plotted in Fig. 5a. These results are of typical brittle rocks from different regions and were obtained from specimens having a diameter of 50 mm and *H/D* ratio of 2, which are the same as U3-2. The transparent material has similar deformation trends with those of natural brittle rocks. It has similar magnitude of elastic modulus of charcoal gray granite and westerly granite, but slightly higher than that of the Beishan granite. However, the UCS of fused silica is only 74% (Beishan granite) and 50% (charcoal gray granite and westerly granite) of the natural brittle rocks.

Uniaxial compression strength

Weilbull statistics have been widely used to characterize the statistical variation in the fracture strength of materials such as rocks [2, 23], ceramics [31, 35], fibers [3, 25], and concrete [32, 46]. This method is based on a "weakest link" theory, which indicates that the most serious flaw in the material will control the strength.

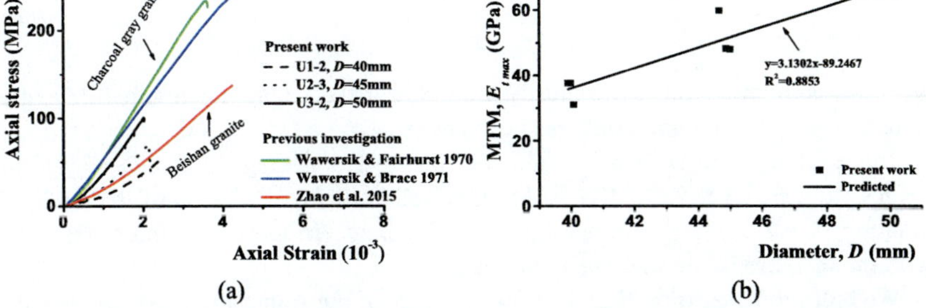

Fig. 5.5 **a** Comparison of the axial stress-axial strain curves of the specimens with different diameter and the previous investigation under uniaxial compression test; **b** curve of maximum tangent modulus (MTM) versus diameter of specimens

According to Weibull [38], the distribution of fracture strength with three parameters is

$$P = 1 - \exp\left[-\frac{1}{V_0}\int\left(\frac{\sigma - \sigma_u}{\sigma_0}\right)^m dV\right] \tag{5.2}$$

where P is the probability of failure of a specimen at σ; σ is the stress applied to the specimen; V is the volume of the specimen; V_0 is the reference volume; σ_0, σ_u and m are the scale parameter, location parameter and shape parameter (Weibull modulus), respectively; σ is not less than σ_u. σ_u is taken as the threshold stress below which failure does not occur and can be set as zero for brittle materials such as rocks and concrete [9]. This assumption allows Eq. (5.2) to be simplified as:

$$P = 1 - \exp\left[-\frac{1}{V_0}\int\left(\frac{\sigma}{\sigma_0}\right)^m dV\right] \tag{5.3}$$

For a constant *H/D* ratio (*H/D* = 2), Eq. (5.3) becomes:

$$P = 1 - \exp\left[-\left(\frac{l}{l_0}\right)^3\left(\frac{\sigma}{\sigma_0}\right)^m dV\right] \tag{5.4}$$

By taking the logarithm twice, Eq. (5.4) can be rewritten in a linear form:

$$\ln[-\ln(1 - P)] = 3\ln l - 3\ln l_0 + m\ln\sigma - m\ln\sigma_0 \tag{5.5}$$

Linear regression analysis is the widely used method to determine Weibull distribution parameters. The measured strength data are ranked in ascending order, and then, a probability of failure P_i is assigned to each strength value. The plotting position that is commonly used for P_i is given by a ranking estimator [9]:

$$P_i = \frac{i}{N + 1} \tag{5.6}$$

where P_i is the probability of failure for the ith ranked strength datum and N is the sample size, i.e., the number of specimens tested.

Table 5.1 lists the UCS for all the specimens. The average UCS of each group ranges from 63.64 MPa to 109.70 MPa, which is 44.80 ~ 74.20% of the UCS of granite (98 ~ 245 MPa). From Table 5.1, an increase in the UCS can be observed with the increase in the specimen diameter.

Weibull plots based on Eq. (5.5) for the UCS of the transparent specimens with different sizes are shown in Fig. 6a. In general, the strength of rocks is scattered in these plots. Several statistical methods were proposed to identify the "outlier", such as the least median square (LMS) method and the reweighted least squares (RLS) method [29], and these statistical methods were compared by Das and Basudhar [10,

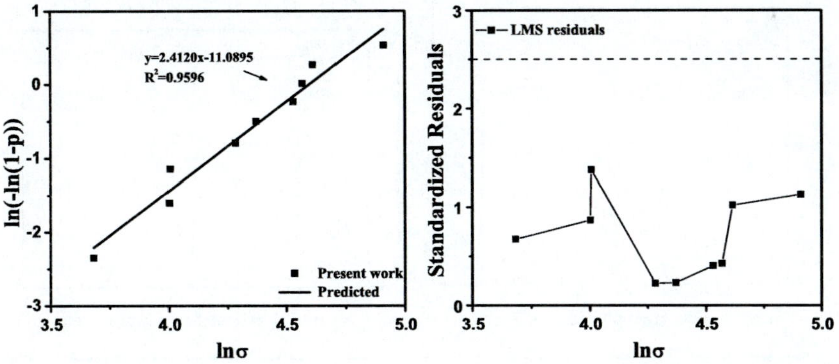

Fig. 5.6 Weibull statistical analysis of UCS: **a** Weibull plots of transparent specimens; **b** standardized residuals of UCS

11]. To evaluate whether there are "outliers" in the experimental data, the standardized residuals were calculated according to Das and Basudhar [10, 11], and the results are shown in Fig. 5.6b. It was noted that there were no "outliers" in these test results. Then, a linear regression analysis was performed on these data. The R^2 coefficient is 0.9596, which indicates a good degree of linearity. The Weibull modulus m is 2.4120, and the scale parameter σ_0 is 235.027.

5.3.2 *Brazilian Tensile Test*

5.3.2.1 The Deformation Behavior

The Brazilian tensile test is a simple indirect testing method used to measure the tensile strength of brittle materials such as concrete and rocks. Figure 5.7 shows the tensile stress–strain curves for all the test specimens. Notably, all the test specimens show similar stress–strain patterns, but with different peak vertical stresses at varying terminated vertical strains. This may be the same reason for the diversity of the compression stress–strain relationships as mentioned above.

Figure 5.8 presents the evolutions of the stress, AE counts and cumulative number of AE counts versus time for B3-2 and B3-3. The acoustic activity was detected at the beginning of the test. In the subsequent loading, almost no AE events were recorded. A sudden increase of AE counts can be noticed when the loading stress reaches approximately 90% of the peak stress, and then the failure of the specimen was obtained. Based on the characteristics of the AE response, the cumulative number of AE counts curves shows an 'S' shape. The stress of the first turning point of the cumulative number of AE counts curve matches the stress at the end of plastic deformation. It proved that local damage did occur.

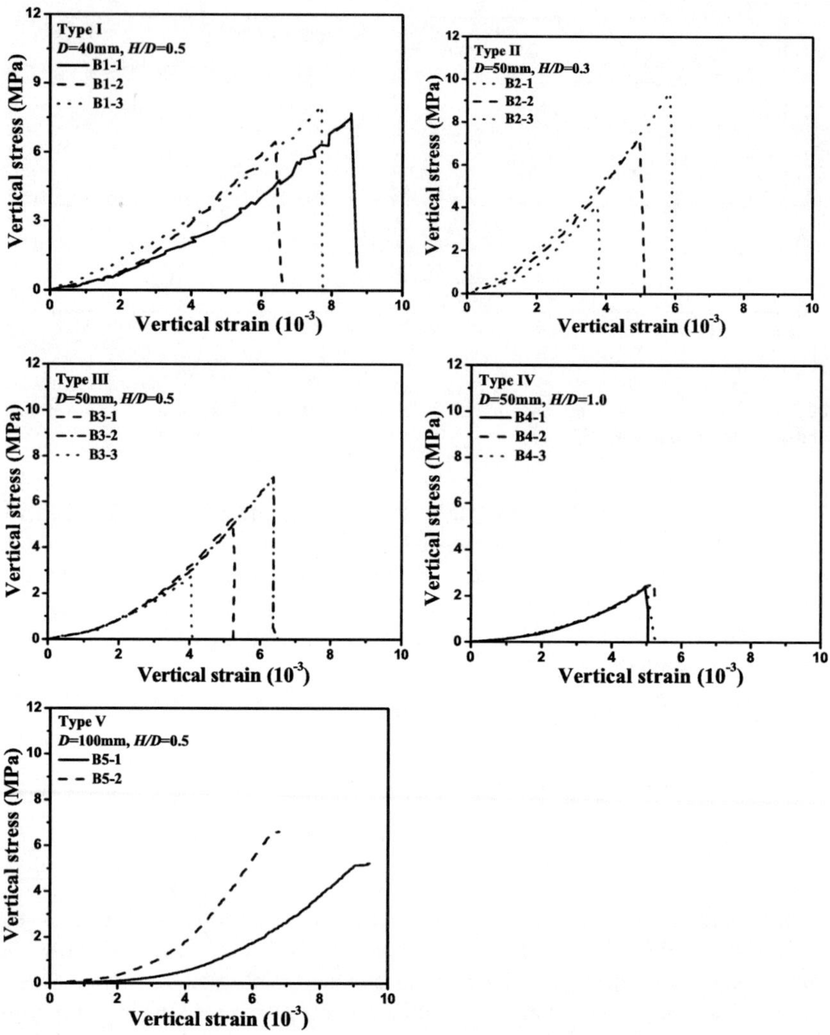

Fig. 5.7 Curves on stress versus strain under Brazilian tensile test

Although, the elastic modulus (*E*) cannot be reliably determined from the Brazilian tensile test due to the uneven stress distribution in the specimen, the stress–strain relation of the specimen can provide some insights into the specimen stiffness. Figure 9a plots the stress–strain relation of the specimens with diameters of 50 mm and different *H/D* ratios. The stiffness decreases about 60% with the *H/D* ratio of specimens increasing from 0.3 to 1.0. The peak vertical stress of the specimens decreases with the *H/D* ratio increasing. This part will be discussed further in the next section. The stress–strain relation of the specimens with the same *H/D* ratio

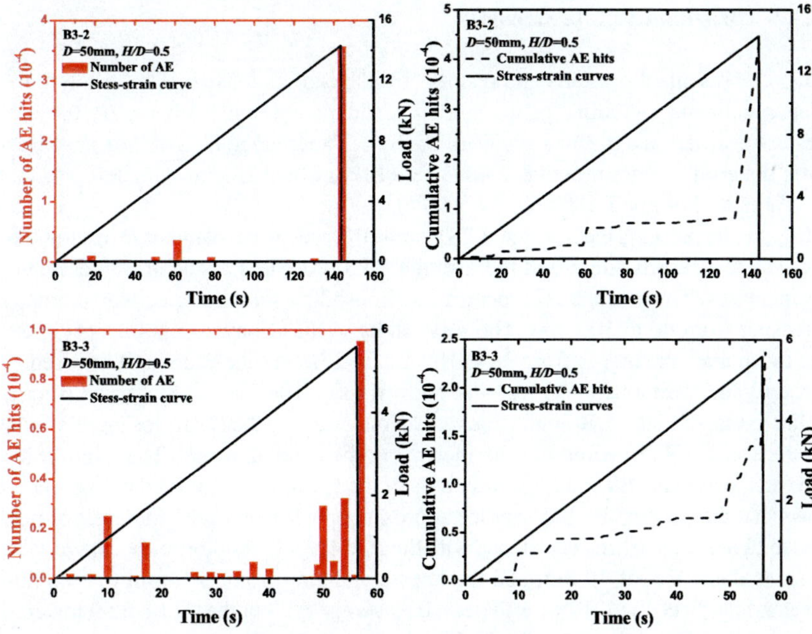

Fig. 5.8 Relations of strain with stress and AE counts under Brazilian tensile test

of 0.5, but different diameters are shown in Fig. 9b. It shows that the stiffness of specimens increases as the diameter increases and the peak vertical stress of the specimens decreases as the diameter increases.

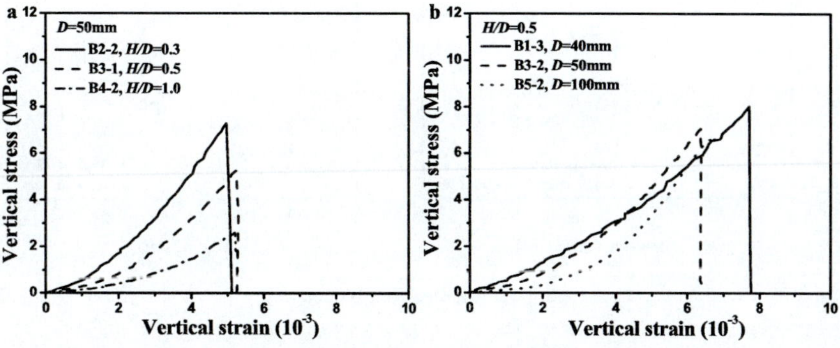

Fig. 5.9 Comparison of the stress–strain curves of the specimens with different diameter and the previous investigation under Brazilian tensile test

5.3.2.2 Brazilian Tensile Strength

Table 5.2 presents the details and results of the Brazilian tensile tests for the transparent specimens, including the specimen size (diameter *D* and thickness *H*), the BTS of each specimen, and the mean value of the BTS. The average BTS of the transparent brittle material is determined as 2.48 ~ 7.40 MPa, which is approximately 30.20 ~ 36.15% of that of granite (6.86 ~ 24.5 MPa).

Figure 10a presents the average BTS against the specimen diameter *D* in a log–log plot. The test results show that the average BTS decreases as the specimen diameter increases. The average BTS decreases almost 20% with the specimen diameter increasing from 40 to 100 mm. The test results of the Dionysos marble specimens with the diameters ranging from 38 to 100 mm and *H/D* ratios equal to 0.5 presented by Kaklis and Vardoulakis [19] are also shown in Fig. 10a. The BTS of fused silica is similar to that of marble. Both the transparent material and the Dionysos marble show that the average BTS decreases with increasing specimen diameter. The relationship between the average BTS and the thickness of the specimens is plotted in Fig. 10b. It shows that the average BTS decreases approximately linearly with increasing specimen thickness. This is in accordance with the results for granodiorite specimens with the same diameters of 49 mm and thicknesses ranging from 22.3 mm to 60.6 mm presented by Yu et al. [42]. Nevertheless, the average BTS of the transparent material are lower than that of the granodiorite. The average BTS of the fused silica specimens decreases by almost 65% with the thickness increasing from 15 to 50 mm. However, it only decreases 25% for granodiorite with the thicknesses increasing from 22.3 mm to 60.6 mm.

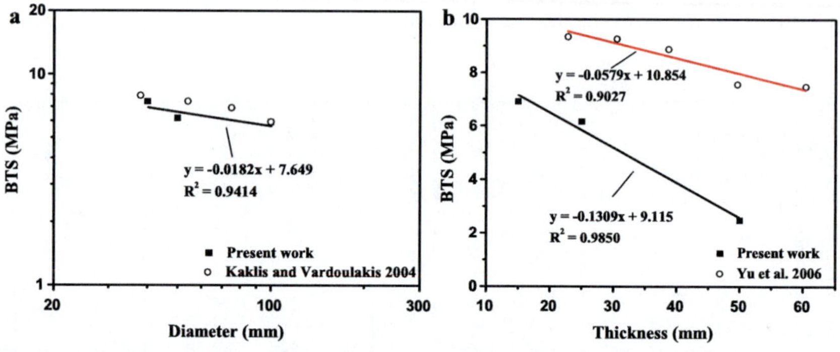

Fig. 5.10 Brazilian tensile strength versus: **a** diameter; **b** thickness

5.4 Conclusions

A transparent material that is suitable to represent brittle rocks or concrete is presented in this study. The strength (UCS and BTS) and deformation properties are studied through uniaxial compression tests and Brazilian tensile tests. The size effects on the strength and deformation properties are also investigated. Based on the analysis of the experimental data, the following conclusions can be made:

(1) The stress–strain relations from the compressive tests of the transparent fused silica are similar to those of brittle rocks. The elastic modulus increases with increasing diameter of the specimens. The results from Brazilian tensile tests show that specimens with a larger diameter or *H/D* ratio may have a lower stiffness.
(2) The UCS of the specimens range from 63.64 MPa to 109.70 MPa, which is 44.80 ~ 74.20% of that of the natural granite. The average tensile strength ranges from 2.48 MPa to 7.40 MPa, which is 30.20 ~ 36.15% of that of granite. The Unconfined compressive strength of the specimens increases with increasing diameter. However, the Brazilian tensile strength decreases as the specimen diameter and *H/D* ratio increase.

References

1. Adams M, Sines G (1978) Crack extension from flaws in a brittle material subjected to compression. Tectonophysics 49(1–2):97–118
2. Amaral PM, Fernandes JC, Rosa LG (2008) Weibull statistical analysis of granite bending strength. Rock Mech Rock Eng 41(6):917–928
3. Andersons J, Sparnins E, Joffe R, Wallstrom L (2005) Strength distribution of elementary flax fibres. Compos Sci Technol 65(3–4):693–702
4. Anderson O, Grew P (1977) Stress corrosion theory of crack propagation with application to geophysics. Rev Geophys Space Phys 15(1):77–104
5. Bahat D (1979) Theoretical considerations on mechanical parameters of joint surfaces based on studies on ceramics. Geol Mag 116(2):81–166
6. Bieniawski ZT (1968) The effect of specimen size on compressive strength of coal. Int J Rock Mech Min Sci 5:325–335
7. Brace WF (1963) A note on brittle crack growth in compression axis posed to occur when the most highly stressed the magnitude suggest that faults and extension fractures in brittle usually studied experimentally. J Geophys Res 68(12):3709–3713
8. Brace WF, Paulding BW, Scholz C (1966) Dilatancy in the fracture of crystalline rocks. J Geophys Res 71(16):3939–3953
9. Danzer R (1992) A general strength distribution function for brittle materials. J Eur Ceram Soc 10(6):461–472
10. Das SK, Basudhar PK (2006) Comparison study of parameter estimation techniques for rock failure criterion models. Can Geotech J 43(7):764–771
11. Das SK, Basudhar PK (2009) Comparison of intact rock failure criteria using various statistical methods. Acta Geotech 4(3):223–231
12. Germanovich LN, Dyskin AV (2000) Fracture mechanisms and instability of openings in compression. Int J Rock MechMin Sci 37(1–2):263–284

13. Germanovich LN, Salganik RL, Dyskin AV, Lee KK (1994) Mechanisms of brittle fracture of rock with pre-existing cracks in compression. Pure Appl Geophys 143(1–3):117–149
14. Haeri H, Shahriar K, Marji MF, Moarefvand P (2014) Experimental and numerical study of crack propagation and coalescence in pre-cracked rock-like disks. Int J Rock MechMin Sci 67:20–28
15. Hallbauer DK, Wagner H, Cook NGW (1973) Some observations concerning the microscopic and mechanical behaviour of quartzite specimens in stiff, triaxial compression tests. Int J Rock Mech Min Sci Geomech Abstr 10:713-726
16. Hoek E, Bieniawski ZT (1965) Brittle fracture propagation in rock under compression. Int J Fracture 1:137–155
17. Hudson J, Crouch S, Fairhurst C (1972) Soft, stiff and servo-controlled testing machines: a review with reference to rock failure. Eng Geol 6(3):155–189
18. ISRM (1978) Suggested methods for determining tensile strength of rock materials. Int J Rock Mech Sci Geomech. Astr 15(3):99–103
19. Kaklis K, Vardoulakis I (2004) An experimental investigation of the size effect in indirect tensile test on dionysos marble. National Congress on Mechanics
20. Keller A (1998) High resolution, non-destructive measurement and of fracture apertures. Int J Rock Mech Min Sci 35(8):1037–1050
21. Ketcham R (2010) Three-dimensional measurement of fractures in heterogeneous materials using high-resolution X-ray computed tomography. Geosphere 6(5):499–514
22. Koerner RM, McCabe WM, Lord AE (1981) Overview of acoustic emission monitoring of rock structures. Rock Mech Rock Eng 14(1):27–35
23. Lai Y, Li S, Qi J, Gao Z, Chang X (2008) Strength distributions of warm frozen clay and its stochastic damage constitutive model. Cold Reg Sci Technol 53(2):200–215
24. Martin CD, Chandler NA (1994) The progressive fracture of Lac du bonnet granite. Int J Rock Mech Min Sci 31(6):643–659
25. Okabe T, Takeda N (2002) Size effect on tensile strength of unidirectional CFRP composites-experiments and simulation. Compos Sci Technol 62(15):2053–2064
26. Palchik V (2010) Mechanical behavior of carbonate rocks at crack damage stress equal to uniaxial compressive strength. Rock Mech Rock Eng 43:497–503
27. Parameswaran VR, Jones SJ (1975) Brittle fracture of ice at 77 K. J Glaciol 14(71):305–315
28. Pratt HR, Black AD, Brown WS, Brace WF (1972) The effect of specimen size on the mechanical properties of unjointed diorite. Int J Rock MechMin Sci 9(4):513–529
29. Rousseeuw PJ (1998) Robust estimation and identifying outliers. In: Wadsworth HM (ed) Handbook of statistical method for engineers and scientists. McGraw-Hill, New York
30. Sui WH, Qu H, Gao Y (2015) Modeling of grout propagation in transparent replica of rock fractures. Geotech Test J 38(5):765–773
31. Thompson GA (2004) Determining the slow crack growth parameter and Weibull two-parameter estimates of bilaminate disks by constant displacement-rate flexural testing. Dent Mater 20(1):51–62
32. Tumidajski PJ, Fiore L, Khodabocus T, Lachemi M, Pari R (2006) Comparison of Weibull and normal distribution for concrete compressive strengths. Can J Civ Eng 33(10):1287–1292
33. Voorn M, Exner U, Barnhoorn A, Baud P, Reuschle T (2015) Porosity, permeability and 3D fracture network characterisation of dolomite reservoir rock samples. J Petrol Sci Eng 127:270–285
34. Wachtman JB (1974) Fracture of ceramics and glass. J Am Ceram Soc 57(12):509–519
35. Wang H, Aboushelib MN, Feilzer AJ (2008) Strength influencing variables on CAD/CAM zirconia frameworks. Dent Mater 24(5):633–638
36. Wawersik WR, Fairhurst C (1970) A study of brittle rock fracture in laboratory compression experiments. Int J Rock MechMin Sci 7:561–575
37. Wawersik WR, Brace WF (1971) Post-failure behavior of a granite and diabase. Rock Mech 3(2):61–85
38. Weibull W (1951) A statistical distribution function of wide applicability. J Appl Mech 18(1):293–297

39. Wong RHC, Huang ML, Jiao MR, Tang CA (2003) Crack propagation in brittle solid containing 3D surface fracture under uniaxial compression. Chin J Rock Mech Eng 22(sup1):2145–2148 (in Chinese)
40. Wong RHC, Lin P, Tang CA (2006) Experimental and numerical study on splitting failure of brittle solids containing single pore under uniaxial compression. Mech Mater 38(1–2):142–159
41. Yang SQ, Jing HW, Wang SY (2012) Experimental investigation on the strength, deformability, failure behavior and acoustic emission locations of red sandstone under triaxial compression. Rock Mech Rock Eng 45(4):583–606
42. Yu Y, Yin JM, Zhong ZW (2006) Shape effects in the Brazilian tensile strength test and a 3D FEM correction. Int J Rock MechMin Sci 43(4):623–627
43. Zabler S, Rack A, Manke I, Thermann K, Tiedemann J (2008) High-resolution tomography of cracks, voids and micro-structure in greywacke and limestone. J Struct Geol 30(7):876–887
44. Zeng L, Xiao LY, Zhang JH, Gao QF (2019) Effect of the characteristics of surface cracks on the transient saturated zones in colluvial soil slopes during rainfall. B Eng Geol Environ. https://doi.org/10.1007/s10064-019-01584-1
45. Zhao XG, Cai M, Wang J, Li PF (2015) Strength comparison between cylindrical and prism specimens of Beishan granite under uniaxial compression. Int J Rock MechMin Sci 76:10–17
46. Zhou J, Chen X (2013) Stress-strain behavior and statistical continuous damage model of cement mortar under high strain rates. J Mater Civ Eng 25(1):120–130
47. Zhu ZD, Lin HX, Sun YL (2017) An experimental study of internal 3D crack propagation and coalescence in transparent rock. Rock Soil Mech 37(4):913–928 (in Chinese)

Chapter 6
Pore Fluid

Abstract Transparent soil is a mixture of pore fluid and granular silica gel or amorphous silica gel. It becomes transparent because the pore fluid has the same refractive index as the transparent solid particles at the same temperature. Therefore, pore fluid is another important component to manufacture the transparent soil. Up to now, there are a few types of pore fluid available to manufacture transparent soils. In this chapter, the investigations on low viscosity pore fluid and the new pore fluid were presented.

6.1 Introduction

Transparent soil is a mixture of pore fluid and granular silica gel or amorphous silica gel. It becomes transparent because the pore fluid has the same refractive index as the transparent solid particles at the same temperature. Therefore, pore fluid is another important component to manufacture the transparent soil. Up to now, there are a few types of pore fluid available to manufacture transparent soils. Iskander et al. [1] discovered two matching pore fluids. The first pore fluid is a 50:50 blend by weight of a colorless mineral (Drakeol 35) and normal paraffinic solvent (Norpar 12). Its refractive index is 1.447 at 24 °C, and its viscosity is 5.0 cP at 24 °C. The density of the pore fluid is 0.804 g/cm^3. The second matching pore fluid is a mixture of calcium bromide and water. The refractive index and viscosity of the mixture are 1.448 and 3.6 cP, respectively, at 25 °C. Zhao and Ge [2] conducted an investigation on the low viscosity pore fluid. Recently, several new types of pore fluid were developed, such as n-dodecane and white oil, and the Krystol 40 and Puretol 7 (or Life Brand™ unscented Baby oil) solvent mixtures [3]. In this chapter, the investigations on low viscosity pore fluid and the new pore fluid were presented.

H. Zhao et al., *Transparent Soil Modelling Technique and Its Application*, Springer Tracts in Civil Engineering, https://doi.org/10.1007/978-981-19-6825-9_6

6.2 Low Viscosity Pore Fluid

To correctly and properly simulate the behavior of natural soil, the selected pore fluid needs to have the following properties: (1) Kinematic viscosity identical or close to that of water; (2) surface tension close to that of water; (3) incompressibility; (4) affordability; and (5) low and no interaction with silica gel, latex membranes, and the model container. Twelve types of chemical fluids were selected and tested in order to identify and determine the appropriate mixture of pore fluids to manufacture the transparent soil. Viscosity and refractive index tests were carried out on the selected fluids. The interaction of the candidate matching pore fluid with the latex membrane was also investigated.

6.2.1 Temperature Variation of the Viscosity and Refractive Index of the Potential Solvents

Twelve (12) different organic fluids were first selected as the possible fluid. These fluids were selected based on their refractive index and their viscosity. Low viscosity pore fluids with refractive index greater than 1.448 and less than 1.448 were selected. Those with refractive index greater than 1.448 were taken as base pore fluid; those with refractive index less than 1.448 were taken as the compensating pore fluid.

To identify the relationship between the refractive index and temperature, an AR200 Digital Handheld Refractometer manufactured by Reichert Analytical Instruments, Depew, NY, was used. It can measure a wide range of refractive indexes from 1.3300 to 1.5600 to an accuracy of $\pm$ 0.0001 nD. All solvents were put into a temperature chamber, and their refractive indexes were measured at six different temperatures (10, 15, 20, 25, 30, and 35 °C), which cover the possible range of operating temperatures.

The viscosity of the chemical solvents was also measured by Cannon–Fenske viscometer tubes at six different temperatures (10, 15, 20, 25, 30, and 35 °C) to examine their variation against temperature. Two different sizes of tubes were used. Size No. 25 was suitable for the viscosity ranging from 0.5 to 2 cSt, while size No.75 was suitable for the viscosity ranging from 1.6 to 8 cSt. All the tests were carried out following ASTM D445-09 [4]. The variation of refractive index with temperature for the tested pore fluids is shown in Fig. 6.1. The refractive index of all tested pore fluids showed a linear variation with respect to temperature within the tested temperature range but with different slopes. The variation of viscosity with temperature for the tested pore fluid is shown in Fig. 6.2. The viscosity of all the tested pore fluids also showed a linear variation with respect to temperature within the tested temperature range [5].

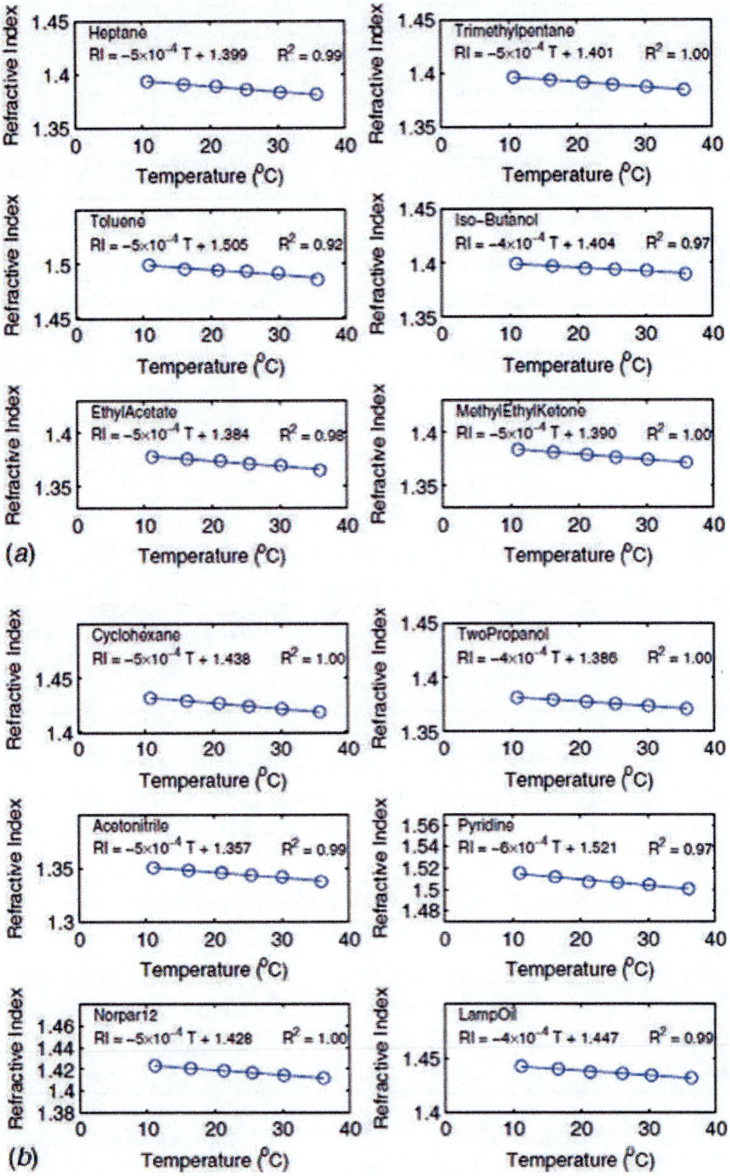

Fig. 6.1 The refractive index variation of the chemical solvents with the temperature

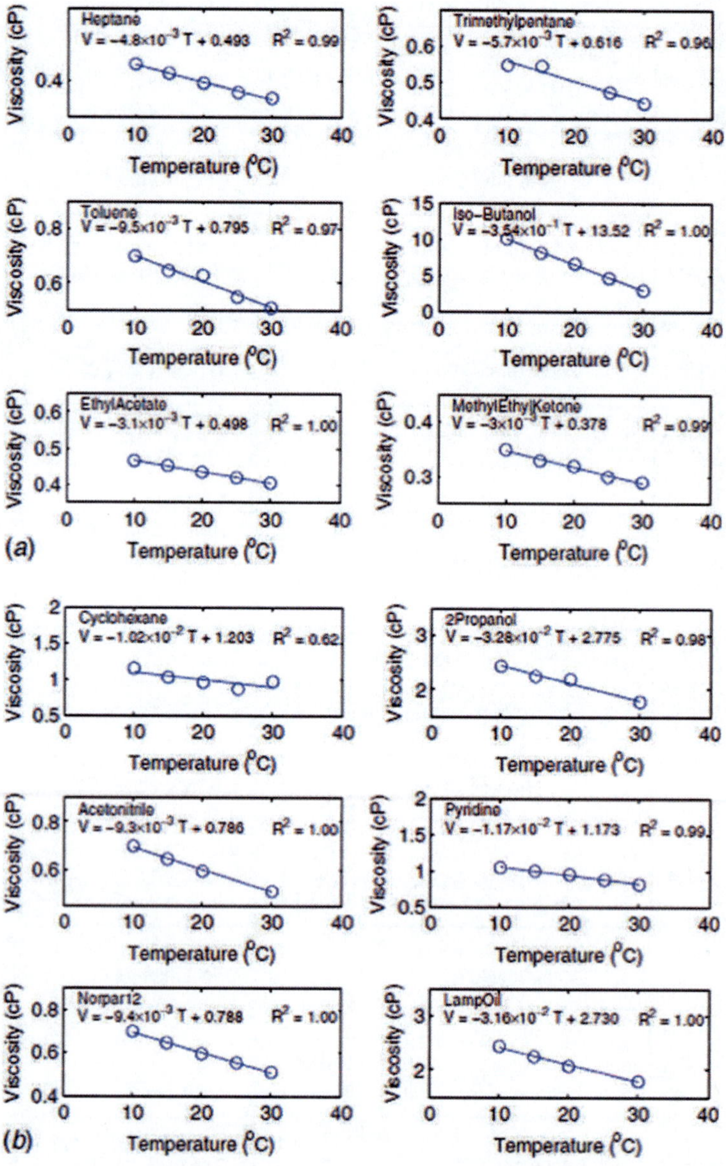

Fig. 6.2 The viscosity variation of the chemical solvents with the temperature

6.2.2 *Determination of the Matching Refractive Index of the Matching Pore Fluid*

The dry silica gel has a refractive index of 1.448 at 25 °C. When the pore fluid has the same refractive index, the silica gel will become transparent. Among those tested candidate pore fluids, those with a refractive index greater than 1.448 were selected as the base pore fluid. Those with a refractive index less than 1.448 were selected as the compensating pore fluid. In order to obtain a pore fluid with the matching refractive index to get the best transparency effects (can be clearly seen through) for the silica gel, 4 mL of the base pore fluid with a refractive index greater than 1.448 (toluene, lamp oil, or pyridine) was first put into the glass tube (diameter of 1.50 cm), and 3–4 g of dry silica gel was added into the tube. Then, the compensating pore fluid (acetonitrile, cyclohexane, ethyl acetate, n-heptane, iso-butanol, methyl ethyl ketone, 2 propanol, 2,2,4-trimethylpentane, and Norpar 12) with a refractive index less than 1.448 was added gradually until the best transparency effect was achieved. The volume ratio between the base pore fluid and compensating pore fluid was then recorded, and its refractive index was measured. During the test, the temperature was kept at 25 °C. Toluene was found to be a good base pore fluid, which a target refractive index can be obtained by mixing with 2 propanol, acetonitrile, ethyl acetate, methyl ethyl ketone, and iso-butanol. However, pyridine cannot make the dry silica gel become transparent when mixed with any other solvent. The pore fluid obtained from pyridine and another pore fluid could not get into the pores of the silica gel.

When the compensating pore fluid was added into the base pore fluid with dry silica gel inside, the interaction between the pore fluid and the silica gel is heat released, and the air bubbles inside the pores are coming out of the pore fluid. The amount of heat released can be felt through the temperature of the testing glass tube. When cyclohexane was added to the silica gel with toluene, air bubbles were observed coming out of the pores of the silica gel but not as quick as when ethyl acetate was added to the silica gel with toluene; no obvious heat was released. The behavior observed when adding the compensating pore fluid to the base solvent toluene was described in Table 6.1. When methyl ethyl ketone was added into the base pore fluid of toluene, a large amount of heat was released; the temperature rise of the tube was felt by the hand. There is an interaction happened between the silica gel and the pore fluid. This does not influence the transparency of the silica gel and the pore fluid. Further investigations need to be performed on the geotechnical properties of the transparent soil made with different types of pore fluid.

The second base fluid investigated in this research was lamp oil, which is a liquid petroleum product designed to burn cleanly in brass and glass oil lamps, torches, and lanterns. Lamp oil is from the same family such as kerosine and has been further processed and refined to reduce the harmful smoke, soot, and other pollutants as kerosine. The lamp oil investigated in this research is 99% pure liquid wax paraffin produced by Lamplight Farms, a Division of the W.C. Bradley Co. It was available at a local supermarket. The test results showed that lamp oil can be mixed with 2 propanol, acetonitrile, 2,2,4-trimethylpentane, ethyl acetate, iso-butanol, and cyclohexane to

Table 6.1 Test results of trial mixing to obtain matching refractive index pore fluid

Mixing fluid	Transparency effect	Volume ratio	Refractive index (25 °C)	Interactions observed
Toluene + cyclohexane	Poor	NA	NA	A moderate amount heat released
Toluene + 2 propanol	Good	1.07:1	1.448	None
Toluene + acetonitrile	Good	1.08:1	1.448	A large amount heat released
Toluene + 2,2,4-trimethylpentane	Poor	NA	NA	None
Toluene + ethyl acetate	Good	0.72:1	1.448	A large amount heat released
Toluene + methyl ethyl ketone	Good	0.95:1	1.448	A large amount heat released
Toluene + heptane	Poor	NA	NA	Turned into yellow color
Toluene + iso-butanol	Good	2.22:1	1.448	None
Toluene + Norpar 12	Poor	NA	NA	None

Note NA = not available

reach the refractive index of dry silica gel. The test results are summarized in Table 6.2. When the dry silica gel was added to the mixture of lamp oil and ethyl acetate, air inside the silica gel pores quickly came out, and a large amount of heat was released. It took about 1 h to make the silica gel become transparent. However, the transparency deteriorated when used to make a 6 cm thick transparent soil mass.

6.2.3 *Investigation on the Interaction Between the Pore Fluid and the Latex Membrane*

The low viscosity pore fluids tested in this research are organic chemical solvents. The degradation of a latex membrane when exposed to these chemical solvents has been a concern during triaxial tests. Iskander et al. [6, 7] observed the deterioration of the latex membrane when using a pore fluid of Drakeol 35 and Norpar 12. To investigate the interaction between a pore fluid and the latex membrane, a series of immersion tests was conducted following ASTM D543-06 [8]. It includes two measurements of changes in (1) weight and dimension and (2) mechanical properties of the latex membrane.

Table 6.2 Mixture of lamp oil and the compensation solvents with dry silica gel

Mixing fluid	Transparency effect	Volume ratio	Refractive index (25 °C)	Interactions observed
Lamp oil + cyclohexane	Poor	NA	NA	None
Lamp oil + 2 propanol	Good	1.0:15	1.434	Heat released
Lamp oil + acetonitrile	Poor	NA	NA	Separation found
Lamp oil + 2,2,4-trimethylpentane	Poor	NA	NA	None
Lamp oil + ethyl acetate	Good	1.28:1	1.434	Heat released; Transparency deteriorates with large sample
Lamp oil + methyl ethyl ketone	Good	5:1	1.434	None; Transparency deteriorates with large sample
Lamp oil + heptane	Poor	NA	NA	None
Lamp oil + iso-butanol	Poor	NA	NA	Heat released
Toluene + Norpar 12	Poor	NA	NA	None

Note NA = not available

Immersion Test—The latex membrane from Humboldt Mfg. Co. was selected as the testing material since it is commonly available at the laboratory for soil testing. The membrane was cut into a specimen size of 2.547.620.0636 cm^3 130.025 in.3 and weighed before immersion. Then, the test specimen was completely immersed into the pore fluid for 7 days, allowing the total surface area to be exposed to the fluid. After 7 days, the specimen was removed from the container and wiped dry; then its weight and dimensions were measured. Observations were made on the appearance of the surface for each specimen after exposure to the pore fluid. The membranes were air dried for 40 h, and then they were weighed and measured to examine any changes. Figure 6.3 shows the membranes after 40 h of air drying and the following observations were made. (1) The pore fluid of toluene and 2 propanol mixture did not change the color of the membrane, which is a sign of low interaction. (2) A moderate deterioration in color of the membrane immersed into the mixed fluid of toluene and the methyl ethyl ketone, mixed fluid of toluene and acetronitrile, mixed fluid of toluene and iso-butanol, and the mixed fluid of lamp oil and 2 propanol was observed. The membrane immersed into the mixed fluid of toluene and ethyl acetate, lamp oil and methyl ethyl ketone, and lamp oil and ethyl acetate had the strongest color deterioration. (3) The membrane immersed into the pore fluid of lamp oil and 2 propanol, toluene and iso-butanol, lamp oil and methyl ethyl ketone, and lamp oil and ethyl acetate curled. The immersion test data are shown in Table 6.3. The data show that after the immersion, the weight of the membrane greatly increased due to the absorption of the pore fluid. There was also an increase in the length, width, and thickness for the membrane immersed in all the solvents. Meanwhile, there was

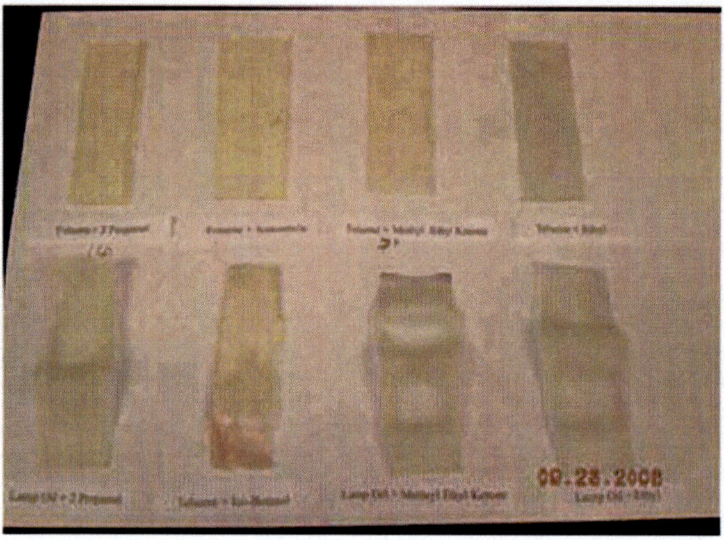

Fig. 6.3 The latex membrane 40 h after the immersion test

an increase in the density for the membrane immersed in the pore fluid of toluene and 2 propanol, toluene and acetronile, toluene and ethyl acetate, toluene and methyl ethyl ketone, and toluene and iso-butanol. However, the density for the membrane immersed in the pore fluid of lamp oil and 2 propanol, lamp oil and ethyl acetate, and lamp oil and methyl ethyl ketone decreased.

Tensile Test—The tensile test was performed on the latex membrane immediately after the immersion test. A low load capacity of 0.045 kN (10 lb) and high sensitivity load cell was used in performing the tensile test because the ultimate tensile strength of the latex membrane was very low. The deflection rate was taken as 0.085 cm/s (2 in./min). The original size of the specimen for the tensile test before the immersion was 3.81 cm (1.5 in.) in width and 76.2 cm (2.5 ft) in length. After the 7-day's immersion test, the specimen was taken out and wiped dry. Then, it was cut into 12.7 cm (5 in.) pieces for the tensile test. The tensile test for the membrane was set up as shown in Fig. 6.4. A piece of sand paper was put between the grip and the membrane to make sure that the grip can hold the tested membrane firmly. The membrane could be stretched to six times of its original length. The detailed information for the tested latex membrane specimens is listed in Table 6.4. The thickness and width of the specimen were measured after the immersion test before the tensile test. The length of the tested specimen was measured after the specimen set up on the testing machines. Because of the different grip length of each specimen, the tested specimen length is different as listed in Table 6.4. During the testing process, sample latex 28 slipped form the grip because the sample became very thin when stretched. Latex 14 was kinked during the test. These conditions do not affect the testing results since the membranes have been stretched to six times of the original size, beyond the range

Table 6.3 7-day immersion testing data on Humboldt membranes

Testing pore fluid	Time	T + 2 pro	T + Ace	T + EA	T + MEK	T + IB	LO + 2 pro	LO + EA	LO + MEK
Mass change (%)	7 days	+185.58	+53.17	+260.92	+220.95	+248.28	+262.96	+278.18	+282.76
Length change (%)	7 days	+35.06	+9.74	+46.75	+38.16	+41.56	+53.85	+57.69	+61.04
Width change (%)	7 days	+32.81	+20.69	+48	+38.46	+48.44	+60.00	+60.78	+67.33
Thickness change (%)	7 days	+53.85	0	+46.15	+46.15	+7.69	+53.85	+53.85	+57.69
Density change (%)	7 days	+3.48	+15.65	+13.70	+14.79	+52.58	−4.16	−3.05	−7.67
Mass change (%)	40 h	+10.26	+5.31	+7.35	+8.19	−1.89	+31.48	+32.73	+43.10
Length change (%)	40 h	0	−1.30	−1.30	0	+1.82	+6.41	+15.38	+12.99
Width change (%)	40 h	+1.56	+4.83	+4.00	0	−1.95	+8.00	+1.96	+5.58
Thickness change (%)	40 h	+7.69	−7.69	0	0	−15.38	0	0	+10.77
Density change (%)	40 h	+0.81	+10.26	+4.58	+8.19	+16.15	+14.41	+12.82	+8.30

Note T + 2 pro = Toluene + 2 propanol; T + Ace = Toluene + acetonitrile; T + EA = Toluene + ethyl acetate; T + MEK = Toluene + methyl ethyl ketone; T + IB = Toluene + iso-butanol; LO + 2 pro = Lamp oil + 2 propanol; LO + EA = Lamp oil + ethyl acetate; LO + MEK = Lamp oil + methyl ethyl ketone

Fig. 6.4 The experimental set up for the tensile test of the latex membrane

of interests. The stress–strain curve for the tested membranes is plotted in Fig. 6.5. Latex 0 was the original Humboldt membrane before any treatment. The test results showed that there was a strong degradation for the latex membranes immersed in the chemical solvents of the mixture of toluene and acetronile (latex 10), lamp oil and 2 propanol (latex 54), lamp oil and methyl ethyl ketone (latex 13), and lamp oil and ethyl acetate (latex 17) in both the tensile strength and elasticity. For the latex membrane immersed in the water (latex 1), the membrane's initial elastic modulus increased. However, the membrane lost most of its elasticity, being easily torn apart. The latex membranes immersed in the mixture of toluene and acetronile (latex 10), lamp oil and 2 propanol (latex 54), lamp oil and methyl ethyl ketone (latex 13), and lamp oil and ethyl acetate (latex 17) were broken at a low stress and also at a small strain level. The latex membranes immersed in the mixture of toluene and methyl ethyl ketone (latex 14) and toluene and 2 propanol (latex 28) only had a slight degradation. For the membranes immersed in the mixture of toluene and ethyl acetate (latex 37), there was a drop in the tensile strength of the membrane, but the elasticity of the membrane remained constant. Therefore, it was concluded that the mixtures of toluene and methyl ethyl ketone, and toluene and 2 propanol would be the best choices for the pore fluid.

The mixture of toluene and ethyl acetate was also recommended to manufacture transparent soil. The lamp oil was not recommended to make the transparent soil since it greatly reduced the tensile strength and elasticity of the latex membrane (latex 13, latex 17, and latex 54), as shown in Fig. 6.5.

Table 6.4 The tested latex membrane size and testing observation for the tensile test

Sample description	Sample number	Thickness (mm)	Width (mm)	Length (mm)	Testing description
L + 2 propanol	Latex 54	0.457	58.0	97.1	Membrane being torn apart
L + MEK	Latex 13	0.508	64.4	103.0	Slipped from the grip, nearly exceeded the testing device capacity
T + ace	Latex 10	0.356	39.0	65.1	Used sand paper to suppress slipping, nearly exceeded the extension limit, aborted
T + ethyl acetate	Latex 37	0.356	38.4	78.7	Sand paper used
T + 2 propanol	Latex 28	0.330	40.6	64.1	Membrane slipped from the grip
L + ethyl acetate	Latex 17	0.305	59.6	68.1	Membrane being torn apart, sand paper used
Water	Latex 1	0.381	38.1	70.1	Sand paper used, membrane slipped from the grip
T + MEK	Latex 14	0.381	39.5	56.3	Sand paper used, membrane kinked, load dropped
Original membrane	Latex 0	0.381	38.6	56.6	Sand paper used, membrane kinked

Note T + 2 pro = Toluene + 2 propanol; L + MEK = Lamp oil + methyl ethyl ketone; T + ace = Toluene + acetonitrile; T + ethyl acetate = Toluene + ethyl acetate; T + MEK = Toluene + methyl ethyl ketone

6.3 New Pore Fluid to Manufacture Transparent Soil

6.3.1 Introduction

Over the last two decades, there have been many attempts to produce transparent soil samples in the laboratory for visualization testing. All the chemicals and mixtures of pore fluids tested are listed in Tables 6.5 and 6.6. Mineral oil and organic chemicals

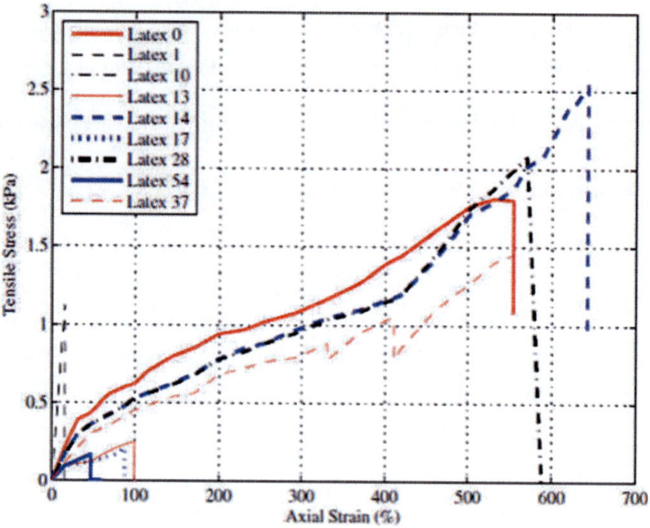

Fig. 6.5 The stress–strain curve for the tested membrane

have been the main type of oil-based pore fluids to satisfy the transparency requirement of the transparent soil sample. The mineral oil (such as, Drakeol 35, and ISO 15) and normal paraffinic (such as, Norpar 10–13) solvent mixtures were identified as pore fluid firstly by Iskander et al. [1], and were also widely used in application [9–17, 27]. The Krystol 40 and Puretol 7 (or Life Brand™ unscented Baby oil) solvent mixtures were also used as pore fluid [3, 11, 18]. Low viscosity of the pore fluid is another important feature in transparent soil manufacturing, since it allows air bubbles to escape the mixture more easily to produce a consistent sample. Two low viscosity pore fluids (toluene versus methyl ethyl ketone mixture, and toluene versus 2 propanol mixture) were introduced for this purpose [2]. However, these oil-based solvent has strong incompatibilities with the latex membranes, and solvent absorption led to swelling and softening of the membranes so that membranes could only be used once and double using in triaxial test [19, 20]. Water-based calcium bromide ($CaBr_2$) solution was also firstly described as pore fluid by Iskander et al. [1], while it is relatively little used in the application [10, 28]. High concentration of $CaBr_2$ solution is easy to precipitate causing the drop transparency of specimens [21], and it has low toxicity which is dangerous to researchers. Water-based sucrose ($C_{12}H_{22}O_{11}$) solution was proposed in recently years, and the mechanical properties were carried out [22, 23]. It shows that $C_{12}H_{22}O_{11}$ solution has high viscosity causing difficulty on large mass of transparent soil manufacture and high cohesion of specimen. Generally the oil-based solutions and $C_{12}H_{22}O_{11}$ solution have much higher viscosity than that of natural soil fluid (i.e. water), and strong incompatibilities with the latex membranes; the water-based calcium bromide solution is easy to precipitate

Table 6.5 All the chemicals which have been tested in previous literature

Chemical name	Organic	Tested by
Drakeol 35	Yes	[1, 6]
Norpar 12	Yes	
Krystol 40	Yes	[19]
Puretol 7	Yes	
Sucrose	Yes	[22, 23]
Toluene	Yes	[2]
Lamp oil	Yes	
Cyclohexane	Yes	
2 Propanol	Yes	
Acetonitrile	Yes	
2,2,4-Trimethylpentane	Yes	
Ethyl acetate	Yes	
Methyl ethyl ketone	Yes	
Heptane	Yes	
Iso-butanol	Yes	
Pyridine	Yes	
Norpar 12	Yes	
15#/3#mineral oil	Yes	[21]
$CaBr_2$	No	[1, 21]
$CaCl_2 \cdot 2(H_2O)$	No	[21]

causing the drop of transparency of specimens, and it is toxic which is hazardous to researchers.

In order to identify the appropriate mixture of pore fluids to manufacture the transparent soil, among the nineteen (19) chemicals/oils/water tested previously, seven (7) of them (4 organic chemical solvents: Heptane (H), Dodecane (D), Lamp oil (LO), 15# mineral oil (15MO) and three chemicals: $C_{12}H_{22}O_{11}$, $CaBr_2$, $CaCl_2$) were selected with another eight (8) new oil and thirty (30) new chemicals have been tested. The eight (8) new oil and thirty (30) new chemicals have been shown in the Tables 6.7 and 6.8 (3 and 4), respectively. Matching RI, low viscosity, and no (or small) interactions with latex membrane and with granular particles are the three key factors for choosing pore fluids. Hence, RI and viscosity versus temperature tests were carried out on the nineteen (19) suitable mixtures of pore fluids (15 oil-based and 4 water-based). Both interactions of the potential matching pore fluids with the latex membrane and with granular particles (silica gel and fused quartz) were also investigated.

Table 6.6 All the fluid mixtures which have been tested in previous literature

Mixture	Transparency	Geotechnical properties tested	Reaction with Membrane[b]	From
Drakeol 35 + Norpar 12	Good	Yes	None test	[6]
[a]Toluene + Methyl ethyl ketone	Good	No	+38.46%	[2]
[a]Toluene + 2 Propanol	Good	No	+32.81%	
Toluene + Acetonitrile	Good	No	+20.69%	
Toluene + Ethyl acetate	Good	No	+48%	
Toluene + Iso-butanol	Good	No	+48.44%	
Lamp oil + 2 Propanol	Good	No	+60%	
Lamp oil + Ethyl acetate	Good	No	+60.78%	
Lamp oil + Methyl ethyl ketone	Good	No	+67.33%	
Toluene + 2,2,4-Trimethylpentane	Poor	No	None test	
Toluene + Cyclohexane	Poor	No	None test	
Toluene + Heptane	Poor	No	None test	
Toluene + Norpar 12	Poor	No	None test	
Lamp oil + Cyclohexane	Poor	No	None test	
Lamp oil + Acetonitrile	Poor	No	None test	
Lamp oil + 2,2,4-Trimethylpentane	Poor	No	None test	
Lamp oil + Heptane	Poor	No	None test	
Lamp oil + Iso-butanol	Poor	No	None test	
Lamp oil + Norpar 12	Poor	No	None test	
Krystol 40 + Puretol 7	Good	Yes	None test	[19]
15# + 3#mineral oil	Good	No	None test	[21]
Sucrose	Good	Yes	None test	[22]

[a] Two suitable pore fluid with low viscosity suggested by Zhao et al. [2]
[b] Width change after 7 days immersion

6.3.2 *Pore Fluids Tested*

The selected pore fluid should have the same RI with granular particles (such as, 1.4475 for silica gel, and 1.4585 for fused quartz, etc.), incompressibility, kinematic viscosity identical or close to that of water, surface tension close to that of water, and low interaction with granular particles, latex membranes, and the model container [20]. Thirty-three (33) different water-based solutions were selected as the possible pore fluids. These pore fluids were selected based on the RI of solute, the solubility, the molecular weight, non-toxic or low-toxic, and affordable following

Table 6.7 The maximum RIs of 33 types water-based saturated pore fluids at temperature from 20 to 33 °C

No	Name	Formula	RI	No	Name	Formula	RI
1	Sucrose	$C_{12}H_{22}O_{11}$	1.5030	18	Ammonium acetate	CH_3COONH_4	1.4064
2	Potassium thiocyanate	KSCN	1.4870	19	Aluminum nitrate	$Al(NO_3)_3$	1.4054
3	Calcium bromide dihydrate	$CaBr_2$	1.4828	20	Sodium metaphosphate	$(NaPO_3)_6$	1.4054
4	Calcium chloride	$CaCl_2$	1.4710	21	Barium acetate	$(CH_3COO)_2Ba$	1.3949
5	Carbon tetrachloride	CCl_4	1.4558	22	Ammonium hydrogen sulfate	$NH_4H_2SO_4$	1.3924
6	Potassium pyrophosphate	$K_4P_2O_7$	1.4360	23	Ammonium chloride	NH_4Cl	1.3877
7	Magnesium chloride hexahydrate	$MgCl_2{\cdot}6H_2O$	1.4312	24	Sodium dihydrogen phosphate	NaH_2PO_4	1.3856
8	Zinc bromide	$ZnBr_2$	1.4296	25	Sodium chloride	NaCl	1.3789
9	Potassium triphosphate	$K_5P_3O_{10}$	1.4286	26	Ammonium sulfate	$(NH_4)_2SO_4$	1.3767
10	Calcium nitrate	$Ca(NO_3)_2$	1.4275	27	Potassium chloride	KCl	1.3752
11	Phosphorous acid	H_3PO_4	1.4265	28	Ammonium dihydrogen phosphate	$NH_4H_2PO_4$	1.3742
12	Potassium phosphate	K_3PO_3	1.4263	29	Sodium carbonate	Na_2CO_3	1.3727
13	Calcium chloride dihydrate	$CaCl_2{\cdot}2H_2O$	1.4244	30	Ammonium iron (III) sulfate	$NH_4Fe(SO_4)_2$	1.3724
14	Potassium carbonate	K_2CO_3	1.4192	31	Sodium thiosulfate	$Na_2S_2O_3$	1.3660
15	Potassium acetate	$C_2H_3KO_2$	1.4164	32	Potassium bicarbonate	$KHCO_3$	1.3579
16	Sodium benzoate	$C_6H_5CO_2Na$	1.4147	33	1-Dodecanesulfonic acid sodium salt	$NaC_{12}H_{25}SO_3$	1.3515
17	L-Sodium glutamate	$C_5H_8NO_4Na$	1.4103				

Note The classification following Lange's Handbook of Chemistry [24]

Table 6.8 Selected organic chemical solvents as pore fluids to manufacture transparent soil

Name	Abbreviation	Formula	Density (g/cm^3)
Ethanol	E	C_2H_6O	0.789
Isopropanol	I	C_3H_8O	0.786
Heptane	H	C_7H_{16}	0.680
Dimethvl silicone oil	DSO	$C_6H_{18}OSi_2$	0.963
Dodecane	D	$C_{12}H_{26}$	0.749
Lamp oil	LO	NA	0.780–0.800
Poly (200)	P	NA	1.125
Paraffin oil	PO	NA	0.877
Heat transfer oil	HTO	NA	1.020–1.080
15# mineral oil	15MO	NA	0.875
3-Propanetriol	3-P	$C_3H_8O_3$	1.261
10# transformer oil	10TO	NA	0.895

Note NA = uncertain

Lange's Handbook of Chemistry [24]. The maximum values of RIs of these water-based saturated solutions at normal laboratory temperatures (from 20 to 33 °C) are shown in Table 6.7. It shows that the RIs of all selected saturated solutions are larger than 1.3515, which is far smaller than those of solute (greater than 1.4500). Hence, in this paper, four (4) types of water-based solutions with relatively high RIs, such as, sucrose ($C_{12}H_{22}O_{11}$), potassium thiocyanate (KSCN), calcium bromide dehydrate ($CaBr_2 \cdot 2H_2O$), and calcium chloride ($CaCl_2$), were selected for measurement. Twelve (12) different organic chemical solvents were selected as the possible solvents. The names (contain the abbreviations), formula and densities details of twelve selected organic chemical solvents are shown in Table 6.8. Sixteen (16) mixture solvents were selected as the possible pore fluids. The details of selected oil-based solutions are shown in Table 6.9.

6.3.3 Apparatus and Procedures

To identify the relationship between the RIs and temperatures, an Abbe Refractometer and temperature control system (Fig. 6.6a) were used to measure the RIs of the pore fluids at 7 different temperatures (20, 25, 30, 35, 40, 45, and 50 °C) to examine their variation against temperatures following ASTM D1218-12 [25]. It can measure a wide range of RIs from 1.3000 to 1.7000 to an accuracy of 60.0001 nD, and can be precisely used in a wide range of temperatures from 0 to 70C. Cannon–Fenske viscometer tubes and temperature control systems (Fig. 6.6b) were used to measure the viscosity of the pore fluids at six different temperatures (20, 25, 30, 35, 40, and 45 °C) following ASTM D445-09 [4], which could cover the possible range of

Table 6.9 Selected oil-based solutions for RIs measures and immersion tests

Mixture	Abbreviation	RIs measures	Immersion tests
Heptane + Paraffin oil	H + PO	Yes	Yes
Dodecane + Paraffin oil	D + PO	Yes	Yes
Lamp oil + Paraffin oil	LO + PO	Yes	Yes
Ethanol + 3-Propanetriol	E + 3-P	Yes	Yes
Isopropanol + 3-Propanetriol	I + 3-P	Yes	Yes
Poly (200) + 3-Propanetriol	P + 3-P	Yes	No
Heptane + 15# mineral oil	H + 15MO	Yes	Yes
Dodecane + 15# mineral oil	D + 15MO	Yes	Yes
Lamp oil + 15# mineral oil	LO + 15MO	Yes	Yes
Dimethvl silicone oil + 10# transformer oil	DSO + 10TO	Yes	Yes
Heptane + 10# transformer oil	H + 10TO	Yes	Yes
Dodecane + 10# transformer oil	D + 10TO	Yes	Yes
Lamp oil + 10# transformer oil	LO + 10TO	Yes	Yes
Dodecane + Heat transfer oil	D + HTO	Yes	Yes
Lamp oil + Heat transfer oil	LO + HTO	Yes	Yes
Heptane + Heat transfer oil	H + HTO	Yes	No

operating temperatures. Six (6) different sizes of tubes were used. The diameters and measuring ranges of tubes are listed in Table 6.10.

Prior to the test, a defined mass of base liquid (RI greater than TRI) was added to the beaker and placed together on the electronic scale at the room temperatures of 25 °C. The electronic scale has an accuracy of ± 0.01 g and can be measured out of the quality of each droplet of liquid. The matching liquid (RI lower than TRI) was added dropwise to the beaker and the refractive index of the mixture was measured by adding a few drops of matching liquid each time until the RI of mixture reaches the TRIs. The quality of the matching liquid can be obtained by recording the change in the total mass of the liquid and the beaker. Then the quality ratios of base liquid to the matching liquid can be obtained at the temperature of 25 °C. The RIs and the viscosity of the pore fluids at seven or six different temperatures can be measured by using temperature control system next.

6.3.4 Results and Discussions

RIs versus concentrations (or quality ratios) with different temperatures

Among those tested candidate water-based solutions (see Table 6.7), four solutions ($C_{12}H_22O_{11}$, KSCN, $CaBr_2{\cdot}2H_2O$, and $CaCl_2$) were relative easy reach to TRI greater than 1.4585 in normal laboratory temperature (range 20–33 °C). The variations of

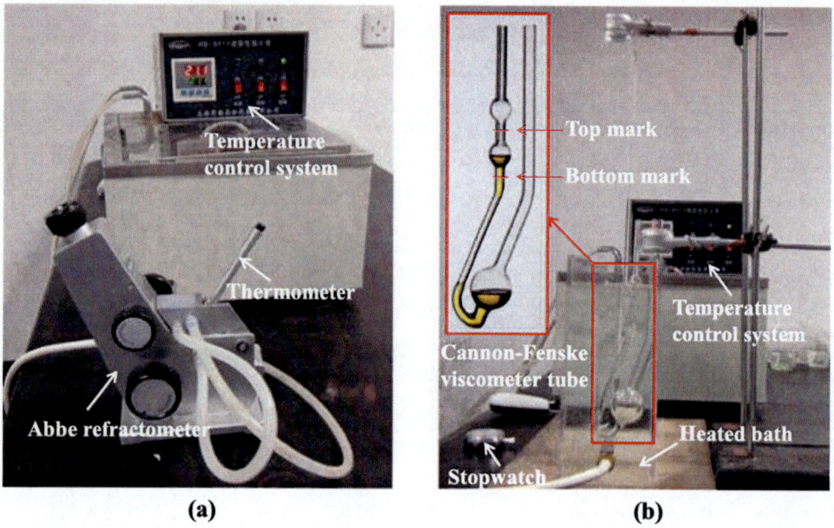

Fig. 6.6 Photos of test components: **a** BM-2WAJ Abbe refractometer; and **b** Cannon–Fenske viscometer tube

Table 6.10 The diameter and range of tubes for viscosity measurement used in this paper

Diameter (mm)	Range (cP)
1.92	240–1200
1.27	50–250
1.01	20–100
0.78	7–35
0.54	1.6–8.0
0.30	0.4–2.0

RIs with concentration for the four tested water-based solutions in room temperature (25 °C) are shown in Fig. 6.7. There are another two solutions shown in the same figure from the test results by Zhang et al. [21]. They are calcium chloride dehydrate ($CaCl_2{\cdot}2H_2O$) and calcium bromide ($CaBr_2$) solutions tested in slightly higher temperature of 30 °C. The RIs of all tested solutions increase fairly linearly with the increasing of solution concentration, though with different gradient. The variations of RIs with temperature for the four tested solutions together with calcium chloride dehydrate ($CaCl_2{\cdot}2H_2O$) and calcium bromide ($CaBr_2$) solutions tested by Zhang et al. [21] are shown in Fig. 6.8. The RIs of all tested solutions decrease fairly linearly with increasing temperature within the tested temperature range (from 15 to 50 °C), though with different gradient.

The variations of RIs with temperature for the twelve tested organic chemical solvents are shown in Fig. 6.9. The test results of three organic chemical solvents (Heptane (H), Dodecane (D), and Lamp oil (LO)) are shown in the same figure from

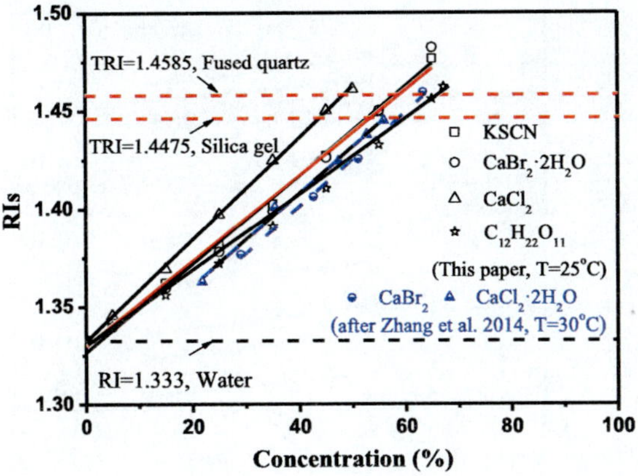

Fig. 6.7 RIs versus concentration for selected water-based solutions (T = 25 °C)

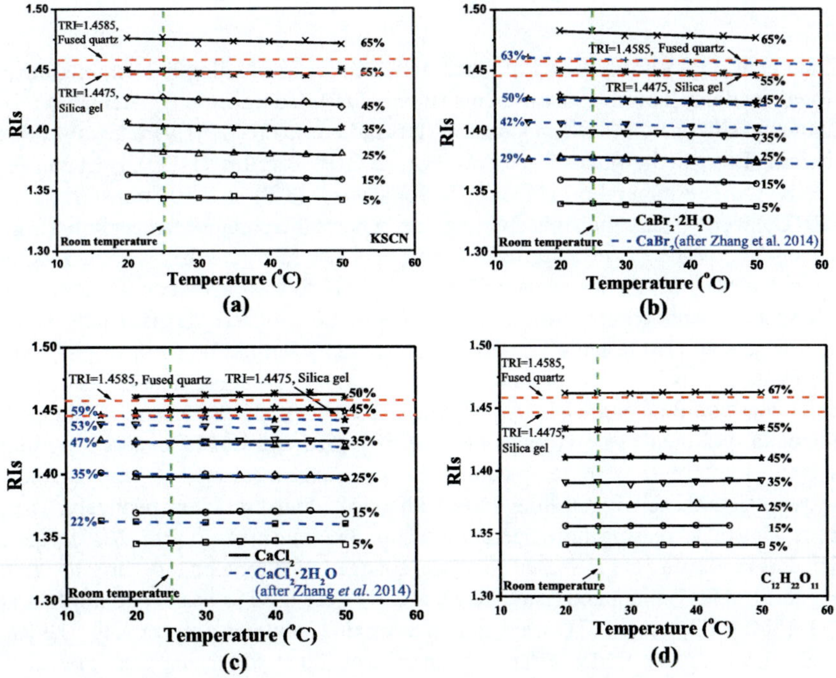

Fig. 6.8 RIs versus temperature of solutions with different concentrations in normal laboratory temperature (from 15 to 50 °C)

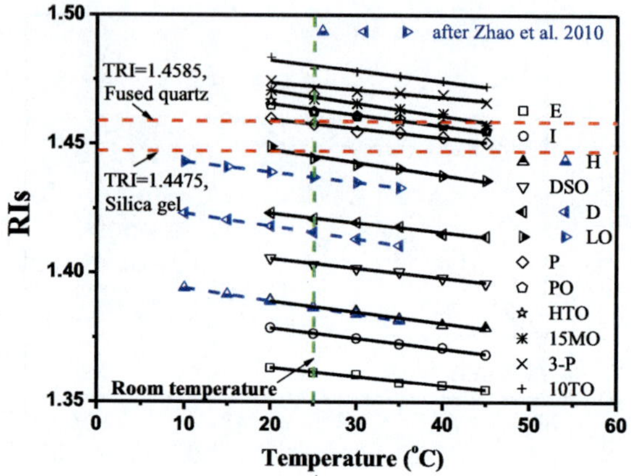

Fig. 6.9 RIs versus temperature of selected organic chemical solvents in normal laboratory temperature (from 10 to 45 °C)

Zhao et al. [2]. Setting the RI of silica gel (1.4475) as dividing line, among those tested candidate solvents, six solvents (Ethanol (E), Isopropanol (I), Heptanes (H), Dimethyl silicone oil (DSO), Dodecane (D), and Lamp oil (LO)) were less than this dividing line,and the other six solvents (Poly-200 (P), Paraffin oil (PO), Heat transfer oil (HTO), 15# mineral oil (15MO), 3-Propanetriol (3-P), and 10# transformer oil (10TO)) were greater than this dividing line in normal laboratory temperature (from 10 to 45 °C). The RIs of all tested solvents decrease fairly linearly with increasing temperature within the tested temperature range (10 to 45 °C), though with different gradient. In room temperature (25 °C), the variations of RIs versus quality ratios for the sixteen (16) tested oil-based solutions are shown in Fig. 6.10. In Fig. 6.10, the relationships are polar plotted, in which angular coordinates are quality ratios and radius coordinates are RIs. Among the sixteen (16) selected oil-based solutions, fifteen (15) of them are easy reaching TRIs of 1.4585 and 1.4475, expect 3-P and P solvents mixture. In room temperature (25 °C), mixture selected organic chemical solvents, lets those RIs reaching TRIs with 1.4585 and 1.4475, respectively; Then increasing or decreasing the temperatures of mixtures and measuring the RIs changes. The RIs of eleven (11) oil-based solutions changed with temperatures are shown in Fig. 6.11. Alkyl compound (contains D, and H) were used as basic pore fluids, and PO, 15MO, HTO, and 10TO were chosen as mixing solutions, respectively. The RIs versus quality ratios under different temperatures (20 to 45 °C) are polar plotted in Fig. 6.12a. Set LO as basic pore fluids, and mixing with PO, 15MO, HTO, and 10TO, respectively. The RIs versus quality ratios under different temperatures (20–45 °C) are polar plotted in Fig. 6.12b. Set 3-P as basic pore fluids, and mixing with I, E, and P, respectively. The RIs versus quality ratios under different temperatures (20–45 °C) are polar plotted in Fig. 6.12c. Similar with single solvent, the RIs of all tested oil-based solutions decrease fairly linearly with increasing temperature within the tested temperature range (20–45 °C), though with different gradient.

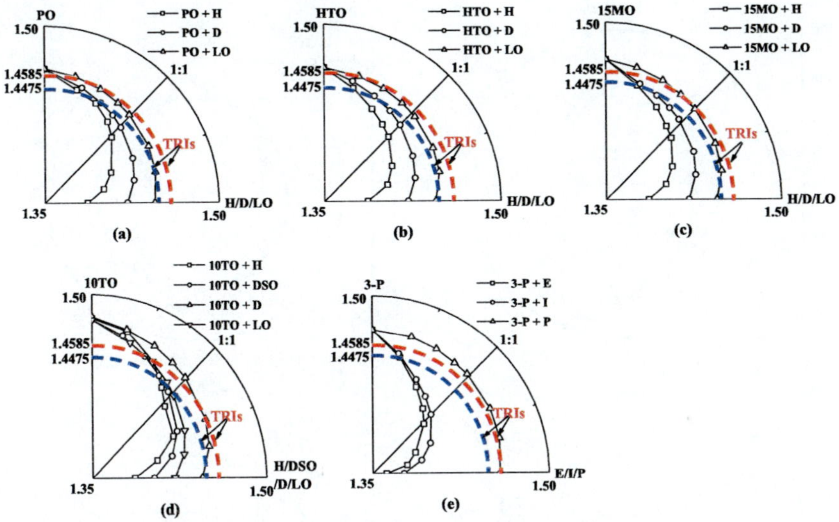

Fig. 6.10 RIs versus quality ratio for selected oil-based solutions (T = 25 °C)

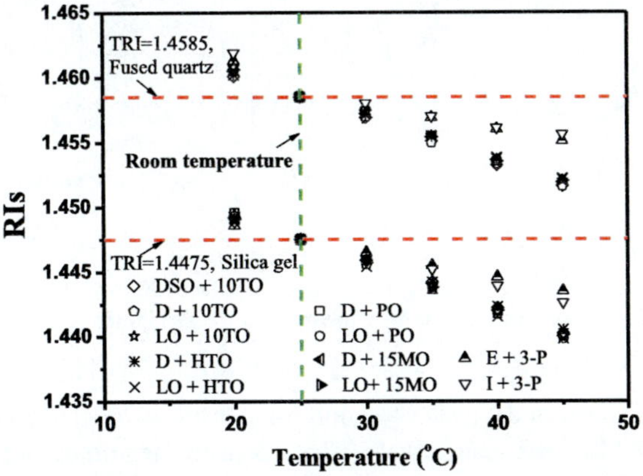

Fig. 6.11 RIs versus temperature for selected oil-based solutions

Viscosities versus concentrations (or quality ratios) with different temperatures

The comparative variations of viscosities with concentrations for the tested four water-based solutions in room temperature (25 °C) are shown in Fig. 6.13. The variations of viscosities with concentrations under six different temperature (20, 25, 30, 35, 40, and 45 °C) for each solution are shown in Figs. 6.14a–d, respectively. Set TRI of solutions with 1.4475 as an example, the viscosity of KSCN and $CaBr_2{\cdot}2H_2O$ solutions with a 1.4475 RI in 25 °C temperature is nearly only 1.2 and 1.9 cP, respectively, which is relative closer to that of water value (0.894 cP in 25 °C temperature);

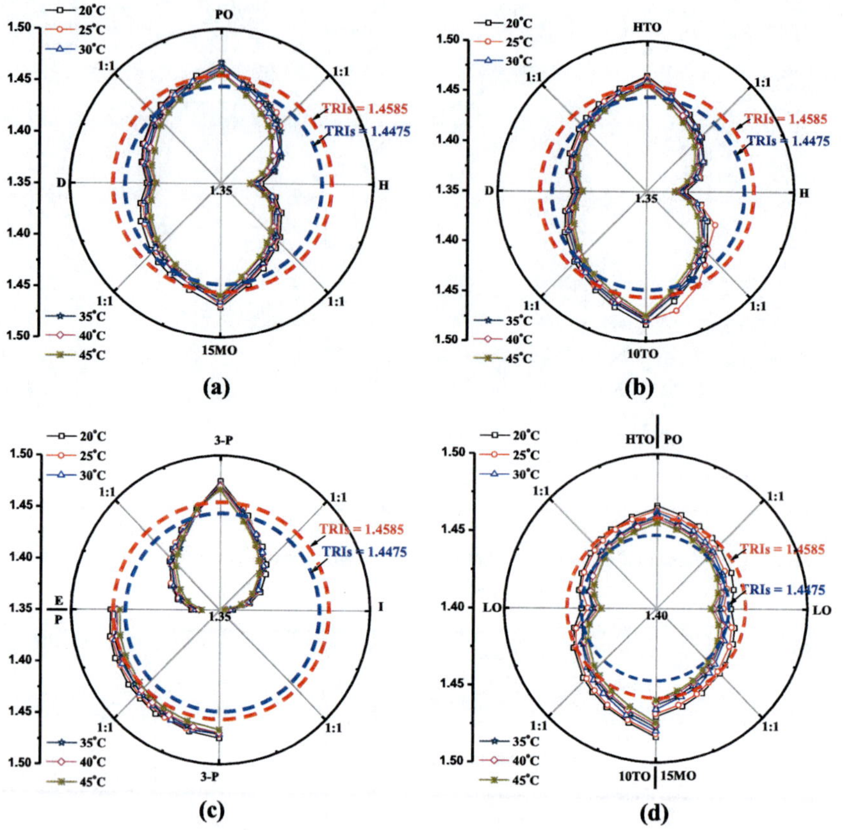

Fig. 6.12 RIs versus quality ratio with temperatures range from 20 to 45 °C for selected oil-based solutions: **a** alkyl compound mixture; **b** LO mixture; and **c** 3-P mixture

however, that of $CaCl_2$ and $C_{12}H_{22}O_{11}$ solution with a 1.4475 RI in 25 °C temperature is nearly 7.2, and 84.5 cP, respectively, which is nearly ten times and hundred times that of water value. The viscosity of three selected water-based solutions (KSCN, $CaBr_2{\cdot}2H_2O$, and $CaCl_2$) showed a non-linear (Eq. 6.1) variation with respect to concentration; while the viscosity of $C_{12}H_{22}O_{11}$ solution showed another non-linear (Eq. 6.2) variation with respect to concentration. The viscosity of solutions will be increased with concentration increasing obviously.

$$y = \frac{a}{1 + bx} \tag{6.1}$$

$$y = ab^{x} \tag{6.2}$$

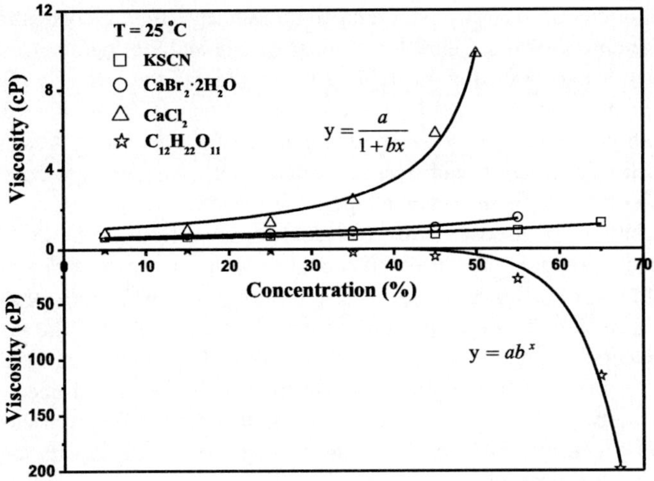

Fig. 6.13 Viscosity versus concentration for selected water-based solutions (T = 25 °C)

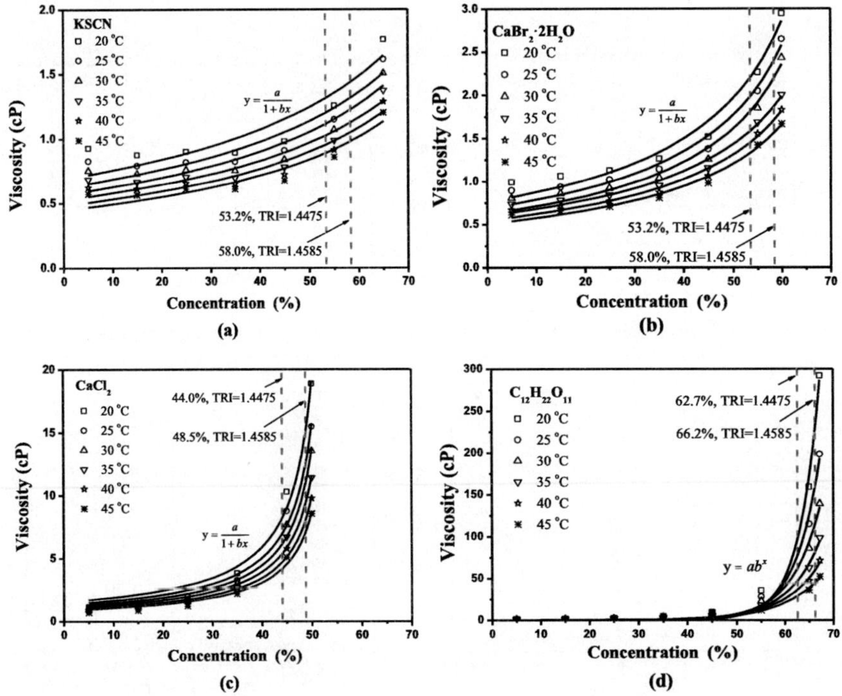

Fig. 6.14 Viscosity versus concentration with temperatures range from 20 to 45 °C for selected water-based solutions: **a** KSCN; **b** $CaBr_2{\cdot}2H_2O$; **c** $CaCl_2$; and **d** $C_{12}H_{22}O_{11}$

The variations of viscosity with temperature under different concentrations for each solution are shown in Fig. 6.15a–d, respectively and the coefficients of the fits for all equations are shown in the Table 6.11. In general, the viscosity of all tested water-based solutions decreased with temperature increasing. It showed nearly a linear variation with respect to temperature under low concentration (less than 45%), and a non-linear variation with respect to temperature under high concentration (larger than 45%) within the tested temperature range.

The variations of viscosity with temperature (range from 20 to 45 °C) for the twelve selected organic chemical solvents are shown in Fig. 6.16. There are two solvents (LO and H) shown in the same figure from the test results by Zhao et al. [2]. They are measured in temperatures range from 10 to 30 °C. The viscosity of majority selected solvents (except 3-P solvent) showed nearly a linear variation with respect to temperature. The viscosity of 3-P solvent showed a non-linear (Eq. 6.3) variation with respect to temperature and the coefficients of the fits are shown in the Table 6.11. In general, the viscosity of solvents can be decreased with temperature increasing with the tested temperature range. Except 3-P solvent, the viscosity of majority solvents is less than 100 cP; the viscosity of these five solvents (D, E, LO, H, and I) is less than 3 cP, especially.

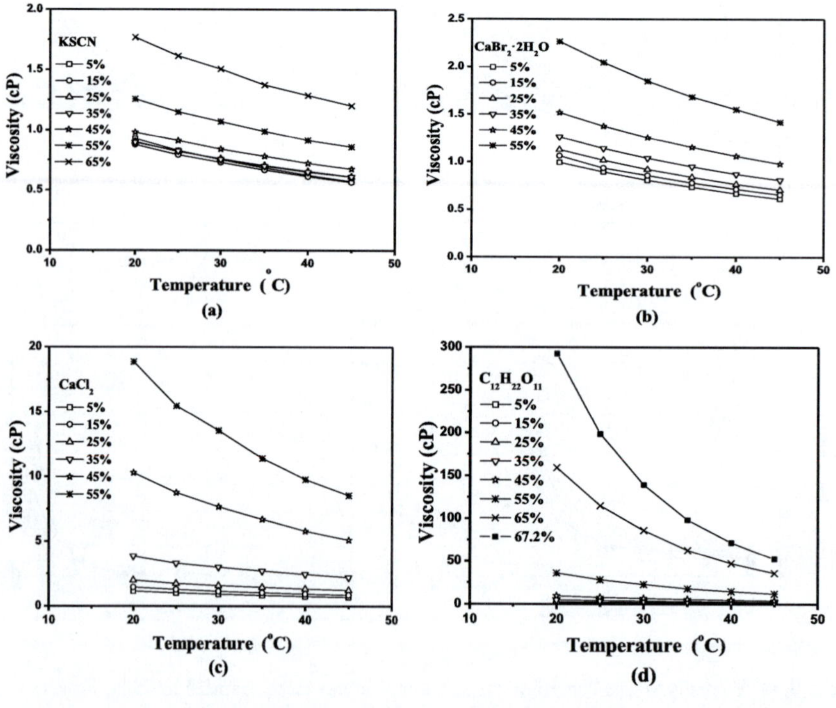

Fig. 6.15 Viscosity versus temperature with different concentrations for selected water-based solutions: **a** KSCN; **b** $CaBr_2 \cdot 2H_2O$; **c** $CaCl_2$; and **d** $C_{12}H_{22}O_{11}$

Table 6.11 The coefficients of the fits, R^2 and SSE for all equations

Figures	Chemicals	Temperature/°C	*a*	*b*	R^2	SSE
Figure 6.13	KSCN	25	0.4828	−0.992	0.8590	0.0085
	$CaBr_2$		0.5533	−0.0115	0.9514	0.0054
	$CaCl_2$		0.9587	−0.0181	0.9825	0.2261
	$C_{12}H_{22}O_{11}$		2.9511E-4	1.2205	0.9908	49.0292
Figure 6.14	KSCN	20	0.6881	−0.0089	0.7847	0.0232
		25	0.6242	−0.0090	0.8197	0.0163
		30	0.5697	−0.0092	0.8308	0.0138
		35	0.5270	−0.0091	0.8492	0.0101
		40	0.4828	−0.0092	0.8591	0.0085
		45	0.4477	−0.0093	0.8763	0.0067
	$CaBr_2$	20	0.7659	−0.0122	0.9639	0.0195
		25	0.6893	−0.0122	0.9683	0.0140
		30	0.6235	−0.0122	0.9677	0.0120
		35	0.5974	−0.0116	0.9770	0.0056
		40	0.5472	−0.0116	0.9785	0.0044
		45	0.5066	−0.0116	0.9811	0.0032
	$CaCl_2$	20	1.5562	−0.0184	0.9818	0.9052
		25	1.3508	−0.0183	0.9797	0.6783
		30	1.2006	−0.0183	0.9816	0.4672
		35	1.0785	−0.0181	0.9802	0.3541
		40	0.9581	−0.0181	0.9825	0.2261
		45	0.8620	−0.0180	0.9839	0.1578
	$C_{12}H_{22}O_{11}$	20	3.7444E-5	1.2660	0.9906	180.1351
		25	2.9463E-4	1.2206	0.9909	49.0292
		30	0.0016	1.1843	0.9923	20.7173
		35	0.0034	1.1647	0.9924	10.2404
		40	0.0063	1.1485	0.9928	5.1796
		45	0.0112	1.1336	0.9936	2.5046
Figure 6.16	3-P	25	−0.0330	3.6650	0.9987	4.8878E-4
	PO		−0.8160	47.3893	0.9603	9.5505
	10TO		−0.5660	32.6760	0.9548	5.2642
	HTO		−1.5556	87.6415	0.9556	38.9971
	P		−2.1957	131.3217	0.9627	64.8841
	15MO		−0.9170	52.5511	0.9542	13.9938
	DSO		−0.7981	69.8596	0.9879	2.7296
	D		−0.0241	2.3997	0.9922	0.0016

(continued)

Table 6.11 (continued)

Figures	Chemicals	Temperature/°C	*a*	*b*	R^2	SSE
	E		−0.0203	1.8946	0.9979	3.0685E-4
	LO		−0.0280	2.6756	0.9954	0.0013
	H		−0.0047	0.6848	0.9973	2.1252E-5
	I		−0.0585	4.0437	0.9843	0.0190

Note a, b—coefficients of the fits; R^2—coefficient of determination; SSE—residual sum of squares

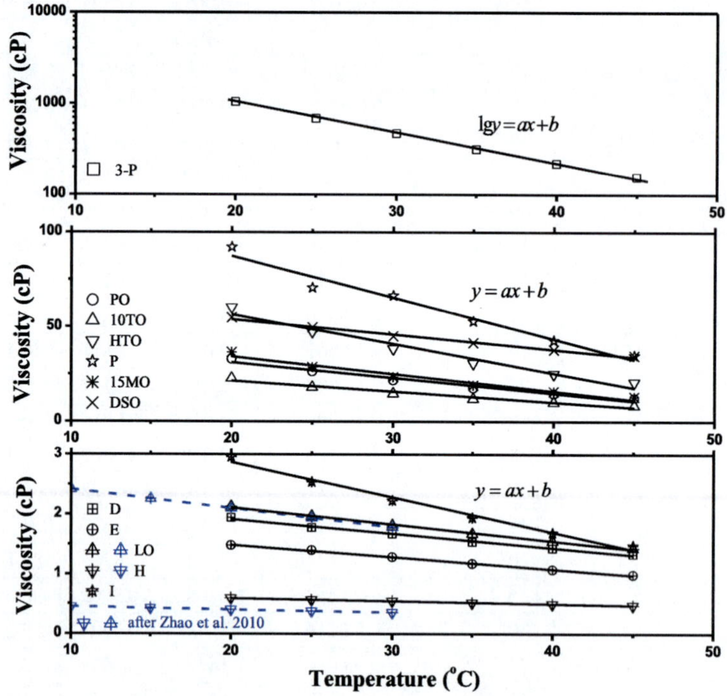

Fig. 6.16 Viscosity versus temperature of selected organic chemical solvents

$$\lg y = ax + b \tag{6.3}$$

Interaction between pore fluids and latex membrane

The degradation of a latex membrane when exposed to these pore fluids has been a concern during triaxial tests. To investigate the interaction between a pore fluid and the latex membrane, a series of immersion tests (contains the changes in both size deformations and strength properties of the latex membrane) was conducted following ASTM D543-14 [26].

The latex membrane from Nanjing Soil Instrument Factory Co., Ltd, China was selected as the testing material since it is commonly available at the laboratory for soil testing. The dimension was measured by Caliper with an accuracy of ± 0.02 mm The membrane was cut into a specimen size 56.4 ± 1.0 mm × 15.2 ± 0.8 mm (length × width) before immersion, and each latex membrane dimension was recorded before and after immersion tests. All the sizes (length × width) of latex membrane before immersion tests are shown in Fig. 6.17. Then, the test specimen was completely immersed into the solutions for 1, 2, 5 and 30 days, respectively; and double latex membrane immersion tests were taken out for each condition. After the selected immersion time, the specimen was removed from the container, and wipe dry. The changes of size deformations (contains length and width) after different immersion time were measured. The areas change ratios of immersed latex membrane influenced by solutions and immersion time are shown in Fig. 6.18. The data show that after the immersion, the size deformation of the latex membrane increased due to the absorption of the solution. Figure 6.18a shows that the maximum area change ratio of latex membrane immersion in selected water-based solutions is approximately 8.5%. Figure 6.18b shows that majority area change ratios of latex membrane immersion in selected oil-based solutions are between 85 and 170%, except two mixtures (E + 3-P, and I + 3-P). The change ratios of latex membrane immersion in the E + 3-P mixtures, and I + 3-P mixtures are only approximately 5%, which reflect small interactions. Large area change ratios of latex membrane immersion in oil-based solutions were also measured by Zhao et al. [2].

The tensile test was performed on the latex membrane immediately after the immersion test. The force was measured by TY10 CNR FT 500 digital dynamometer with an accuracy of ± 0.1 N. The initial stress strength of latex membranes with widths of 15.36 mm and 15.26 mm are equal to 12.72 and 12.32 N/mm, respectively. Hence, the 12.52 N/mm average initial stress strength of latex membranes was used for comparative analysis in the following sections. After selected immersion time,

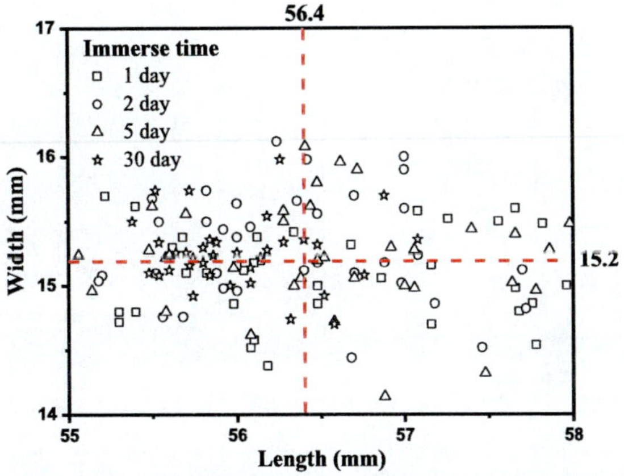

Fig. 6.17 The sizes (length × width) of latex membrane before immersion tests

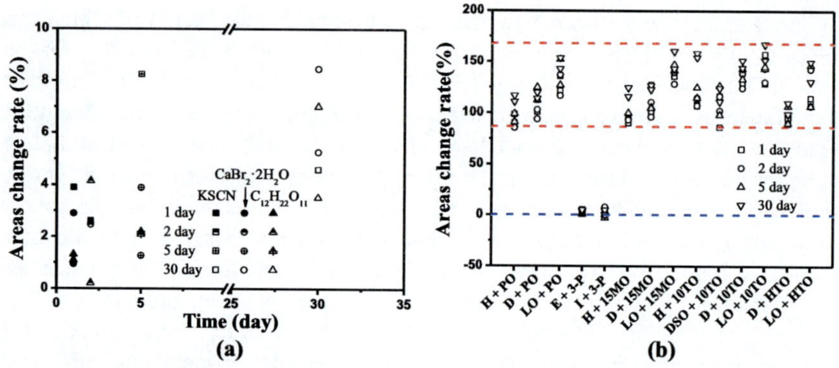

Fig. 6.18 The areas change rate of immersed latex membrane influenced by pore fluids and time: **a** water-based solutions; and **b** oil-based solutions

the specimen was taken out, wiped dry, and the size deformation measure; then, the tensile test was carried out. The stress reduction ratios of immersed latex membrane influenced by water-based solutions and oil-based solutions are shown in Figs. 6.19a and b, respectively. In general, the stress reduction ratios increased with the increasing of immersion time. For water-based solutions, Fig. 6.19a shows that less than 20% stress reduction ratios happened during 5 days immersion time, and the maximum value reaches 45% after 30 days immersion time. For oil-based solutions, Fig. 6.19b shows that more than 90% stress reduction ratios happened for latex membrane immersed in majority oil-based solutions, except two mixtures (E + 3-P, and I + 3-P). For E + 3-P, and I + 3-P mixtures, less than 20% stress reduction ratios happened during 5 days immersion time, except one test point. The stress strengths of latex membrane measured through tensile tests are related to width deformation after immersion.

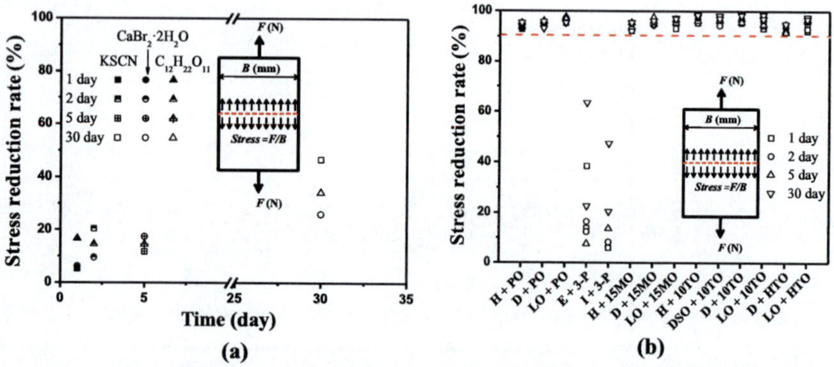

Fig. 6.19 The stress strengths reduction rate of immersed latex membrane influenced by pore fluids and time: **a** water-based solutions; and **b** oil-based solutions

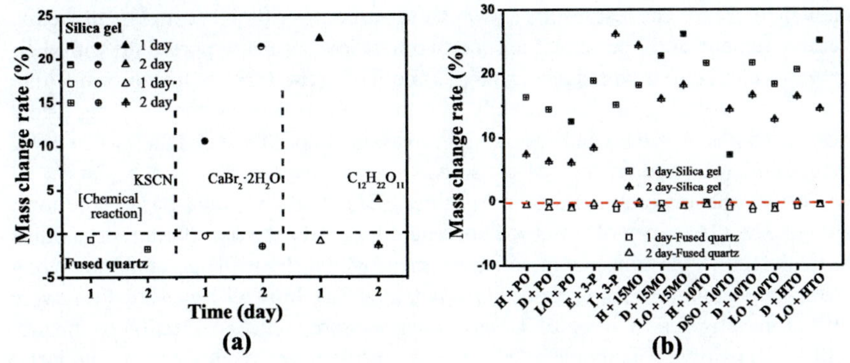

Fig. 6.20 The mass change rate of granular particles influenced by pore fluids and time: **a** water-based solutions; and **b** oil-based solutions

Interaction between pore fluids and granular particles

To investigate the interaction between a pore fluid and the granular particles (silica gel and fused quartz with 0.5–1.0 mm particle sizes), immersion tests was also conducted following ASTM D543-14 [26]. The weight of silica gel and fused quartz was measured first, and then was immersed in selected solutions. After selected immersion time (1 day and 2 days), the specimen was taken out, dry, and then weight measure. The mass change ratios of silica gel and fused quartz influenced by water-based and oil-based solutions are shown in Fig. 6.20a and b, respectively. It shows that the mass of fused quartz immersed in both selected water-based and oil-based solutions were decreased, and the decrease ratios increased with immersion time increasing within the tested immersion time. After 2 days immersion time, the mass change of fused quartz immersed in three selected solutions was less than 2%. Figure 6.20a shows that the mass of silica gel immersed in $CaBr_2{\cdot}2H_2O$ and $C_{12}H_{22}O_{11}$ solutions were increased, and the mass changes were difference with immersion time. Chemical reaction was happened when silica gel and KSCN solution mixture. After 1 day or 2 days immersion time, the maximum mass change of silica gel immersed in $CaBr_2{\cdot}2H_2O$ and $C_{12}H_{22}O_{11}$ solutions was nearly 22%. Figure 6.20b shows that the mass of silica gel immersed in oil-based solutions were increased, and the mass changes were difference with immersion time. After 1 day or 2 days immersion time, the maximum mass change of silica gel immersed in oil-based solutions was approximately between 7 and 27%.

6.4 Summary and Conclusions

Twelve (12) different types of chemical solvents were tested to identify low viscosity pore fluid to manufacture transparent soil together with the silica gel. Eight potential matching pore fluids were further investigated to examine their interactions with the latex membrane. The test results show that the mixtures of toluene and methyl ethyl ketone, toluene and 2 propanol are the two best low viscosity pore fluid with their very low interactions with the latex membrane. These two types of low viscosity pore fluids are harmful to the human body, and the protective measures (wearing safety masks and plastic gloves and operating in a venting hood) must be taken when using them to manufacture the transparent soils.

Further the maximum RI of thirty-three (33) types of water-based solutions, and sixteen (16) types of oil-based solvents to identify the appropriate pore fluids to manufacture the transparent soil were investigated. Four (4) potential matching water-based solutions, and fourteen (14) potential matching oil-based solutions were further investigated to measure RI versus temperature and concentration (or quality ratio), viscosity versus temperature and concentration (or quality ratio). The interactions' results of potential matching pore fluid with the latex membrane, and with granular particles (silica gel and fused quartz) show that water-based $CaBr_2{\cdot}2H_2O$, $C_{12}H_{22}O_{11}$ and $CaCl_2$ solutions are suitable pore fluids to manufacture transparent soil; KSCN solution is another new suitable water-based solution with low viscosity, easy reaching TRIs, and very low interactions with the latex membrane and the fused quartz granular particles; Ethanol versus 3-Propanetriol (E + 3-P) mixture, and Isopropanol versus 3-Propanetriol (I + 3-P) mixture are identified as two new suitable oil-based solutions with minimum interaction with latex membranes. Further research on the mechanical properties of transparent soil manufactured by potential pore fluid (KSCN, E + 3-P mixture, and I + 3-P mixture) and each granular particle (silica gel or fused quartz), and the difference with using other pore fluids is greatly encouraged.

References

1. Iskander M, Lai J, Oswald C, Mainnheimer R (1994) Development of a transparent material to model the geotechnical properties of soils. Geotech Test J 17(4):425–433
2. Zhao HH, Ge L, Luna R (2010) Low viscosity pore fluid to manufacture transparent soil. Geotech Test J 33(6):463–468
3. Peters SB, Siemens G, Take WA (2011) Characterization of transparent soil for unsaturated applications. Geotech Test J 34(5):1–12
4. ASTM D445-09 (2009) Standard test method for kinematic viscosity of transparent and opaque liquids (and calculation of dynamic viscosity). In: Annual Book of ASTM Standards, ASTM International, West Conshohocken, PA
5. Mallinckrodt Baker Inc. (2009) Summary of key physical data for baker analyzed hplc solvents. http://www.mallbaker.com/conversion/solventphydata.htm (Last accessed Sept. 19, 2009)
6. Iskander M, Liu J, Sadek S (2002) Transparent silica gels to model the geotechnical properties of sand. Int J Phys Modell Geotech 2(4):27–40
7. Iskander M, Liu J, Sadek S (2002) Transparent amorphous silica to model clay. J Geotech Geoenviron Eng 128(3):262–273
8. ASTM D543-06 (2006). Standard practices for evaluating the resistance of plastics to chemical reagents. In: Annual Book of ASTM Standards, ASTM International, West Conshohocken, PA
9. Song Z, Hu Y (2009) Plate anchor keying under inclined pullout in clay: observation and estimation. Alexandria, Millpress Science Publication, Alexandria, Egypt, Proc. XVII IC-ISMGE, pp 708–711

10. Welker AL, Bowders JJ, Gilbert RB (1999) Applied research using a transparent material with hydraulic properties similar to soil. Geotech Test J 22(3):266–270
11. Ezzein FM, Bathurst RJ (2014) A new approach to evaluate soil-geosynthetic interaction using a novel pullout test apparatus and transparent granular soil. Geotext Geomembrances 42:246–255
12. Hover ED, Ni Q, Guymer I (2013) Investigation of centerline strain path during tube penetration using transparent soil and particle image velocimetry. Geotech Lett 3(2):37–41
13. Kong GQ, Cao ZH, Zhou H, Sun XJ (2015) Analysis of piles under oblique pullout load using transparent soil models. Geotech Test J 38(5):725–738
14. McKelvey D, Sivakumar V, Bell A, Graham J (2004) Modelling vibrated stone columns in soft clay. Proc ICE—Geotech Eng 157(3):137–149
15. Stanier SA, Black JA, Hird CC (2012) Enhancing accuracy and precision of transparent synthetic soil modelling. Int J Phys Modell Geotech 12(4):162–175
16. Sun JL, Liu JY (2014) Visualization of tunneling-induced ground movement in transparent sand. Tunn Undergr Space Technol 40:236–240
17. Toiya M, Hettinga J, Losert W (2009) 3D imaging of particle motion during penetrometer testing. Granul Matter 9(5):323–329
18. Siemens GA, Peters SB, Take WA (2013) Comparison of confined and unconfined in transparent porous media. Water Resour Res 49(2):851–863
19. Ezzein FM, Bathurst RJ (2011) A transparent sand for geotechnical laboratory modeling. Geotech Test J 34(6):590–601
20. Zhao HH, Ge L (2014) Investigation on the shear moduli and damping ratios of silica gel. Granul Matter 16(4):449–456
21. Zhang YP, Li L, Wang SZ (2014) Experimental study on pore fluid for forming transparent soil. J Zhejiang Univ (Eng Sci) 48(10):1828–1834
22. Guzman IL, Iskander M, Suescun-Florez E, Omidvar M (2014) A transparent aqueous-saturated sand surrogate for use in physical modeling. Acta Geotech 9(2):187–206
23. Guzman I, Iskander M (2013) Geotechnical properties of sucrose-saturated fused quartz for use in physical modeling. Geotech Test J 36(3):448–454
24. McGraw-Hill Education (2005) Lange's Handbook of Chemistry, 16th edn. McGraw-Hill Education LLC
25. ASTM D1218-12 (2012) Standard test method for refractive index and refractive dispersion of hydrocarbon liquids. In: Annual Book of ASTM Standards, ASTM International, West Conshohocken, PA
26. ASTM D543-14 (2014) Standard practices for evaluating the resistance of plastics to chemical reagents. In: Annual Book of ASTM Standards, ASTM International, West Conshohocken, PA
27. Lehane BM, Gill DR (2004) Displacement fields induced by penetrometer installation in an artificial soil. Int J Phys Model Geotech 1(1):25–36
28. Gao Y, Sui WH, Liu JY (2015) Visualization of chemical grout permeation in transparent soil. Geotech Test J 38(5):774–786

Part II
Transparent Soil Imaging and Image Processing

Chapter 7
Laser Speckle Effect

Abstract When a laser sheet shines into the transparent soil model, its interaction with transparent solids will generate a laser speckle effect. This laser speckle effect is the basis of soil deformation measurement. Characteristics of laser speckle effect and digital laser speckle effect were presented in this chapter.

7.1 Introduction

When the laser shines on the surface of an object with diffuse properties, according to the Huygens principle, each point of the surface of the object can be regarded as a point light source, the light reflected from the surface of the object is superimposed in space, it will interfere with each other throughout the space, forming a random distribution of bright and dark spots, called laser spots. For some natural surface with strong texture, itself can be used as a speckle image, called natural speckle. For some surfaces with weak texture, manually made speckle images are required. In summary, there must be a rough plane that can scatter, and the depth of the rough plane must be greater than the wavelength of light to form a scattered spot [1].

Water itself is a transparent liquid. If there are small grains or particles inside water with diameter larger than the wave length of light, when light passes through water vapor it reflects and refracts to form speckle pattern. Similarly, when a light is cutting through the transparent soil, it will introduce a laser speckle effect when the laser is interacting with the solid particles of transparent soil. The laser speckle pattern was recorded through a digital camera, through analyzing the speckle pattern during the deformation and movement of soil particles, the displacement information can be obtained through image pattern matching technique such as digital image correlation [2]. Figure 7.1 shows a typical laser speckle pattern under the green laser light inside the transparent soil model. This speckle pattern provides the basis for the displacement measurement of transparent soils when disturbed.

H. Zhao et al., *Transparent Soil Modelling Technique and Its Application*, Springer Tracts in Civil Engineering, https://doi.org/10.1007/978-981-19-6825-9_7

Fig. 7.1 Laser speckle pattern of transparent soil

7.2 Characteristics of Laser Speckle Field

According to the scalar theory of light field diffraction, the propagation process (including imaging) of a monochromatic light field can be characterized by a simple superposition integral. If we know the complex amplitude of a monochromatic light field is $A_0(x_0, y_0)$ on the plane of (x_0, y_0), the complex amplitude on the plane of (x, y) can be expressed as the following,

$$A(x,y) = \iint\limits_{\Sigma} A_0(x_0, y_0)h(x, y; x_0, y_0)dx_0dy_0 \quad (7.1)$$

where $h(x, y; x_0, y_0)$ is the power function for the propagation process from the plane of (x_0, y_0) to the parallel plane of (x, y); the integration area of $\sum$ is decided by the distribution area of the light field on the plane (x_0, y_0) in the free space; under imaging conditions, it is determined by the width of the point spread function of the imaging system [3].

The generation, propagation and receiving of light is a multiple random process. In general, it assumes that the object surface is a plane surface, its reflection index is $R(x_0, y_0) = r(x_0, y_0)\exp[j\Phi(x_0, y_0)]$, the light field is $A_L(x_0, y_0)$, therefore the scattering light field on the surface of the object can be expressed as the following,

$$A_0(x_0, y_0) = R(x_0, y_0)A_L(x_0, y_0) = r(x_0, y_0)A_L(x_0, y_0)\exp[j\Phi(x_0, y_0)] \quad (7.2)$$

$r(x_0, y_0)$ and $\Phi(x_0, y_0)$ are random variables related with specific scattering primitives. Because the illumination light field is a slowly changing quantity in space, the scattering light field is mainly determined by the reflection characteristics. Previous experiments indicate that the scattering light field on the optical rough surface has the following statistical features:

(1) The complex amplitude of light field $A_0(x_0, y_0)$ is independent with the phase angle $\Phi(x_0, y_0)$. The complex amplitudes of light fields scattered by different scattering primitives are statistically independent of each other.
(2) The measured surface is rough, that is the standard deviation of its surface relief height is much larger than the light wave, so that $\Phi(x_0, y_0)$ can be considered to distribute uniformly in $[-\pi, \pi]$, its probability density function is as the following,

$$P_\phi(\phi) = \begin{cases} \frac{1}{2\pi} & (-\pi < f\phi < \pi) \\ 0 & (other) \end{cases} \tag{7.3}$$

(3) When the scattering primitives of the measured surface are very fine, which is much smaller compared with the illumination area and the effective area covered by point spread function formed by measurement system on the surface, but large enough compared with the light wave length. The self-correlation function of the light field scattered from the measured surface can be expressed as the following,

$$I_{A0}(r_{02} - r_{01}) = < A_0(r_{01})A_0^*(r_{02}) \geq \langle I_0(r)\rangle\delta(r_{02} - r_{01}) \tag{7.4}$$

where, operator $\langle * \rangle$ refers to the average computation of the ensemble, $\delta(*)$ is two-dimensional Dirac function, $\langle I_0(r_0)\rangle$ is the spatial slowly varying intensity function determined by the macroscopic reflection characteristics of illumination light field and the object surface; the vector $\boldsymbol{r}$ is the abbreviation coordinates (x, y). This equation indicates that after scattering, the light field of the object surface is not the spatial interference field, it changes into strictly non-interference spatial field.

7.3 Digital Image of Laser Speckle

With the development of imaging technology, the charge coupled device (CCD) and complementary metal oxide semiconductor (CMOS) were developed as two important image sensors and widely used in industry. When the light is focused through the lens to form an image, CCD will convert the optical signal of the image into electrical signal (voltage). The greater the amount of light, the more electrons will be released, the stronger the electrical signal, and the brighter the pixel display will be. The smallest unit constituting the picture on the CCD is called pixel. The number and size of pixels in each CCD are directly related to the image quality. The higher the pixel, the higher the output image quality. The obtained images are called digital images.

There are four types of digital images: binary image, gray image, color image and index image. A binary image composed of l's and 0's is then obtained which is a display of all the points in the image where its gradient is larger than the value of the threshold that is applied. The black and white image is the binary image.

Gray scale digital image is an image with only one sampling color per pixel. Such images are usually displayed in grayscale from the darkest black (0) to the brightest white (255), although theoretically this sampling can be different shades of any color, or even different colors in different brightness, which is an image expressed with two-dimensional array, or it can be viewed as a collection of pixels. For each pixel in a gray image, a value between 0 and 255 is assigned to express the gray scale.

For an image with M × N pixels in order for the computer image processing, each image is expressed by an M × N array usually, as shown in Eq. (7.5).

$$\begin{bmatrix} g(0,0) & g(0,1) & \dots & g(0,N-1) \\ g(1,0) & g(1,1) & \dots & g(1,N-1) \\ \vdots & \vdots & & \vdots \\ g(M-1,0) & g(M-1,1) & \dots & g(M-1,N-1) \end{bmatrix} \tag{7.5}$$

An index image is an image that directly uses the pixel value as the subscript of the RGB palette. Index images can "directly map" pixel values to palette values. An index image contains a data matrix and a palette matrix map. The data matrix can be uint8, uint16 or double precision, while the palette matrix is always an M × 3 double precision matrix. The palette is usually stored with the index image, and when the image is loaded, the palette is automatically loaded with the image.

Color image refers to the image composed of R, G and B components per pixel, in which R, G and B are described by different gray levels. An RGB image is an M × N × 3 array of color pixels. M, N represents the number of rows and columns of the image, and the three two-dimensional matrices of m × n represent the R, G and B color components of each pixel. It can also be regarded as a "pile" formed by three gray images. When it is sent to the red, green and blue inputs of the color monitor, a color image is formed. Figure 7.2 illustrates the above concepts classically.

Fig. 7.2 The matrix form of color image

When the laser speckle pattern is recorded by a CCD camera or a CMOS camera (digital camera), we thus obtain a digital image of laser speckle. With the fast development of image processing technology, the digital images of laser speckle can be used to measure strain, displacement, crack, and stress distribution of solid surface, which has become a common method used daily in wide applications [4].

References

1. Knopoff L (1964) Scattering of elastic waves by small inhomogeneities. J Acoust Soc Am 36(2):338
2. Liu J, Iskander MG (2010) Modelling capacity of transparent soil. Can Geotech J 47(4):451–460
3. Jin G, Li J (1978) Laser surveying. China Science Publishing & Media Ltd. (CSPM)
4. Yoneyama S, Kitagawa A, Kitamura K, Kikuta H (2006) In-plane displacement measurement using digital image correlation with lens distortion correction. JSME Int J Ser A Solid Mech Mater Eng 49(3):458–467

Chapter 8
2D Transparent Soil Imaging and Digital Image Cross-Correlation

Abstract Transparent soil is used to model natural soils in geotechnical research. The laser speckle effect generated when a laser light goes through the transparent soil were used to provide the displacement information of soil particles. Digital image cross-correlation is usually adopted to process the laser speckle images of the transparent soil.This chapter presented a typical experimental set up for 2D transparent soil modelling, 2D and 3D digital image cross correlation methods, and influences of particles size on digital image deformation measurement.

Transparent soil is used to model natural soils in geotechnical research. The laser speckle effect generated when a laser light goes through the transparent soil were used to provide the displacement information of soil particles. Digital image cross-correlation is usually adopted to process the laser speckle images of the transparent soil.

8.1 2D Transparent Soil Model and Imaging

In 2D transparent soil model, a typical setup is shown in Fig. 8.1. It consists of a transparent soil model, investigated structures (such as pile, anchor, tunnel), a loading system, a laser, and a digital camera. The laser is used to produce a laser sheet inside the transparent soil model. The digital camera was used to record the images of the transparent soil and the structure model during the loading process of the structures. Continuous green wave laser (wave length 532 nm, 1.4–3.0 W) are usually used in the 2D transparent soil modeling [39]. Currently, the size of the transparent soil model can have a dimension of 60 cm × 50 cm × 30 cm (Length × height × width). Although the transparency is decreasing with the increase of the model width, but with the increase of the transparency of the transparent soil, the width of the model is expected to increase in the future and a larger scale model can be possible.

H. Zhao et al., *Transparent Soil Modelling Technique and Its Application*, Springer Tracts in Civil Engineering, https://doi.org/10.1007/978-981-19-6825-9_8

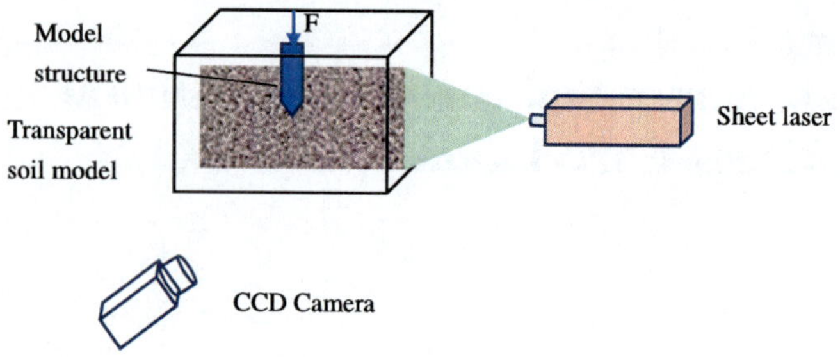

Fig. 8.1 2D typical set-up of transparent soil model

8.2 Digital Image Correlation (DIC)

Digital image correlation (DIC) is a non-contact, full field displacement optical measurement technique. It is most often used in the following applications: material characterization, sample testing (displacement and strains), applications where displacement or deformation measurements are needed, high speed/frequency applications (i.e. crash testing, vibration). Now DIC has been extended from 2D DIC to 2.5D DIC [1] and stereo-DIC [2]. The in-plane 2D DIC technology is still the most widely used method for noncontact surface deformation measurement.

The first step in DIC measurement is to prepare a speckle pattern on the surface of the testing sample, the speckle pattern serves as a carrier of deformation information. The speckle pattern may be natural or artificially made. In the transparent soil, the speckle pattern is due to the interaction of laser light and transparent soil particles. DIC processes two successive images obtained during the loading stages. The images are divided into small sub-windows, through tracking the corresponding position between matching sub-windows in the reference/undeformed images and the target/deformed images, the displacement/deformation of the materials can be obtained. Figure 8.2 shows the basic principles of the standard sub-window based DIC method.

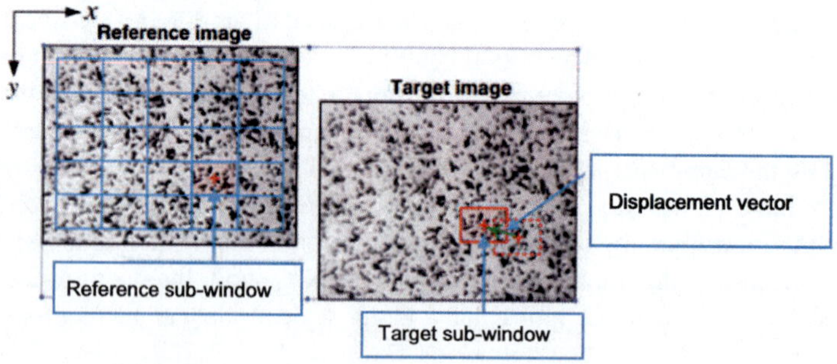

Fig. 8.2 Schematic illustration of the undeformed sub-window and the corresponding deformed sub-window in 2D-DIC method

The cross-correlation coefficient between the two sub-windows is usually calculated through the following Eq. (8.1),

$$C(s,t) = \sum_{m=0}^{M-1}\sum_{n=0}^{N-1} I_1^{i,j}(m,n) \times I_2^{i,j}(m-s, n-t) \tag{8.1}$$

where I_1(i, j) is the intensity of the sub-window number (*i, j*) in the first image, I_1; and $I_2(i, j)$ is the intensity of the sub-window number (*i, j*) in the second image, I_2. When this expression reaches its peak value, the two sub-windows will most likely match each other. The sub-window has moved *s* and *t* units in *i* and *j* directions, respectively.

In order to assess the similarity of intensity between the reference sub-window and the target sub-window, there are different computation algorithm developed except Eq. (8.1). The zero-mean normalized sum-of-square difference criterion (ZNSSD) and zero-mean normalized cross-correlation criterion (ZNCC) have been widely used in DIC analysis, which have been proved to be more robust methods. The ZNSSD and ZNCC are defined as follows:

$$C_{ZNSSD}(p) = \sum_{i=-M}^{M} \sum_{j=-M}^{M} \left[\frac{f(x,y) - f_m}{\sqrt{\sum_{i=-M}^{M}\sum_{j=-M}^{M}[f(x,y)-f_m]^2}} - \frac{g(x*,y*) - g_m}{\sqrt{\sum_{i=-M}^{M}\sum_{j=-M}^{M}[g(x,y)-g_m]^2}} \right] \tag{8.2}$$

$$C_{ZNCC}(p) = \sum_{i=-M}^{M} \sum_{j=-M}^{M} \left[\frac{[f(x,y)-f_m] \times [g(x*,y*)-g_m]}{\sqrt{\sum_{i=-M}^{M}\sum_{j=-M}^{M}[f(x,y)-f_m]^2} \times \sqrt{\sum_{i=-M}^{M}\sum_{j=-M}^{M}[g(x*,y*)-g_m]^2}} \right] \tag{8.3}$$

where f(x, y) is the gray level intensity at coordinates (*x, y*) in the reference sub-window of the reference image I_1; g(x*, y*) is the gray level intensity at coordinates (x*, y*) in the current sub-window of the current image I_2; and f_m and g_m are the mean gray level intensity values of the reference and deformed sub-window, respectively, as shown in the following:

$$f_m = \frac{1}{(2M+1)^2} \sum_{i=-M}^{M}\sum_{j=-M}^{M} [f(x,y)] \tag{8.4}$$

$$g_m = \frac{1}{(2M+1)^2} \sum_{i=-M}^{M}\sum_{j=-M}^{M} [g(x^*,y^*)] \tag{8.5}$$

The major step of implementing DIC is to use a nonlinear optimization scheme to achieve sub-pixel accuracy with a better match, and a specific interpolation is undertaken subsequently. Interpolation methods such as forward additive Newton–Raphson (FA-NR) algorithm, three-point gaussian interpolation, high-order B-spline interpolation algorithm etc. are frequently used [3–5].

The relationship between the ZNCC and ZNSSD is expressed as:

$$C_{ZNCC}(p) = 1 - 0.5 \times C_{ZNSSD}(p) \tag{8.6}$$

In the application of DIC method, the reference image is usually taken as the first image captured during the experiment, but during large deformation, it is easy to become decorrelated, therefore, in order to avoid the error caused by this decorrelation during large deformation, the reference image can be updated according to a setting threshold value [6]. In addition, in order to achieve better correlation data, the processed data usually went through a series of filtering operations. Gaussian smoothing is widely used a pre-processing tool to remove the high-frequency components (e.g., noise and details) from images in DIC analysis [7, 8]. A Gaussian low-pass filter is expressed as

$$g(x, y) = \frac{1}{2\pi\sigma^2} e^{-(x^2+y^2)/2\sigma^2} \tag{8.7}$$

where x and y are the distances from the origin in the horizontal and vertical axes, respectively; σ is the standard deviation (SD) of the Gaussian function. Zhou et al. [9] recommended Butterworth filter as its flexibility of passband selection and maximal preservation of the allowed frequencies, Binomial filter is also recommended if computational cost is a critical issue.

8.3 Main Error Sources in 2D-DIC Measurement

Errors in 2D-DIC measurement may come from several resources such as image quality, image distortion, correlation algorithm, subpixel interpolation error etc. Two major error sources are from DIC analysis algorithm and the image acquisition process.

During obtaining the digital images in the experiment, it needs to avoid the disturbances due to the existence of adverse factors affecting measurement precision. The potential errors can be summed up as follows: (i) the quality of the speckle pattern; (ii) image distortion caused by geometric distortion of the imaging lens, motion of the sensor, and current camera operating status; (iii) testing environment; and (iv) Out-of-plane motion of the test sample surface.

(i) **the quality of the speckle pattern**

The quality of the speckle pattern is the most important factor that influences the 2D-DIC measurement. The grey intensity distribution of artificially made speckle pattern may vary with different people and different methods. The gray intensity distribution of the natural material may be too weak to be observed. The size of the speckle maybe too small or too large. The gradient of the grey intensity may be too small to give any differences. Therefore, it needs an appropriate method to assess the quality of the speckle pattern. A few methods now available to assess the quality of the speckle pattern.

The mean speckle size as a simple and global index to assess the quality of the whole image speckle pattern for DIC analysis [10]. Hua et al. [11] proposed the concept of the mean subset fluctuation, which is the mean sum of a pixel and its eight adjacent pixels form the subset to reflect gradient of the speckle subset. Yu et al. [12] proposed the second derivative of mean intensity ω_f to assess the quality of the speckle image. Su et al. [13] recently proposed a new global parameter called the root mean square error as a quality index. Some local parameters that can reflect the local quality of the speckle pattern based on individual subsets are also proposed by some researchers.

$$\omega_f = \frac{\sum_{i=1}^{W}\sum_{j=1}^{H}\left|\nabla^2 f(x_{ij})\right|}{W \times H} \tag{8.8}$$

where $\nabla^2 f(x_{ij}) = \sqrt{f_{xx}(x_{ij})^2 + f_{yy}(x_{ij})^2}$), W and H are the width and height of the image in pixels.

(ii) **image distortion**

The image distortion mainly caused by the camera hardware. Lens distortion is one of the main causes of image distortion. In high-precision DIC measurement systems, the image distortion caused by lens distortion is strongly related to the quality of the imaging equipment. Lenses can introduce distortion due to manufacturing accuracy and deviations in the assembly process, resulting in distortion of the original image. The distortion of the lens is divided into radial distortion and shear distortion. To deal with the problem of image distortion to restore fidelity of image, several proved methods for 2D-DIC measurement have been presented [14–17].

Pan et al. [15, 16] proposed a convenient method for systematic errors in two-dimensional digital image correlation due to lens distortion. The in-plane displacement *u, v* in x- and y-directions are correlated with the u_i', v_i' measured by DIC through the following equations:

$$\begin{aligned} u_i' &= u + a_1 x_{di}' + b_1 y_{di}' - k_1[x_{di}'(x_{di}'^2 + y_{di}'^2) - x_{di}(x_{di}^2 + y_{di}^2)] \\ v_i' &= v + a_2 x_{di}' + b_2 y_{di}' - k_1[y_{di}'(x_{di}'^2 + y_{di}'^2) - y_{di}(x_{di}^2 + y_{di}^2)] \end{aligned} \tag{8.9}$$

where k_1 is the lens distortion coefficient, a_1, b_1, a_2, b_2 are parameters accounting for the possible small out-of-plane motion and in-plane motion, x_d, y_d is the distorted image coordinates of the considered image point in reference image, $x_d' = x_d + u'$, $y_d' = y_d + v'$. Then, the seven unknown parameters can be resolved using conventional linear least-squares method:

$$p = (X^T X)^{-1} X^T d \tag{8.10}$$

where p is the parameter vector, $p = (u, a_1, b_1, v, a_2, b_2, k_1)^T$,

$$X = \begin{bmatrix} 1 & x'_{d1} & y'_{d1} & 0 & 0 & 0 & x'_{d1}(x'^2_{d1} + y'_{d1}) - x_{d1}(x^2_{d1} + y^2_{d1}) \\ 1 & x'_{d2} & y'_{d2} & 0 & 0 & 0 & x'_{d2}(x'^2_{d2} + y'^2_{d2}) - x_{d2}(x^2_{d2} + y^2_{d2}) \\ \vdots & \vdots & \vdots & \vdots & \vdots & \vdots & \vdots \\ 1 & x'_{dn} & y'_{dn} & 0 & 0 & 0 & x'_{dn}(x'^2_{dn} + y'^2_{dn}) - y_{dn}(x^2_{dn} + y^2_{dn}) \\ 0 & 0 & 0 & 1 & x'_{d1} & y'_{d1} & y'_{d1}(x'^2_{d2} + y'^2_{d2}) - y_{d1}(x^2_{d1} + y^2_{d1}) \\ 0 & 0 & 0 & 1 & x'_{d2} & y'_{d2} & y'_{d2}(x'^2_{d2} + y'^2_{d2}) - y_{d2}(x^2_{d2} + y^2_{d2}) \\ \vdots & \vdots & \vdots & \vdots & \vdots & \vdots & \vdots \\ 0 & 0 & 0 & 1 & x'_{dn} & y'_{dn} & y'_{dn}(x'^2_{dn} + y'^2_{dn}) - y_{dn}(x^2_{dn} + y^2_{dn}) \end{bmatrix}$$

$$d = [u_1, u_2, \cdots u_n, v_1, v_2, \cdots v_n]^T \tag{8.11}$$

This lens distortion correction method using rigid body, in-plane translations is not only simple in principle, but also very practical for real 2D-DIC applications.

In addition, self-heating is a very common phenomenon for the cameras in working state. After operating for 1–2 h, the internal temperature will become steady. The increase in temperature caused by self-heating not only brings darker current noise but also causes thermal expansion of various supports for connecting camera sensor, resulting in further small movement of the sensor current and subsequent image expansion and translation. This can cause several and complex image distortion. In the case of the long-time operation of the digital camera, it is necessary to preheat the digital camera to the thermal balance stage before experiment.

(iii) **testing environment noises**

The speckle images are often affected by various environmental noises, mainly due to illumination fluctuations and acquisition hardware, such as thermal noise, readout noise, and cutoff noise. These noises will directly affect the accuracy and reliability of image registration and subpixel interpolation. In general, using high quality low-noise camera is a straightforward and effective way to decrease or suppress the effect of image noise. Another simple approach is to average the image data in each step measurement. Typically, an average of 100 images is sufficient to suppress the noisy signals of the image. Now before calculate the correlation coefficient, presmoothing the speckle images using Gaussian low-pass filter is one effective method to remove the noise [7].

The illumination fluctuations are another environmental noise that may occur in the actual experiment. It is generally known that the ZNSSD and ZNCC are insensitive to fluctuations in white light illumination. Jerabek et al. [18] proposed a concrete way for preventing overexposure of the optimal image contrast corresponding to the maximum light intensity, which is determined by compromising two characteristic parameters: the mean value of strain standard deviations (SSDs) and the standard deviation of mean strain values.

(iv) **Out-of-plane motion of the test sample surface**

The out-of-plane motion of the sample surface is one of the most important sources of potential error. The sample surface may move relative to the imaging sensor plane, which can include out-of-plane forward, backward translation, and out-of-plane rotation during loading. Small out-of-plane motion alters the compliance with the principle of the pinhole imaging model and then results in additional errors. It is usually considered that the higher (lower) the displacement gradients, the higher (lower) the sensitivity to out-of-plane motion, especially in strain measurements. Haddadi and Belhabib [19] had shown that an out-of-plane translation of 1 mm brought strain measurement error up to 2×10^{-3} mm^{-1} when the relative distance between the surface of the test sample and the camera was about 21 cm. Recent work by Pan et al. [7] confirmed that the bilateral telocentric lens was scarcely affected by the out-of-plane motion of the sample surface. This is validated by the fact that the longer focal lengths indirectly cause a smaller error in out-of-plane motion. Hoult et al. [20] illustrated five potential solutions for the correction of strain error caused by out-of-plane motion, denoting that it is possible to control the mean strain error within 5 $\mu\varepsilon$. The distance between the camera and the testing sample can be appropriately increased to limit the error caused by the out-of-plane motion, and using a telocentric lens can be considered to be the same as the above principle.

8.4 Particle Image Velocimetry (PIV)

Particle image velocimetry (PIV) is the most powerful modern technique to measure extended velocity fields [21, 22] in experimental fluid mechanics. This relies on seeding the flow-volume with small tracer particles which are illuminated with pulsed laser light and the digital cameras captured images of their displacement with time. By recording two images of two laser shots in a short time interval Δt, the velocity is subsequently extracted through digital image correlation with the specialized software. A typical PIV system is shown in Fig. 8.3. Now the PIV has evolved into Dual plane Stereoscopic PIV [23], 2D-3C PIV [24]), stereoscopic PIV [25] and Tomographic PIV [26] over the last twenty years. Micro-PIV has also been developed to measure liquid flows in micro channels [27].

The seeding particles' properties have a great impact on the domain flow topology. Ideally, the properties should be identical to the actual fluid to obtain the exact flow physics in the domain. The correct choice of the seeding material primarily depends on the physical problem investigated. An appropriate concentration of the seeding particles was required to provide the good cross-correlation between the images captured. As a rule, seeding particle's diameter should be large enough to scatter the significant light and small enough to follow the fluid flow. The Lorenz-Mie scattering theory is generally applied to estimate the diameter of spherical seeding particles [28]. When the seeding particles are small with a low magnification factor, Eq. (8.11) can be used, and in the case of larger particles or larger magnification,

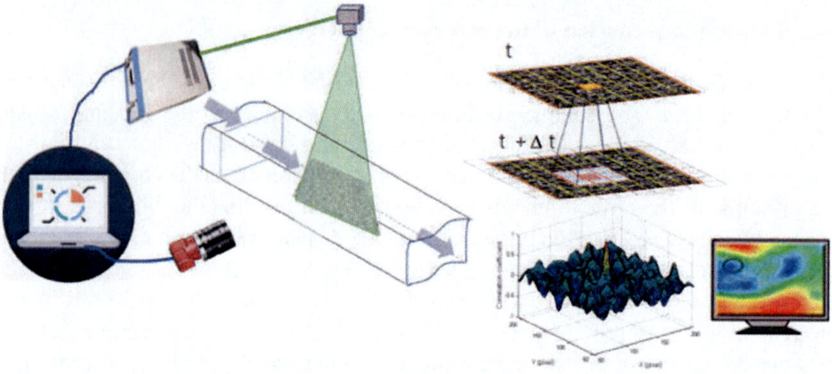

Fig. 8.3 A typical PIV system set-up

Eq. (8.12) can be applied to calculate the diffraction diameter of particles. In the case of large particles, geometric imaging becomes more dominant [28, 29].

$$d_{diff} = 2.44 f(M+1)\lambda \tag{8.12}$$

$$d_{\tau} = \sqrt{(Md_p)^2 + d_{diff}^2} \tag{8.13}$$

where f, M, λ, d_{diff}, and d_p are the f-number, ratio between the focal length and aperture diameter, the magnification factor, the wavelength of light, diffraction diameter, and the particle diameter, respectively.

PIV is basically using the similar cross-correlation function as in DIC. But with the specific applications in fluid mechanics, some improved calculation algorithms are developed. Scarano and Riethmuller [30] proposed iterative multigrid approach in PIV image process with discrete widow offset (WIDIM). In this method, first splitting each of the two pictures into sets of interrogation areas, applying a predictor for the displacement field δ_p(x, y) according to the flow distribution pattern, then the cross-correlation is applied to the two images to give the correction displacement δ_c(x, y). The result is validated through two threshold values of ε_{div} and ε_{mean} which can be used as the input for a new interrogation based on a finer grid or as the output. The iterative form for this method is as the following,

$$\delta^k(\mathrm{x}, \mathrm{y}) = \delta_p^k(x, y) + \delta_c^k(x, y) \tag{8.14}$$

ε_{div} and ε_{mean} are used to check the validity of a measured velocity vector.

(1) ε_{div}, a threshold value for the local divergence, it limits the maximum level to the divergence evaluated as

$$\nabla \cdot V = \left(\frac{u_{i,j+1} - u_{i,j}}{\Delta y} \right) - \left(\frac{v_{i,j+1} - v_{i,j}}{\Delta x} \right) \le \varepsilon_{div} \quad (8.15)$$

(2) ε_{mean}, a threshold value, it limits the maximum level for the ratio between the deviation from the mean and the root mean square (rms). It is calculated as follows:

$$\frac{u_{i,j} - mean(u)}{rms(u)} \le \varepsilon_{mean} \quad (8.16)$$

The quality of a PIV analysis can be assessed through the signal-to-noise ratio (SNR). It is estimated as the ratio between the first peak of the cross-correlation function and the second peak due to noise [31]. A good image is the prerequisite for a good SNR. And to achieve a reliable SNR, each interrogation window (IW) should contain a certain amount of tracers (seeding particles): the optimal tracers' density has to be ≈ 5–10 per IW [32].

White and Bolton [33] first applied the PIV technique to investigate the installation of a displacement pile. White et al. [34] applied this technique to study the instantaneous velocity field at failure of a skirted strip foundation under an eccentric vertical load in the centrifuge test. They later developed a commercial software Geo-PIV to measure soil deformation.

Since for the transparent soil model, a laser sheet is used to generate speckle pattern inside the soil mass, it operates much more like a PIV system. Therefore, the available PIV technique in theory can be applied to transparent soil model to get the velocity and displacement information of the transparent solids. Zhao [35] applied PIV technique to measure the dynamic soil deformation with transparent soil. Currently PIV has become an important tool to measure the soil and rock deformation.

8.5 Influences of Fused Quartz Grain Size on the Displacement by DIC

Transparent soil has been applied in the study of pile penetration and other geotechnical problems. A large number of experiments show that the displacement measurement of DIC method is closely related to the quality of obtained laser speckle image, and the accuracy of transparent soil model test depends on the target displacement amplitude and query window size [36]. Transparent particles of different material characteristics and pore fluids of different components are used in transparent soil modelling, skeleton materials such as silica gel and fused quartz often have different particle sizes, but the characteristics of laser speckle image of transparent soil, and whether it affects the accuracy of the DIC and the implementation of the DIC algorithm (such as whether different sizes of query windows are required), has not yet been found to publish the relevant results. In this section, fused quartz of four different

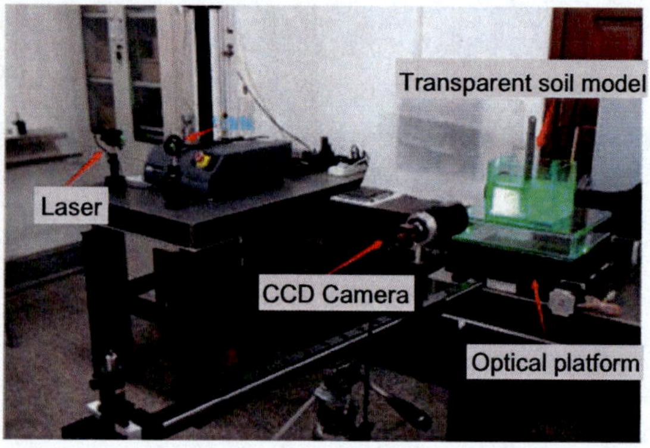

Fig. 8.4 Experimental testing system set up

particle sizes were used as skeleton materials, matching calcium bromide aqueous solution to make a transparent soil model for research on effects of size.

8.5.1 Experimental Program

The test consists of 4 transparent soil models, consisting of fused quartz of four particle sizes (medium fine sand 0.1–0.3 mm, medium sand 0.3–0.5 mm, coarse sand 0.5–2 mm, and fine gravel 2–5 mm). Transparent soil model of dry fused quartz with a solution of calcium bromide with a refractive index of 1.4582 is made in a glass box of 16 cm × 12 cm × 12 cm (length, width and height) and 5 mm thick. Optical displacement platform (MODEL No.: STEL0606 SP100, SERIAL No.: 10040, PACIFIC PRECISION LABS, INC.) can provide 0.125 mm per step movement (32 steps in total). The green laser (15 mW, 532 nm) creates a 4–5 mm thick laser sheet through a cylinder mirror, forming a laser speckle pattern in the model. The high-speed camera captures the image (768 pixels × 1024 pixels) after each movement of the displacement platform. The test system is shown in Fig. 8.4.

8.5.2 Influences of Different Sized Fused Quartz on Displacement Measurement

The grayscale histogram is a function of grayscale, indicating that the number of pixels in an image that reflects the relative frequency of grayscale distribution. The image grayscale is randomly distributed, the speckle effect of particles with different sizes is different, and the grayscale distribution of the image is different. Take 4 images from Fig. 8.5 and cut out the transparent soil area of each image to calculate the grayscale histogram of the different sized fused quartz (see Fig. 8.6). The grey intensity of speckle image of 0.1–0.3 mm particles is uniform, the grayscale value

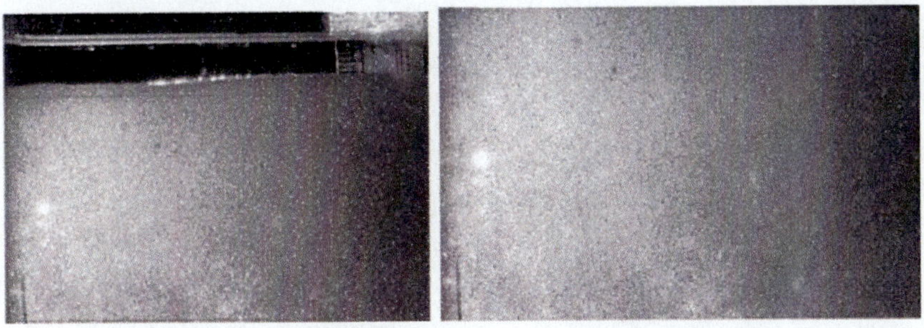

(a) Average image at 4.0 mm (b) Cutting image

Fig. 8.5 The cutting of speckle area from the captured image (i.e., 0.1–0.3 mm silica sand)

of most pixels is close to the grayscale median, higher or lower pixels are less. The result from the DIC calculation is good. The speckle image of 0.3–0.5 mm particle size is dark, the grayscale value of most pixels is low. The results from DIC calculation are also not good. The speckle image of 0.5–2.0, 2–5 mm particle size is very dark, the mean pixel grey scale is very low, the reason may be due to that when the laser intensity remains the same, with the melting quartz particle size increases, the laser entering into the model becomes weak, the speckle intensity distribution is not uniform since the laser is entering from one side of the model. For the two transparent soil models with larger particle sizes, using higher-power, thinner laser light sources, will help to obtain better laser speckle images.

8.5.3 Selecting the Query Window Based on Average Gray Gradient

From previous studies, the measurement error of displacement by DIC is usually considering random noises, interrogation window size and the interpolation error [37, 38]. Here, the influences of fused quartz particle size on interrogation window size were investigated.

The 32 image sequences taken by each transparent soil model are processed by DIC method to compare the mean error SEM and standard deviation Std of the error field between the displacement field and the pre-added displacement field obtained by different size query windows.

$$SEM = u_{\text{mean}} - u_{\text{imp}} \tag{8.17}$$

where $u_{\text{mean}} = \frac{1}{n}\sum_{i=1}^{n} u_i$, the average of the calculated displacement field results, u_{imp} is the pre-added displacement; SEM is the error level.

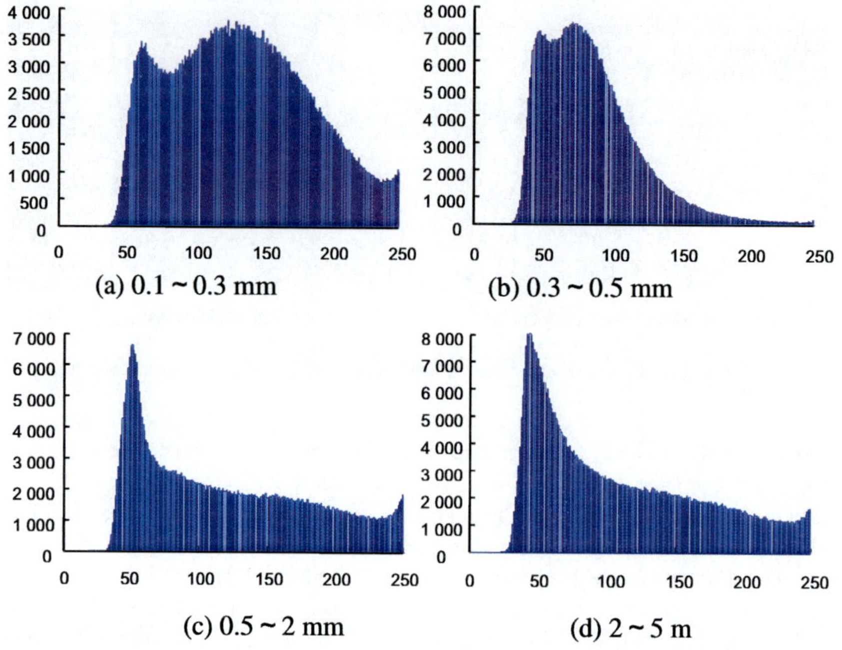

Fig. 8.6 The gray histograms of four transparent soil models

$$Std = \sqrt{\frac{1}{n-1}\sum_{i=1}^{n}(u_i - u_{\text{mean}})^2} \tag{8.18}$$

where, Std is the dispersion of the calculation results to SEM (Fig. 8.6).

Because the displacement measurement of any point is based on calculating the distribution of the speckle pattern in a sub-window around that point on the image, the size of the sub-window directly affects the accuracy of the displacement field measurement. If the sub-windows are too large, because there are points with similar grayscale values in the sub-window and should be error information misidentified into the displacement field, or the sub-windows are too small, and may be because the actual displacement is larger than the size of the sub-window and miss the displacement vector.

The math derivation by [38] indicates that using the one-dimensional strain of linear interpolation and the two-dimensional strain field using bilinear interpolation, the resulting sub-window parameter mean error SEM (u_e) is a biased estimate and is a function of the interpolation error h (i, j), white noise standard difference σ^2, sub-pixel displacement τ_x and square sum of the sub-window gray value after the change as follows

$$SEM(u_e) \cong \frac{\sum_{i=1}^{N}\sum_{j=1}^{N}\left[-h_x(x_{ij}) * f_x(x_{ij})\right] + (1 - 2\tau_x) * N^2 * \sigma^2}{\sum_{i=1}^{N}\sum_{j=1}^{N} f_x^2(x_{ij})} \tag{8.19}$$

where $h_x(x_{ij}) = g(\xi_{ij}) - f(x_{ij})$, $\xi_{ij} = x_{ij} - u$, they are the coordinates of pixel point x_{ij} in the deformed image I_2.

In addition, [37, 38] showed that Std(u_e) of measured displacement is the function of the image noise and the sum of squares of the grey gradient in the image sub-window,

$$f_x^2(x_{ij}) \tag{8.20}$$

It can be seen that the sum of squares of gray gradients in the image sub-window can be used as an effective parameter for the quality of speckle pattern in the sub-window, and can be used for the adaptive selection of the size of image sub-window, but it can not be directly applied to the whole speckle pattern. Pan et al. [37] then proposed the global parameter mean intensity gradient from the sum of gray squares of local parameter image subregions δ_f is used to evaluate the quality of the whole speckle pattern, which is defined as:

$$\delta_f = \frac{\sum_{i=1}^{W}\sum_{j=1}^{H}\left|\nabla f(x_{ij})\right|}{W \times H} \tag{8.21}$$

where, W and H (in pixles) are the image width and height, $\left|\nabla f(x_{ij})\right| = \sqrt{f_x^2(x_{ij}) + f_y^2(x_{ij})}$ is the modulus of gray intensity gradient vector of each pixel, $f_x(x_{ij})$ and $f_y(x_{ij})$ are the gray intensity derivative in x and y directions, determined by gradient operator by the central difference method. Comparing SEM and Std we know that the speckle image with larger mean intensity gradient can get a better measurement accuracy/precision.

Pan et al. [37] proved that the sum of squares of intensity gradients in the x direction of the sub-window in the formula (8.19) and (8.20) can be approximated as the function of the sub-window size N (pixels) and the mean intensity gradients of the whole image δ_f:

$$\sqrt{\sum_{i=1}^{N}\sum_{j=1}^{N} f_x^2(x_{ij})} \cong N \times \delta_f \tag{8.22}$$

As can be seen from Eq. (8.22), for different gray intensity gradients δ_i, by adjusting the size of the sub region N_i, the accuracy of the displacement field results obtained can be guaranteed.

8.5.4 *Influences of Fused Quartz Grain Size on the Query Window Size in DIC*

Obviously, different speckle fields have different speckle sizes. In order to study the influence of fused quartz particle size on the size of query window, randomly select the average images of speckle images taken at the set displacement of 3.375 and 3.500 mm of four models with different grain size. Query windows with different sizes of n = 8–64 (step size of 2) are used respectively, and they are iterated three times according to the overlap ratio of 0.5, the displacement field is obtained and the SEM and Std of the error between the obtained displacement and the pre applied displacement are calculated (Fig. 8.7).

It can be seen from Fig. 8.7 that for the query window of the same size, the larger the particle size, the smaller the measurement mean error, the higher the accuracy, the smaller the standard deviation and the higher the stability. It can be seen from Fig. 8.8 that when n = 16, the displacement field of 0.1–0.3 mm fused quartz model obtained by DIC method is compared with the pre-added displacement field, the error is the smallest (the smallest Std, closest to 0) and the stability is the best. Therefore, in the selected four query windows, query window of n = 16 has the highest accuracy. Similarly, by comparing Figs. 8.8, 8.9, 8.10 and 8.11, it can be obtained that the query window sizes with the highest accuracy of displacement field obtained by digital image correlation processing of 0.3–0.5 mm, 0.5–2 mm and 2–5 mm fused silica models are N = 32, N = 16 and N = 32, respectively. It shows that in digital image correlation processing, applying a query window size to the correlation calculation of different speckle patterns without analysis cannot

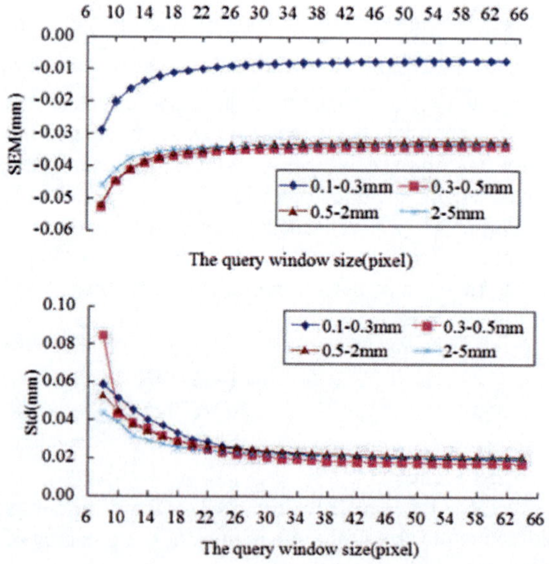

Fig. 8.7 Variation of *SEM* and *Std* with different query window sizes

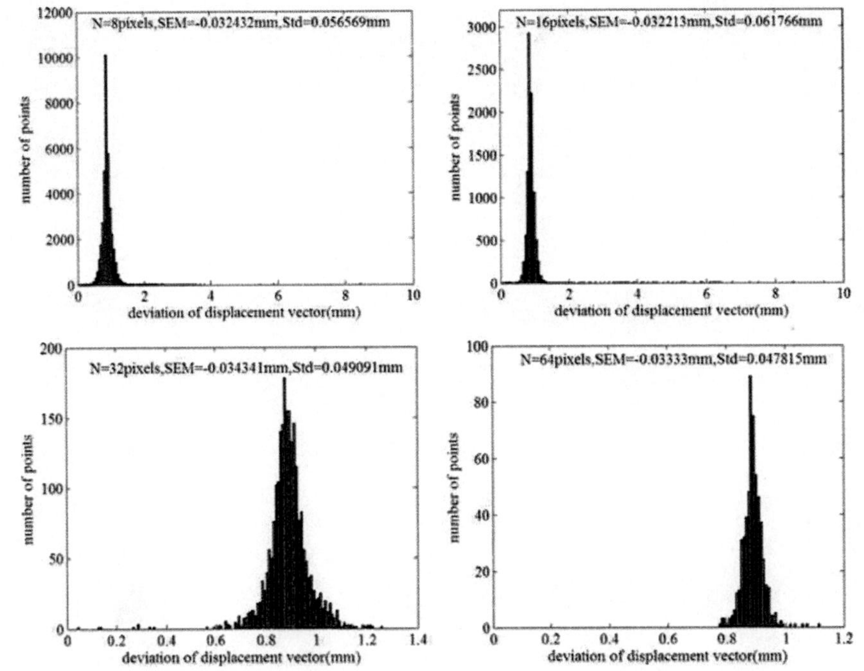

Fig. 8.8 Histograms of displacement error for transparent soil model of 0.1–0.3 mm fused quartz using query window size $N = 8, 16, 32, 64$

ensure the measurement accuracy and the comparability of the displacement field obtained by processing the speckle patterns of transparent soil model with different fused quartz particle sizes. Different sizes of query windows are suitable for different speckle fields in digital image correlation processing. Before correlation calculation of each group images, it is necessary to select the size of query window.

8.5.5 Translation Test

Equivalent sub-window size

The accuracy analysis of displacement measurement is carried out through digital translation test and physical translation test. In order to make the analysis equivalent and ensure that the displacement field obtained by digital correlation calculation of transparent soil models with different particle sizes has equivalent accuracy, it is necessary to select the query window size N, so as to make N* δf equivalent [formula (8.21)] of different digital images. Randomly select the digital speckle image group obtained when the displacement platform moves 4.0 mm, average them respectively, calculate the average gray gradient of the processed speckle image, and adjust the

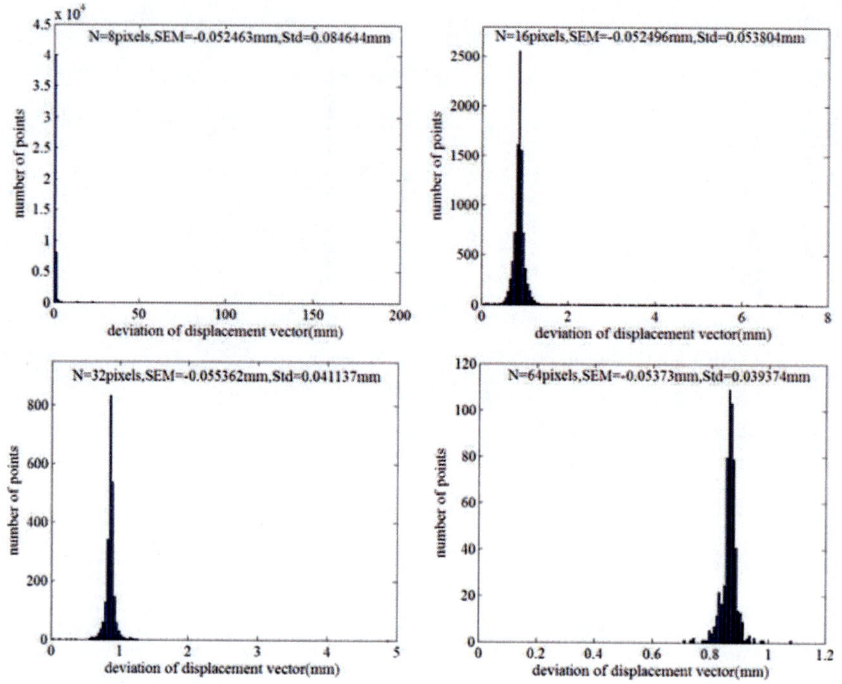

Fig. 8.9 Histograms of displacement error for transparent soil model of 0.3–0.5 mm fused quartz using query window size $N = 8, 16, 32, 64$

size of the query window. The size of the query window is shown in Table 8.1. It can be seen that the large-size fused silica model needs a relatively large query window, so it also has a large average speckle field.

Digital translation test

In this study, the influence of fused quartz models with different particle sizes on the accuracy of the measured displacement field in the pre added displacement field without considering the random error of the test system is studied. Select the average image of the images taken at the pre displacement of 4.0 mm, and use the Fourier transform phase shift theorem to digitally translate each image in the horizontal direction, 0.05 pixels per step and total 40 times. The two adjacent images of each group were correlated with the $n = 16$ equivalent sub-window obtained in Table 8.1 and used in the 0.1–0.3 mm particle size model to obtain SEM and STD (Fig. 8.12). The displacement field is calculated using the obtained query window size (Table 8.1).

It can be seen from Fig. 8.12 that SEM and Std are functions of the sub-pixel displacement, and they are approximately distributed in a straight line. This distribution has been proved by Scherier et al. [40] and is caused by the sub-pixel interpolation algorithm used in the phase shift theorem (in this experiment, it is determined by the

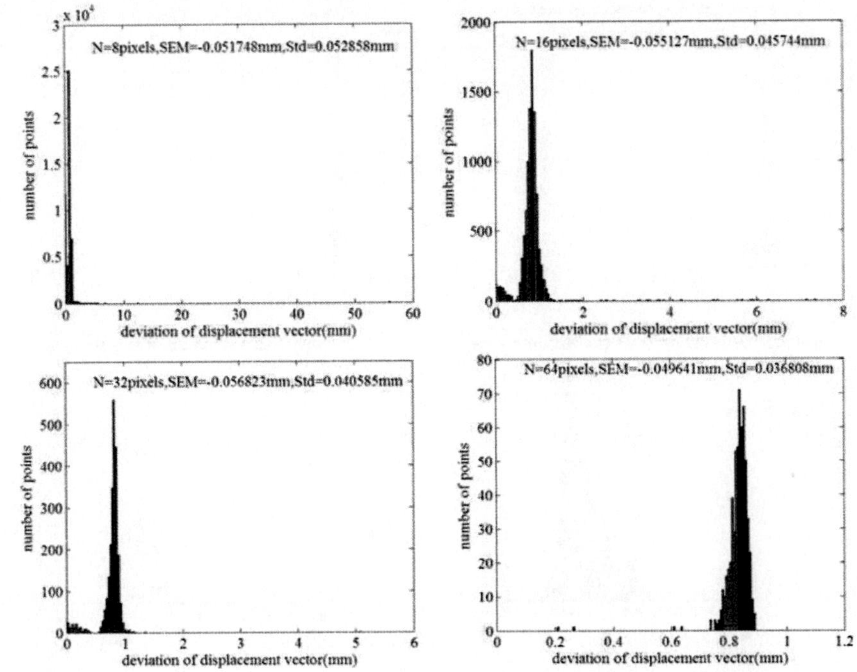

Fig. 8.10 Histograms of displacement error for transparent silica sand model of 0.5–2 mm fused quartz using a query window size $N = 8, 16, 32, 64$

interpolation algorithm used by the sine function called by MATLAB program); The SEM and Std distribution interval after window adjustment is almost independent of the fused quartz particle size, but with the increase of the fused quartz particle size, the data jitter tends to be obvious, and the jitter of the 0.5–2 mm particle size model curve is more severe than that of the 2–5 mm particle size model. It shows that the accuracy and stability of displacement field of large particle sized fused quartz model is not as good as that of small particle sized fused quartz model.

Physical translation test

Image groups (6 in each group) of 0.1–0.3 mm fused quartz model with pre-displacement of 0–4 mm (step size 0.125 mm) at 1, 2, 3, 4 mm were selected for average processing, then we obtained 33 average translation images for the cross-correlation calculation. The query window of corresponding size adjusted as shown in Table 8.1 is used for correlation processing between the two adjacent frames to obtain SEM and Std.

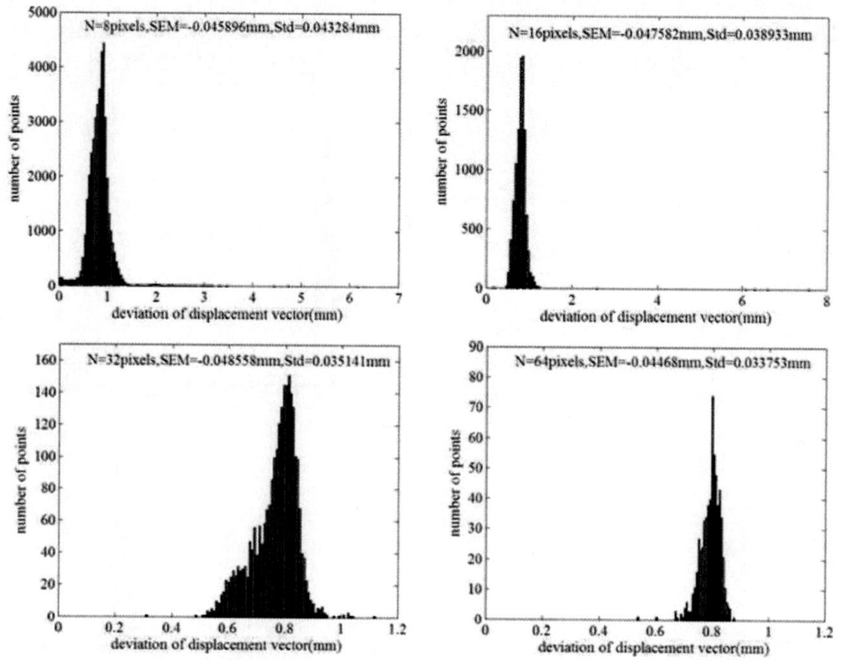

Fig. 8.11 Histograms of displacement error for transparent soil model of 2–5 mm fused quartz using query window size *N*

Table 8.1 The calculated query window sizes for different grain size group to achieve the equivalent measurement precision (at 4 mm displacement)

Grain size group	0.1–0.3 mm	0.3–0.5 mm	0.5–2 mm	2–5 mm
Average gray gradient	21.2824	17.9236	17.2448	17.2187
Equivalent sub-window size (pixel)	16	20	20	20

Do the same for 0.3–0.5, 0.5–2 and 2–5 mm fused quartz models, and the results are shown in Fig. 8.13. It can be seen from the figure that the distribution intervals of Std and SEM are almost the same in different particle size groups. Although 0.1–0.3 mm model of small particle size group has better transparency than 0.3–0.5 mm model, in this group of tests, 0.3–0.5 mm model has better displacement measurement accuracy, but is not as stable as 0.1–0.3 mm model. The 0.5–2 and 2-5 mm models showed poor displacement measurement stability, but the overall measurement accuracy was not much inferior to that of the 0.1–0.3 and 0.3–0.5 mm models. This indicates that, on the premise of fully evaluating the stability of the data, the measurement of the displacement field of the large particle size fused quartz model is possible to ensure a good measurement accuracy.

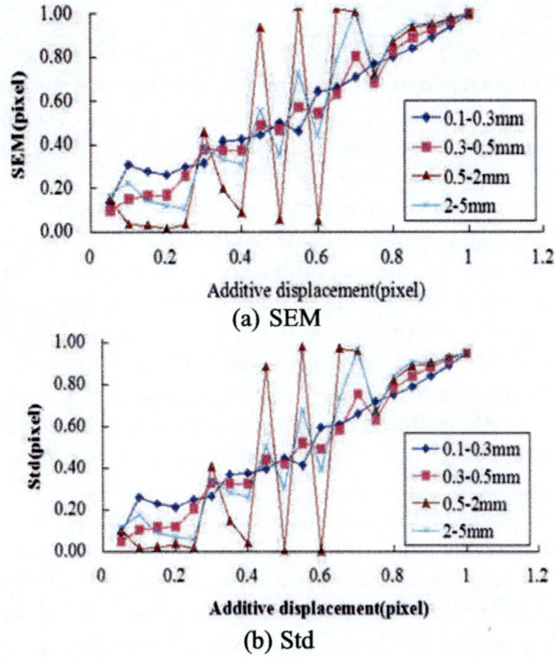

Fig. 8.12 The measurement precision of displacement field computed using digital image correlation method, compared with the displacement field added using phase shift theorem of Fourier transform: **a** SEM, **b** Std

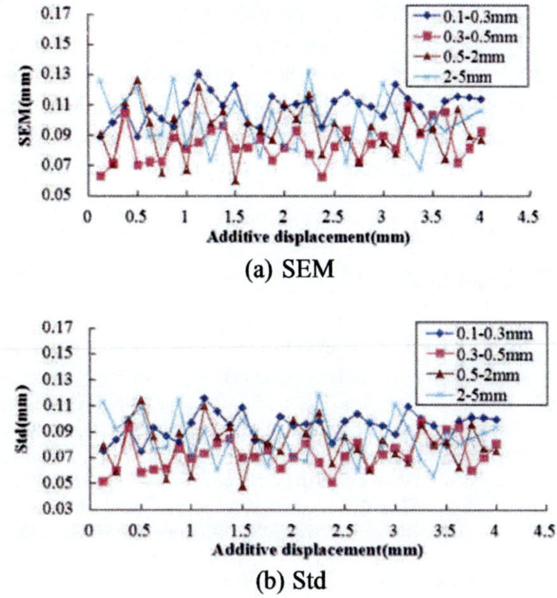

Fig. 8.13 The measurement precision of displacement field computed using digital image correlation method, compared with the displacement field added using optical displacement platform: **a** SEM, **b** Std

8.6 Summary

The principle of deformation measurement of transparent soil imaging is explained in this section. It is based on the DIC method. The computation algorithm of DIC and PIV are briefly introduced in this section. The factors which affect the accuracy of computation is also reviewed.

As an important natural sand simulation material for transparent soil model test, the fused quartz sand is the direct cause of the formation of the laser speckle field, which has an important influence on the quality of speckle pattern used in the DIC processing, and therefore determines the accuracy of displacement field obtained from the speckle pattern. In this chapter, the grain size influence of transparent soil made of fused quartz is investigated. The comparison and analysis are carried out from the measurement system, related processing procedures and processing parameters, the statistical histogram of speckle pattern, and measurement accuracy of the digital and physical pre-added displacement field. It is found that the fused quartz model with large particle size can obtain a certain measurable region under the condition of increasing the laser power and ensuring the speckle recognizable on the one side of the model. In order to ensure the measurement accuracy and comparability of the displacement fields of the speckle patterns formed by quartz particles with different particle sizes, it is necessary to select the window sizes of each group of images before DIC calculation. The displacement field accuracy of the correlation algorithm for the fused quartz model with large particle size is less stable than that of the fused quartz model with small particle size.

In addition, according to the equivalent window size obtained from the global variable of average gray level, it can be seen that the speckle size formed by the fused quartz model with large particle size has a large average speckle size, but due to the small number of test window sizes, it is not enough to get a universal conclusion. This conclusion can be further tested in more experiments.

References

1. Brachman RWI, McLeod HA, Moore ID, Take WA (2010) Three-dimensional ground displacements from static pipe bursting in stiff clay. Can Geotech J 47(4):439–450
2. Tang Z, Liang J, Xiao Z, Guo C (2012) Large deformation measurement scheme for 3D digital image correlation method. Opt Lasers Eng 50(2):122–130
3. Bruck HA, McNeill SR, Sutton MA et al (1989) Digital image correlation using Newton-Raphson method of partial differential correction. Exp Mech 29:261–267. https://doi.org/10.1007/BF02321405
4. Nobach H, Honkanen M (2005) Two-dimensional Gaussian regression for sub-pixel displacement estimation in particle image velocimetry or particle position estimation in particle tracking velocimetry. Exp Fluids 38(4):511–515
5. Chen Y, Li Y, Zhao Y (2016) Sub-pixel detection algorithm based on cubic B-spline curve and multi-scale adaptive wavelet transform. Optik 127(1):11–14. https://doi.org/10.1016/j.ijleo.2015.09.144

6. Pan B, Dafang W, Yong X (2012) Incremental calculation for large deformation measurement using reliability-guided digital image correlation. Opt Lasers Eng 50(4):586–592
7. Pan B (2013) Bias error reduction of digital image correlation using Gaussian pre-filtering. Opt Lasers Eng 51(10):1161–1167
8. Nixon SM, Aguado SA (2002) Feature extraction and image processing. Newnes
9. Zhou Y, Sun C, Song Y et al (2015) Image pre-filtering for measurement error reduction in digital image correlation. Opt Lasers Eng 65:46–56
10. Lecompte D, Smits A, Bossuyt S et al (2006) Quality assessment of speckle patterns for digital image correlation. Opt Lasers Eng 44(11):1132–1145
11. Hua T, Xie H, Wang S, Hu Z, Chen P, Zhang Q (2011) Evaluation of the quality of a speckle pattern in the digital image correlation method by mean subset fluctuation. Opt Laser Technol 43:9–13
12. Yu H, Guo R, Xia H, Yan F, Zhang Y, He T (2014) Application of the mean intensity of the second derivative in evaluating the speckle patterns in digital image correlation. Opt Lasers Eng 60:32–37
13. Su Y, Zhang Q, Xu X, Gao Z (2016) Quality assessment of speckle patterns for DIC by consideration of both systematic errors and random errors. Opt Lasers Eng 86:132–142
14. Yoneyama S, Kitagawa A, Kitamura K, Kikuta H (2006) In-plane displacement measurement using digital image correlation with lens distortion correction. JSME Int J Ser A Solid Mech Mater Eng. 49(3):458–467
15. Pan B, Yu L, Wu D, Tang L (2013) Systematic errors in two-dimensional digital image correlation due to lens distortion. Opt Lasers Eng 51(2):140–147
16. Pan B, Yu L, Wu D (2013) High-accuracy 2D digital image correlation measurements with bilateral telecentric lenses: error analysis and experimental verification. Exp Mech 53(9):1719–1733
17. Yoneyama S, Kikuta H, Kitagawa A, Kitamura K (2005) Lens distortion correction for digital image correlation by measuring rigid body displacement. Opt Eng 45(2):409–411
18. Jerabek M, Major Z, Lang RW (2010) Strain determination of polymeric materials using digital image correlation. Polym Testing 29(3):407–416
19. Haddadi H, Belhabib S (2008) Use of rigid-body motion for the investigation and estimation of the measurement errors related to digital image correlation technique. Opt Lasers Eng 46(2):185–196
20. Hoult NA, Take WA, Lee C, Dutton M (2013) Experimental accuracy of two dimensional strain measurements using digital image correlation. Eng Struct 46:718–726
21. Raffel M, Willert EC, Wereley TS, Kompenhans J (2007) Particle image velocimetry, a practical guide. Springer
22. Scharnowski S, Khler CJ (2020) Particle image velocimetry—classical operating rules from today's perspective. Opt Lasers Eng:106185
23. Tanahashi M, Hirayama T, Taka S, Miyauchi T (2008) Measurement of fine scale structure in turbulence by time-resolved dual-plane stereoscopic PIV. Int J Heat Fluid Flow 29(3):792–802
24. Yamada S, Nakamura H (2016) Construction of 2D-3C PIV and high-speed infrared thermography combined system for simultaneous measurement of flow and thermal fluctuations over a backward facing step. Int J Heat Fluid Flow 61(Part A):174–182
25. Mattern P, Babi M, Kriegseis JA (2017) Plane-to-plane comparison of common averaged vs. POD patterns of time-resolved stereo-PIV data within a pump. Euro J Mech B/Fluids 61(2):321–329
26. Martins JWAF, Foucaut J-M, Stanislas M, Azevedo AF (2019) Characteriza-tion of near-wall structures in the log-region of a turbulent boundary layer by means of conditional statistics of tomographic PIV data. Exp Thermal Fluid Sci 105:191–205
27. Sinton D (2004) Microscale flow visualization. Microfluid Nanofluid 1:2e21
28. Melling A (1997) Tracer particles and seeding for particle image velocimetry. Meas Sci Technol 8(12). https://doi.org/10.1088/0957-0233/8/12/005
29. Tropea C, Yarin A, Foss JF (2007) Springer handbook of experimental fluid mechanics. Springer-Verlag, Berlin

30. Scarano F, Riethmuller ML (1999) Interactive multigrid approach in PIV image processing with discrete window offset. Exp Fluids:513–523
31. Charonko JJ, Vlachos PP (2013) Estimation of uncertainty bounds for individual particle image velocimetry measurements from cross-correlation peak ratio. Meas Sci Technol 24(6):065301
32. Adrian RJ, Westerweel J (2011) Particle image velocimetry, no 30. Cambridge University Press
33. White DJ, Bolton MD (2004) Displacement and strain paths during plane-strain model pile installation in sand. Geotechnique 54(6):375–397
34. White DJ, Randolph M, Thompson B (2005) An image-based deformation measurement system for the geotechnical centrifuge. Int J Phys Model Geotech 5(3):01–12
35. Zhao H (2007) Development and assessment of transparent soil and particle image velocimetry in dynamic soil-structure interaction. University of Missouri, Rolla
36. Sadek S, Iskander GM, Liu JY (2003) Accuracy of digital image correlation for measuring deformations in transparent media. J Comput Civ Eng 17(2):88–96
37. Pan B, Wu D-F, Xia Y (2010) Study of speckle pattern quality assessment used in digital image correlation. J Exper Mech 25(2):120–129
38. Wang YQ, Sutton MA, Bruch HA et al (2009) Quantitative error assessment in pat tern matching: effects of intensity pat tern noise, interpolation, strain and image contrast on motion measurement. Strain 45:160–178
39. Sang Y, Wang Z, Yu S, Zhao H (2019) The loading test on the singe pile with pile cap in transparent soil model. Geotech Test J 42(2):385–406
40. Schreier HW, Sutton MA (2002) Systematic errors in digital image correlation due to undermatched subset shape functions[J]. Experimental Mech 42(3):303–310

Chapter 9
Camera Calibration Based on Neural Network Method

Abstract Camera calibration is an important step in soil deformation measurement with digital image cross-correlation method. Its purpose is to transform the measured soil deformation in image plane into real space. This chapter presented a three-layer, feed-forward neural network for camera calibration which achieves very good performance.

9.1 Camera Calibration

In digital image processing, one first important step is the camera calibration, especially when the camera is not perfectly perpendicular to the measurement field, which occurs frequently in the laboratory model test. The goal of a camera calibration is to link a point P(X, Y) in the object coordinate system to the corresponding point p(u, v) in the image plane coordinate system, as depicted in a pinhole model in Fig. 9.1.

The simplest way to do camera calibration is through the pinhole camera method, which is based on the principle of collinearity [1]. A linear relation between an object point and its image point is assumed. Due to the fact that camera lens distort images and also because the object is commonly behind a refractive window, a simple relation between the object points and the image points does not normally exist. Taylor et al. [2] first applied close range photogrammetry in centrifuge modeling. In their camera calibration model, nine parameters are needed, including 2 orthogonal offsets between the center of the CCD array and the intersection of the optical axis of the camera with the array, the focal length of the camera, 3 polynomial coefficients representing radial geometric lens distortions, 2 polynomial coefficients corresponding to tangential geometric lens distortions, and a differential scale parameter in the image to account for the non-square pixels. This nine-parameter model can correct image distortions such as fisheye and barreling. Paikowsky and Xi [3] used a steel plate with 121 evenly distributed 4.76-mm diameter points, 50 mm apart on as the calibration object points. A polynomial function was used to fit points in the object plane to the image plane. The absolute error could reach 0.4 mm. Recently, [4] have implemented the 14-parameter model of [1] to perform camera calibration for their image-based deformation measurement system in centrifuge experiments.

H. Zhao et al., *Transparent Soil Modelling Technique and Its Application*, Springer Tracts in Civil Engineering, https://doi.org/10.1007/978-981-19-6825-9_9

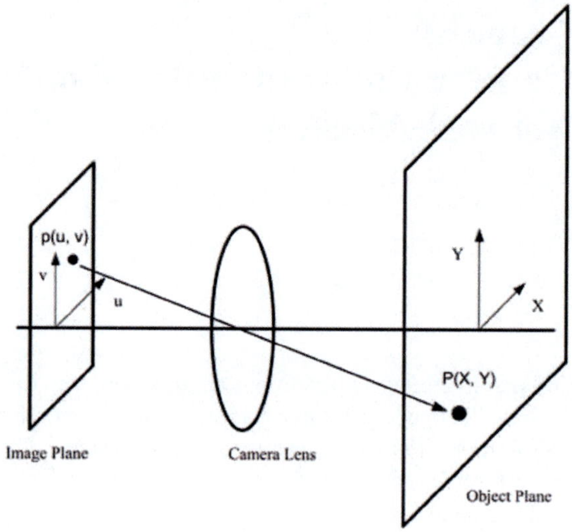

Fig. 9.1 Concept of camera calibration

The Camera Calibration Toolbox for Matlab developed by Bouguet [6] at Caltech was used as a framework for the current research (Fig. 9.2). The features of the toolbox include extracting image corners, running the calibration engine, displaying the results, controlling accuracies, etc. A neural network-based calibration method was proposed and implemented into the Toolbox. The method was then examined by the calibration data for measuring the planar deformation of soils. An example of this calibration algorithm integrated with PIV code MatPIV [5] was illustrated with a rectangular footing model. The image acquiring system used in this research is the Light Wise LW-1.3-G-1394 digital camera from the Image Solution Group, which has a resolution of 1280 × 1024 pixels. It is noted, however, that this method does not give images containing both the soil under observation and also the calibration points. If the camera or test model container moves between calibration and the experiment, the calibration is invalid. This can be improved by using a method with concurrent soil and control point data [2, 4].

9.2 Neural Network Calibration Method

The neural network trained in this study is a three-layer, feed-forward neural network (2–15–2 NN). The input layer consists of n1 = 2 input nodes, and the output layer consists of n2 = 2 output nodes. The hidden layer consists of 15 neurons. Input data are the image points *p*(u, v) obtained from the extraction of the corner points. The target data are the corner points *P*(X, Y) from the calibration plane, and the training data are from half of the calibration points. The other half of the calibration data is used as testing data to test the neural network [7]. A total of 2500 data points were

Fig. 9.2 Coordinate system with 2.5 mm spacing grids for corner points

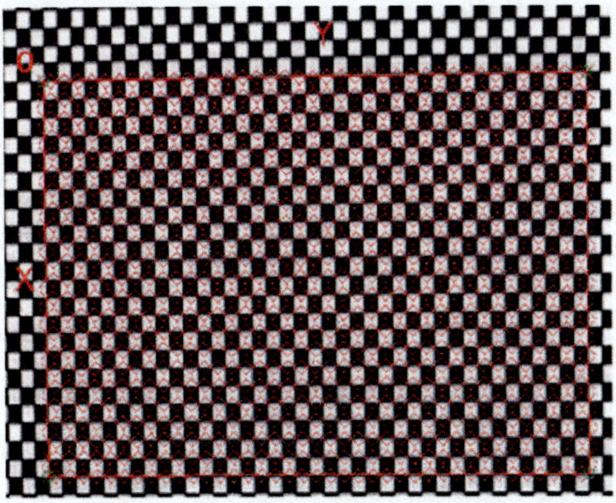

used in this study with 2.5 mm separation distance between two neighbor points. This gives a field of view (FOV) of 145 mm × 155 mm. The coordinates were set on the top left corner of the calibration board as shown in Fig. 9.2.

The neural network training algorithm used was a back-propagation function, which updates the weight and bias values according to Levenberg–Marquardt optimization method. In the training process, the input layer takes an input vector of size n1 = 2 and passes it to the hidden layer. Next, the hidden layer maps this input vector to a vector of size n3 = 2 by a tangent sigmoid transfer function tansig (ζ):

$$a = \frac{2}{1 + e^{-2\zeta}} - 1 \tag{9.1}$$

where ζ is an activation function, the sum of the weighted inputs and biases. This vector from the hidden layer is then mapped by a pure linear transfer function to an output layer. The performance of the neural network model is defined by the normalized square error MSE, expressed as where Q is the number of outputs, N is the total number of data sets, d_{qh} is the set of target data; $x_{\mathrm{out,qh}}$ is the set of network output data. To determine the appropriate number of neurons in the hidden layer, different numbers of neurons (5, 10, 15, and 20) were tested. It was found that 15 neurons in the hidden layer gave the best performance.

$$MSE = \frac{1}{2QN}\sum_{q=1}^{Q}\sum_{h=1}^{N}\left(d_{qh} - x_{out,qh}\right)^2 \tag{9.2}$$

The proposed neural network calibration model was verified through the testing data, which were not used in the training process. The maximum absolute residual errors for X and Y coordinates were 0.011 mm (1/13182 of FOV) and 0.014 mm (1/11071 of FOV), respectively. The residual errors for X and Y are shown in Figs. 9.3

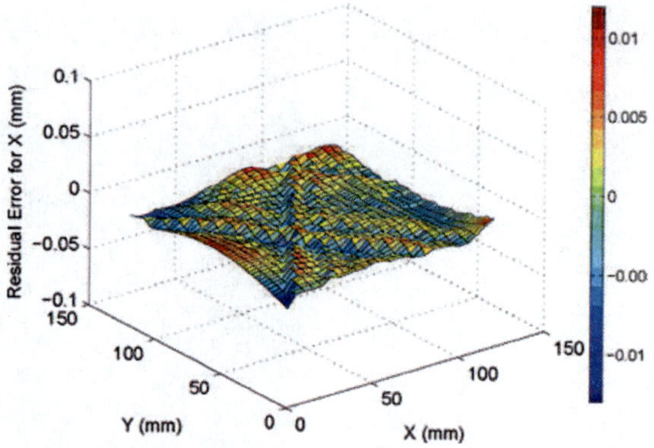

Fig. 9.3 Residual error for X from the Neural Network model

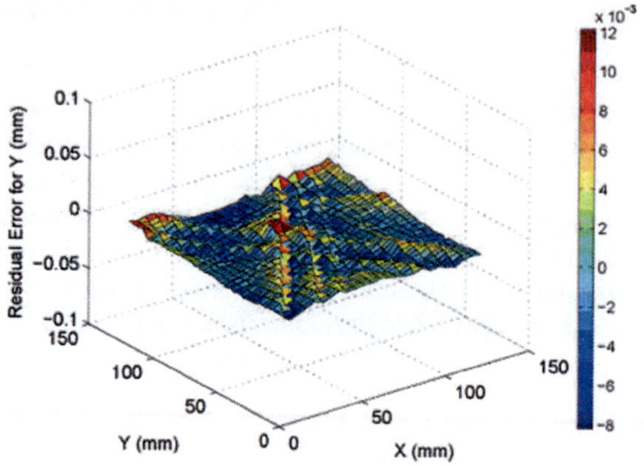

Fig. 9.4 Residual error for Y from the Neural Network model

and 9.4, respectively. It is clear from Figs. 9.3 and 9.4 that the residual error is small and distributed uniformly throughout the entire calibration area.

To examine the effectiveness of this neural network model, the linear and second order polynomial calibration algorithms were also studied in this research, which are expressed as in Eqs. (9.3) and (9.4):

$$\begin{cases} X = a_1 + a_2 u + a_3 v \\ Y = b_1 + b_2 u + b_3 v \end{cases} \tag{9.3}$$

$$\begin{cases} X = a_1 + a_2 u + a_3 v + a_4 u^2 + a_5 v^2 + a_6 uv \\ Y = b_1 + b_2 u + b_3 v + b_4 u^2 + b_5 v^2 + b_6 uv \end{cases} \tag{9.4}$$

where X, Y are the coordinates of point P in the object plane; u, v are the coordinates of corresponding image point p in the image plane; and ai, bi are the coefficient for the model. The coefficients ai, bi were determined from the normalized least square method using the same training data as used for the neural network model.

These models were also tested using the same testing data to examine their prediction accuracy.

Figure 9.5 shows the residual error for X from the linear calibration model. The residual error is small in the center of the calibration plane and becomes large towards the edges of the calibration plane. Figure 9.6 shows the residual error for Y from the linear calibration model. The residual error for Y has a convex distribution.

Along X = 50 mm and 100 mm, the residual error is small. The residual error increases as it moves away from these two lines with the maximum residual error appearing at the edge and center area parallel to the X axis. Figures 9.7 and 9.8 present the residual error distribution for X and Y over the calibration plane from the second-order polynomial model. The residual error for X is the same as that for Y when rotating by 90°. Both have a larger error at the four corners of the plane and a smaller error at the center area of the calibration plane. The maximum absolute residual error is 0.672 mm (1/215 of FOV) for X and 0.685 mm (1/266 of FOV) for Y from the linear calibration algorithm. The maximum absolute residual error is 0.030 mm (1/4833 of FOV) for X and 0.032 mm (1/4844 of FOV) for Y from the

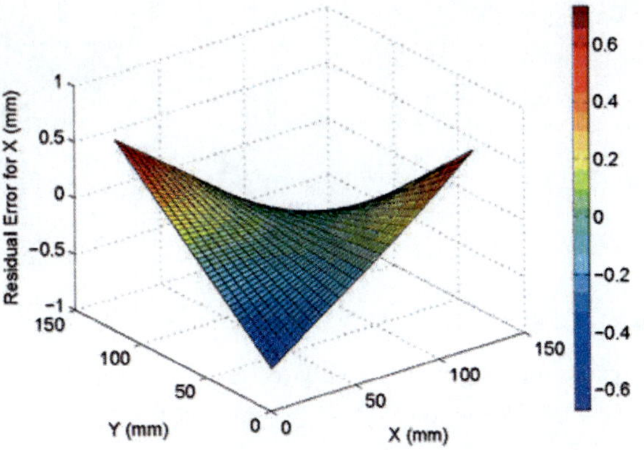

Fig. 9.5 Residual error for X from the linear calibration model

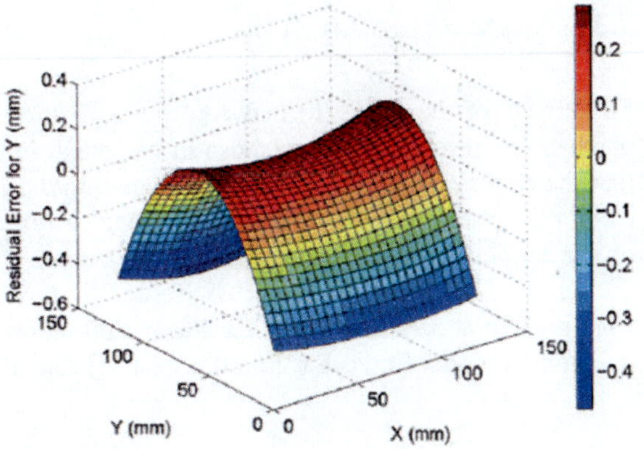

Fig. 9.6 Residual error for Y from the linear calibration model

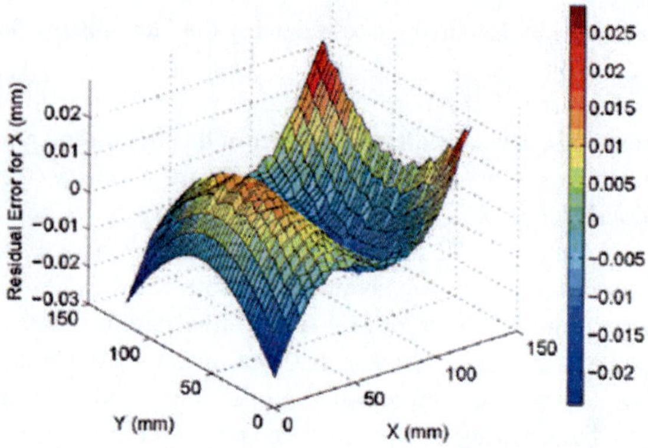

Fig. 9.7 Residual error for X from the second-order polynomial model

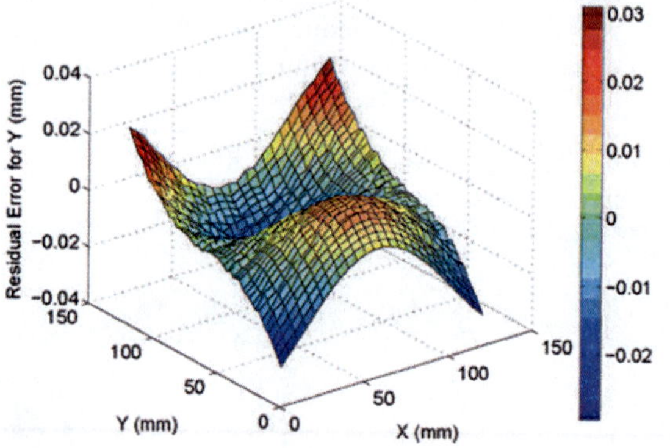

Fig. 9.8 Residual error for Y from the second-order polynomial model

second-order polynomial algorithm. These values are much higher than those errors from the neural network calibration algorithm.

9.3 Angle Error Analysis

The effects of the angle (θ) between the calibration plane and the image sensor plane on the residual error were also investigated. In the setup for this experiment, first the calibration plane was closely aligned to the image sensor plane; then, the calibration plane was rotated to different angles (θ), as shown in Fig. 9.9. Images of the calibration plane were taken by the camera while keeping the position of the camera fixed during this process. Table 9.1 shows the residual errors obtained from the neural network model, the linear calibration model, and the second-order polynomial model for each calibration plane. From Table 9.1, it is evident that the linear algorithm gives the largest absolute residual error among the three different calibration methods.

When the angle, θ, is small, the second-order polynomial model can predict the data well. However, the error increases as the angle becomes larger. The neural network calibration algorithm was found to perform the best, because it gives the minimum absolute residual error, and also the error does not vary much as the angle increases.

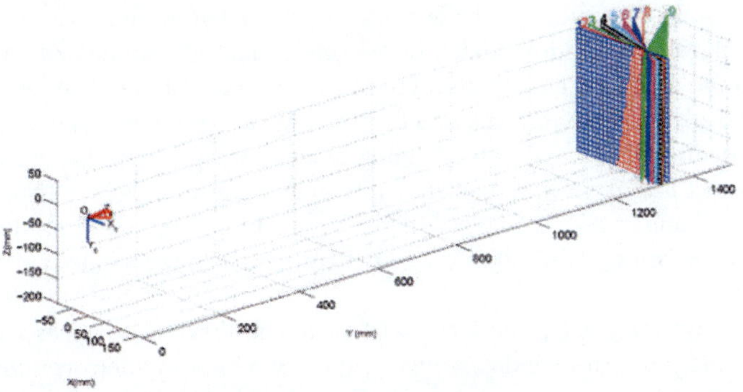

Fig. 9.9 Different calibration planes for angle error analysis

Table 9.1 Absolute error for nine planes from three calibration algorithms

Plane no	Angle (°)	Maximum absolute residual error					
		Linear model		Second-order polynomial model		Neural network model	
		mm	Fraction of FOV	mm	Fraction of FOV	mm	Fraction of FOV
1	7.62	0.672	1/215	0.030	1/4833	0.014	1/11071
2	13.61	1.499	1/97	0.044	1/3295	0.025	1/5800
3	17.97	1.932	1/75	0.057	1/2544	0.016	1/9063
4	25.60	1.932	1/75	0.095	1/1526	0.039	1/3718
5	30.87	2.348	1/43	0.130	1/1115	0.038	1/3816
6	36.49	3.944	1/37	0.199	1/729	0.014	1/10357
7	39.43	4.427	1/33	0.202	1/718	0.016	1/9063
8	50.66	4.482	1/32	0.274	1/529	0.056	1/2589
9	53.93	4.722	1/31	0.303	1/479	0.062	1/2339

9.4 Application in DIC and Particle Image Velocimetry (PIV)

A small-scale wood block (6.35 cm × 7.62 cm × 15.88 cm) was placed on top of the sand in a model container (35.56 cm × 6.60 cm × 35.56 cm), as shown in Fig. 9.10. The height of the sand model was 18.92 cm (7.45 in.), and a loading frame was used to apply the vertical load to the foundation. The camera was set up to take images of the soil below the footing during the loading. The loading rate was set to 2.54 mm/min (0.1 in/min). The test setup is illustrated in Fig. 9.11. A long steel plate was put on the loading platen of the frame, the sand box rested on the steel plate, and the digital camera was set about 2 m from the sand box at the other end of the steel plate. This setup ensured that the digital camera was moving together with the loading platen during the loading process. Images were taken every 30 s. The proposed neural network camera calibration algorithm was implemented into MatPIV, a computer code initially developed by [5] for PIV, to process the images. The principles behind DIC (PIV) image pattern matching techniques are presented in Chap. 8.

Assume two images (I_1 and I_2) are obtained, and these two images are divided into smaller regions (sub-windows, interrogation-windows, or interrogation regions). Then, each sub-window in the image, I_1, is compared with the corresponding sub-window in the image, I_2. The cross-correlation between the two sub-windows is defined as the following

$$R(s,t) = \sum_{m=0}^{M-1}\sum_{n=0}^{N-1} I_1^{i,j}(m,n) \times I_2^{i,j}(m-s,n-t) \tag{9.5}$$

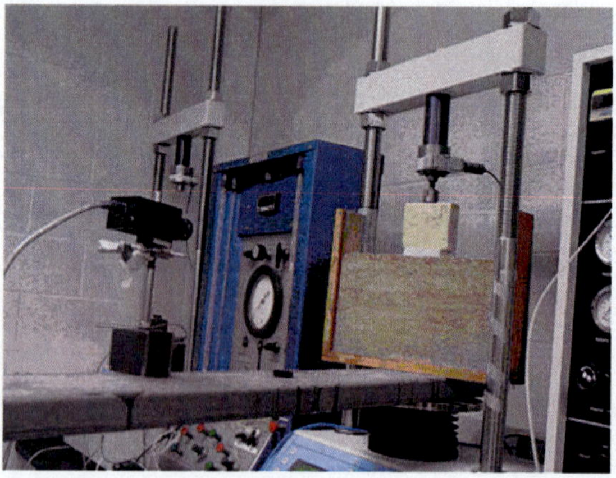

Fig. 9.10 Experimental setup for the rectangular footing model

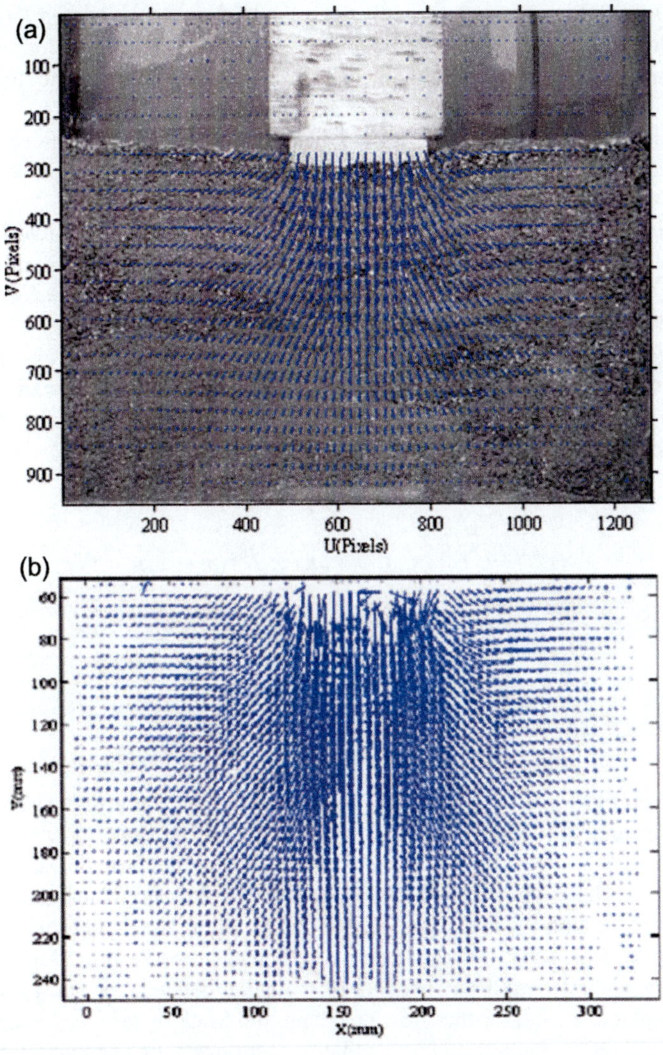

Fig. 9.11 **a** Deformation vector field of sand under the rectangular footing in the image space (t = 6 min, P = 0.92 kN). **b** Deformation vector field of sand under the rectangular footing in the object space (t = 6 min, P = 0.92 kN)

where is I_1(i, j) the sub-window number (*i, j*) in the first image, I_1; and I_2(i, j) is the sub-window number (i, j) in the second image, I_2. When this expression reaches its peak value, the two sub-windows will most likely match each other. The sub-window has moved s and t units in *i* and *j* directions, respectively. By a centroid fit, a Gaussian fit, or a parabolic fit to the region close to the peak, the system can achieve sub-pixel resolution. The smallest window size used in this analysis was 36 pixels × 36 pixels in order to cover the large displacement of sand particles.

The digital image correlation method was applied to analyze the two images to obtain the full-field deformation of the soil surrounding the footing. Figure 9.11a, b show the displacement vector field of the soil under the footing at t = 6 min, and P = 0.92 kN (20.6 lb), which are similar to cylindrical cavity expansion.

Therefore, Fig. 9.12a, b show circular contours of displacement. From the experimental results, no Terzaghi's failure mechanism was observed, i.e. the vertical movement extends to too great a depth to match Terzaghi's mechanism, and there is no heave beside the footing as predicted by Terzaghi.

$$R(s,t) = \sum_{m=0}^{M-1}\sum_{n=0}^{N-1} I_1^{i,j}(m,n) \times I_2^{i,j}(m-s,n-t) \tag{9.6}$$

9.5 Summary and Conclusions

A three-layer back-propagation neural network calibration algorithm was developed for camera calibration in an image-based measurement system. This algorithm was compared with the linear calibration and second-order polynomial calibration algorithms. When the angle between the image plane and object plane became large, the linear calibration algorithm and second-order polynomial calibration algorithm gave a considerable error. It was found that the neural network calibration model will give a very accurate result independent of the angle between the image plane and the object plane. This algorithm was integrated into MatPIV software. This fast and accurate calibration method should facilitate image correction in situations when the camera is not aligned with the model.

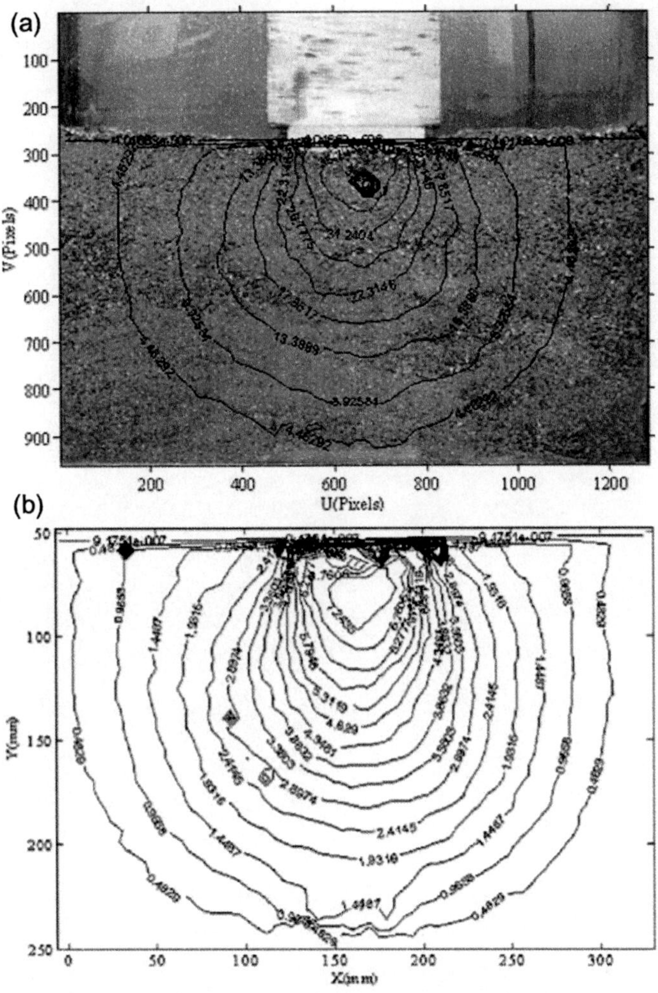

Fig. 9.12 **a** Magnitude contour of displacement vector for sand under the rectangular footing in the image space (contour label in unit of pixels, t = 6 min, P = 0.92 kN). **b** Magnitude contour of displacement vector for sand under the rectangular footing in the object space (contour label in unit of pixels, t = 6 min, P = 0.92 kN)

References

1. Heikkila J, Silven O (1997) A four-step camera calibration procedure with implicit image correction. In: IEEE computer society conference on computer vision and pattern recognition (CVPR 1997), San Juan, Puerto Rico, pp 1106–1112
2. Taylor R, Grant R, Robson S, Kuwano J (1998) An image analysis system for determining plane and 3-D displacements in soil models. In: Proceedings of Centrifuge 1998, Tokyo, pp 73–78
3. Paikowsky S, Xi F (2000) Particle motions tracking utilizing a high-resolution digital CCD

camera. ASTM Geotech Test J 23(1):123–134
4. White D, Take W, Bolton M (2003) Soil deformation measurement using particle image velocimetry (PIV) and photogrammetry. Geotechnique 53(7):619–631
5. Sveen J (2004) An introduction to MatPIV, v.1.6.1. Eprint No.2, Department of Math. University of Oslo, ISSN 0809-4403
6. Bouguet J-Y (2022) Camera calibration toolbox for Matlab (version 1.0) [computer software]. CaltechDATA. https://doi.org/10.22002/D1.20164
7. Zhao H (2007) Development and assessment of transparent soil and particle image velocimetry in dynamic soil-structure interaction. University of Missouri, Rolla

Chapter 10
Three-Dimensional Transparent Soil Imaging and Processing

Abstract Non-intrusive observation of the spatial deformation field of the soil has been a difficult problem for laboratory model test testing. Based on the advantage that the visualization test method of transparent soil can observe the two-dimensional (2D) deformation inside the soil, this chapter described a new automatic tomographic scanning measuring device to observe the three-dimensional (3D) spatial deformation of the soil inside the transparent soil. A series of 2D laser speckle images of different vertical cross sections before and after deformation were obtained, and an improved 3D reconstruction algorithm was used to reconstruct the 3D displacement field of soil after deformation. Its applications in pile penetration are reported.

10.1 Introduction

In soil-structure interaction, many problems are three dimensional in nature such as tunnel excavation, slope failure, pile-soil interaction etc. This requires to investigate the problem from the three-dimensional point of view. In recent years, many research efforts have been devoted to the development of non-destructive three-dimensional (3D) deformation measurement technology. Typically, stereo particle image velocimetry (stereo-PIV) was a more advanced optical measurement method based on the PIV technique and a binocular vision model. Two cameras capture images from different angles, then the 3D displacement field was calculated using the paired images and reconstruct algorithm [1–4]. Recently, in order to reduce hardware investment and eliminate the need of precise camera synchronization, single-camera stereo-PIV technique was realized by placing an additional optical device [5]. However, thus far, the stereo-PIV technique could only provide three-dimensional surface information. Additionally, the use of advanced non-intrusive techniques, such as X-ray computed tomography (CT) scanning and magnetic resonance imaging, in geotechnical small-scale model have been applied to produce visual images for not only in cross section but also the images in three dimension [6–8]. Nevertheless, routine application of these techniques was limited by their high costs and small experimental operation space.

H. Zhao et al., *Transparent Soil Modelling Technique and Its Application*, Springer Tracts in Civil Engineering, https://doi.org/10.1007/978-981-19-6825-9_10

Three-dimensional transparent soil imaging provides an important and convenient tool to carry this type of research. The first three-dimensional transparent soil imaging system was developed by [9], in which the soil model box is moving in parallel to the laser sheet. Zhao et al. [26] developed another transparent soil imaging system, in which the laser sheet and the camera were moving synchronously while the transparent soil model does not move. In this section, the later 3-D transparent soil imaging system will be introduced.

10.2 Transparent Soil Model and Testing Set Up

The granular material used in this experiment is fused quartz with a purity of 99.99%. In order to simulate sandy soil, the grain size of the fused quartz is in the range of 0.1–2 mm. The grain size distribution curve for the fused quartz is shown in Fig. 10.1. Fused quartz has a specific gravity of 2.33, and a drained friction angle in the range of 38.6°–40.3°. More details regarding physical properties of fused quartz can be found in Table 10.1. The mixed pore fluid comprised of nParaffin C12 and White Oil 15 at a volume ratio of 1:4.75. The magnetic agitator (Fig. 10.2a) was used to fully stir the two immiscible liquids, and the refraction index (RI) of the mixed pore liquid was 1.4590 at 20.0 °C, which is the same refractive index as the reported RI of fused quartz [10]. The relationship between temperature and pore fluid viscosity was measured by the Cannon–Fenske viscometer, and the relationship curve is shown in Fig. 10.2b. The blend of pore fluid has a dynamic viscosity of 11 mPas at 20.0 °C (the viscosity of the water is 1.01 mPas). The viscosity is proportional to the contact area and velocity. In the experiment of jacked-pile penetration, the penetration velocity and the contact area between piles and soil are small, and the magnitude of penetration resistance is much larger than that of the viscosity of pore fluid, so the influence of viscosity on penetration resistance can be ignored [11].

In order to prepare the transparent soil model, the fused quartz which meets the requirements of particle size distribution was cleaned with de-aired water and oven dried. After then, the sand pluviation method is used to fill the fused quartz particles into the glass model box with a dimension of 300 mm × 180 mm × 280 mm (length × width × height) through a funnel. In order to prevent the occurrence of stratification, it is necessary to mix and compact the filled fused quartz. Then, the hose was inserted into the bottom of the model box and fill the model box with pore fluid from the bottom to top slowly. Afterward, the transparent soil model is placed into a sealed bucket, a vacuum machine is used to extract air from the bucket to make the transparent soil model become saturated. The saturated transparent soil model is consolidated through the loading system by means of the pre-drilled organic glass plate. The transparent soils sample has excellent transparency as illustrated in Fig. 10.3a.

In the jacked-pile penetration tests, the model piles were made of polymethyl methacrylate, as shown in Fig. 10.3b. The length of the pile was about 125 mm, with a diameter of 22 mm. In order to minimize reflection of light, the surface of all

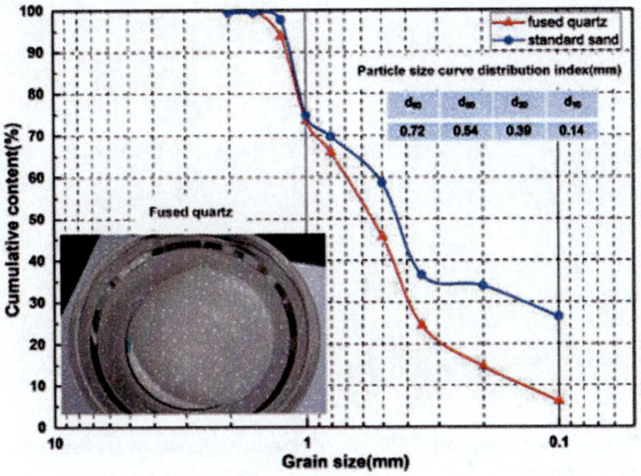

Fig. 10.1 Grain size distribution curve for the fused quartz and Fujian (China) standard sand

Table 10.1 Parameters of the fused quartz

C_u	C_c	ρ_d (g/cm^3)	$\rho_{d\max}$ (g/cm^3)	$\rho_{d\min}$ (g/cm^3)	D_r (%)
5.14	1.51	1.4	1.56	1.18	65

C_u coefficient of ununiformity; C_c coefficient of curvature

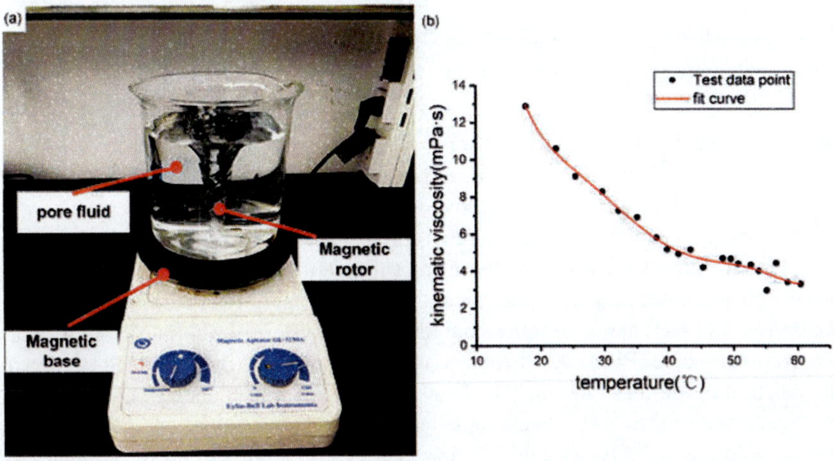

Fig. 10.2 **a** The magnetic agitator. **b** The relationship curve between temperature and pore fluid viscosity

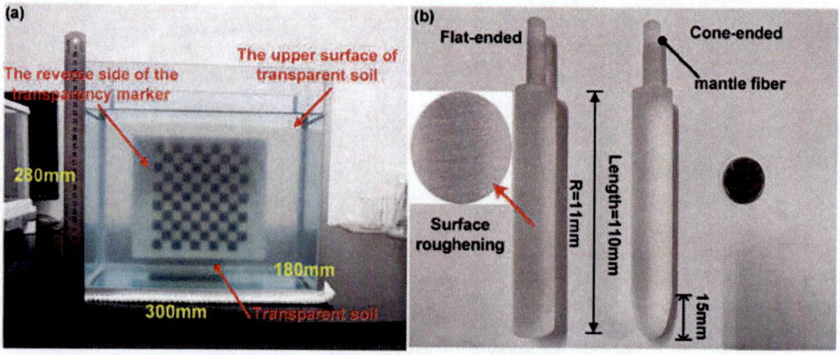

Fig. 10.3 **a** Reference viewed behind a 18 cm thick model made of transparent soil. **b** Two types of the small scale pile models

piles are roughened. Although the corresponding method is not used to measure the specific friction coefficient, the same method is used to rough the model pile, which provides consistent friction conditions for the model test and gives full play to the interaction between the pile and the soil.

10.3 Automatic Tomographic Scanning Measuring Device and Experimental Setup

In order to visualize the three-dimensional spatial deformation of jacked-pile penetration problem, a high-precision motorized linear stages with a CCD camera and a laser scanning illumination system is employed. Figure 10.4 illustrates the custom experimental setup of novel automatic tomographic scanning measuring. It includes an axial loading subsystem and two measuring subsystem. The axial loading subsystem provides a guarantee for the continuous and accurate penetration of the pile model into the transparent soil sample. For the axial loading subsystem, the classic Proportional-Integral-Derivative (PID) closed-loop control algorithm was implemented to control the position of the servo motor. The displacement sensor is connected to the bottom of the servo motor as a feedback signal source for displacement control. The load cell is indirectly fixed by a loading frame mounted on the top of the servo motor as a feedback signal source for the force control. The measuring subsystem is mainly composed of a laser transmitter, a CCD camera, a computer, a set of image acquisition software, two sets of high-precision motorized linear stages: one for the laser device and one for the CCD camera. The two sets of high-precision motorized linear stages are composed of a stepper motor, a couplings, a ball screw, a small size rail and grating ruler. Based on the displacement signal collected by the grating ruler, the displacement deviation is calculated. The PID control algorithm is used to correct the displacement deviation in the computer controller to realize the closed-loop control,

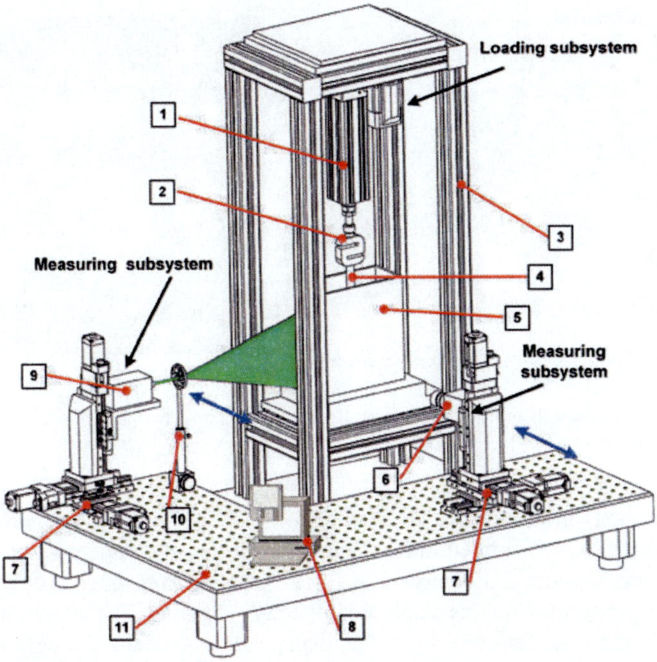

Fig. 10.4 Schematic diagram of experimental setup: (1) Electric servo cylinder, (2) Load cell, (3) Loading frame, (4) Model pile, (5) Transparent soil model box, (6) CCD industrial camera, (7) High-precision motorized linear stages, (8) Computer, (9) Laser device, (10) Optical prism, (11) Optical platform

so that the two stepper motors remotely and accurately controlled by the computer controller are used in the measuring subsystem for driving the motorized linear stages to move, respectively. In image collection, the CCD camera and the laser device move the same distance synchronously to obtain a two-dimensional (2D) image at different locations, and the closed-loop control system is used to ensure high synchronization control accuracy. By referring to the idea of 3D reconstruction of CT scan in medical science, a modified 3D reconstruction method is used to quantitatively calculate 3D deformation field from a series of deformed 2D images at different positions.

The tests were performed in a darkroom. The first step is to make sure that the laser device produced a laser sheet inside the transparent soils sample. The initial position of the laser sheet should be aligned with the axis of model pile. This step is performed manually. Once the laser sheet is in place, we placed the CCD industrial camera (1028 × 1536 pixels) approximately 800 mm away from the model box with its optical axis was adjusted perpendicularly to the laser sheet. Due to the finite thickness of the light sheet, in order to prevent the overlapping effect of adjacent two slices, the spacing between the two laser sheets should be greater than 1.5 times of the maximum particle size of the transparent soil sample. The schematic diagram of core principles of the experimental approach is shown in Fig. 10.5a, which describes

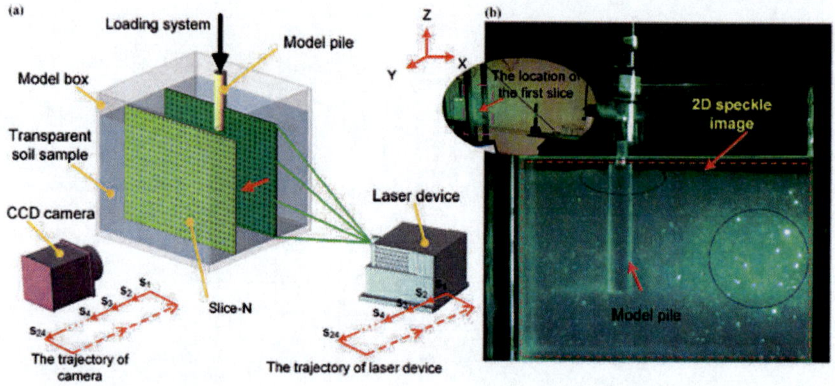

Fig. 10.5 **a** Schematic diagram of experimental approach. **b** The actual model test image captured from the CCD camera

the synchronous movement trajectories of the CCD camera and the laser device. The high-precision motorized linear stages control the simultaneous translation of the laser device and CCD camera along the Y-axis at fixed intervals. The test of model pile penetration adopts displacement control method. The overall process of the experiment is as follows: At the beginning, we started the tomographic scanning measuring device, synchronously control the CCD camera and laser device to move to the designated position in turn (24 target locations were set in this experiment). Each move to a specified location (the spacing between each movement is 3.5 mm), the computer program automatically controls the CCD camera to take photos of the laser bright speckle in the irradiation position of the laser device in the transparent soil model, and the whole tomographic scanning process obtains 24 two-dimensional speckle images. After the CCD camera and laser device return to the initial position, turn off the tomographic scanning measuring device. Then the penetration test was performed at a constant speed of 0.25 mm/s. According to the set penetration displacement, the model pile was penetrated to the specified depth of the transparent soil model by the axial loading subsystem (the penetration experiment takes a penetration depth of 0.5R as a stage). Then the tomographic scanning measuring device was started again and the above operation steps were repeated to obtain the deformed speckle images of all the specified positions after the first stage of the pile penetration test. And so on, we observed the whole pile penetration test through automatic tomographic scanning measuring device, and obtained a series of 2D speckle image. The actual image captured from the CCD camera, as shown in Fig. 10.5b. The transparent soil model is nearly completely transparent, and transparent particles cannot be identified individually in the transparent soil sample. The speckle pattern created by the interaction between transparent soils and laser light, has a well-defined spatial structure [12]. It is a unique signature of the interaction between the transparent soil matrix, impurities, entrapped air, and the laser. The speckle properties typically depend on the roughness, refraction and reflection of the surface. The penetration of the pile disturbs the surrounding transparent soil particles will result in change in the speckle distribution in the slice [9]. From analysis of two consecutive images, the displacement field of soil can be obtained.

10.4 Optimized Particle Image Velocimetry Image Processing Algorithm

White et al. [13] reported that particle image velocimetry (PIV) image processing technology was used as a method to measure soil deformation in geotechnical model tests. Since then, PIV technique has become routine experimental practice to obtain deformation patterns in many geotechnical research laboratories [9, 14]. An important step in PIV technique is to track the deformation information between consecutive speckle images taken before and after deformation. This step first selects seed point in the reference image and defines the target image, and then each image is partitioned into many smaller areas called subsets or sub-window. Each subset of 2 N × 2 N pixels corresponds to one seed point in each image. The similarity between reference subset and target subset is evaluated according to the correlation criterion. The seed points in the target image can be located through searching the peak location of the correlation coefficients. With the aid of the camera calibration parameter (the scale ratio of the object space coordinates to the pixel coordinates), the pixel differences of the location of the reference subset center and the target subset center yield the displacement vector at seed point [15]. To determine the corresponding location of each subset in the deformed image, various correlation functions are defined, such as Cross-Correlation (CC), Sum of Absolute Differences (SAD) [16] and Zero-mean Normalized Sum of Squared Difference (ZNSSD). Generally speaking, the ZNSSD is insensitive to the scale and offset changes in illumination lighting fluctuations, which is highly recommended to use among the many correlation criterion.

To date, the most effective pattern-recognition algorithm is incremental reliability-guided particle image velocimetry (RG-PIV) based on ZNSSD algorithm In the conventional PIV algorithm, the first image is considered as the reference image and all the deformed images are compared with the first image to extract the deformation information. For the large deformation, the displacement cannot be accurately obtained by the conventional PIV algorithm. This is called decorrelation effect. Furthermore, frequently updating reference image would unavoidably accumulate large additional errors in the deformation measurement [14, 17]. In this optimized PIV algorithm, some initial seed points are defined in the original reference image and tracked its corresponding locations in a series of deformed images. If the computed correlation coefficient of initial seed points in a series of image are smaller than predefined threshold, which means no decorrelation effect in the deformed image, the predefined reference image is employed to continue correlation analysis. However, if the ZNSSD coefficient of the current deformed image is larger than the predefined threshold, the image recorded just before the current deformed image is selected as an updated reference image to continue correlation analysis [15]. With the aid of the idea of incremental RG-PIV, the image post-processing program in this study is an improved program based on the existing MATPIV program with the method of automatically updating the reference image. Considering that the penetration problem of jacked piles is a three-dimensional large deformation problem with out-of-plane displacement, the predefined threshold of the ZNSSD coefficient is set to be 0.6 (1 is completely correlated, 0 is completely irrelevant).

10.5 The Calibration Tests

10.5.1 The Calibration Tests of Automatic Tomographic Scanning Measuring Device

Calibration is a necessary step in verifying the accuracy of the measuring device. To obtain accurate measurements when performing tomographic scanning experiments, the special calibration process was designed to achieve two tasks: (1) lens distortion of a camera: rectifying the image distortion; (2) synchronous matching: the camera and laser device move synchronously to avoid pixel equivalent changes. The calibration device is shown in Fig. 10.6a. The calibration plane uses a pattern of 10 × 11 squares, which has 130 corners (Fig. 10.6b). The size of a square is 15 mm × 15 mm. For calibrating camera lens distortion task, a simple and flexible camera calibration technique proposed by [18] was used, which consists of a closed-form solution, followed by a nonlinear refinement based on maximum likelihood criterion. The specific calibration process is as follows: First, take six images of the calibration plane under different orientations by moving either the plane or the camera, which is similar to the one illustrated in Fig. 10.6a. Next, detect the feature points in the images, as shown in Fig. 10.6c. Then, Zhang's calibration algorithm was used to estimate camera parameters. The estimated distortion parameters allow us to correct the distortion in the original images [18].

To achieve the calibration of synchronous matching between camera and laser device, the pixel equivalents of images at different tomographic positions can be calculated by using the calibration board. At the same time, in order to demonstrate the superiority of the synchronous motion design of camera and laser device in the tomographic scanning measuring device, it was verified by contrast test. By setting up two groups of experiments: one is the synchronous movement of the camera and laser; the other set is to fix the camera but allow the laser moving. The calibration plate in the two groups of experiments was placed at the same position overlapping the plane illuminated by the laser, and photographing the calibration plates at 24 different locations to calculate pixel equivalents in images at different locations. Three parallel tests were carried out for each group of tests, and the error lines of pixel equivalent for the two groups of tests were shown in Fig. 10.7. The results of the calibration tests indicate that the pixel equivalent in the synchronous movement test

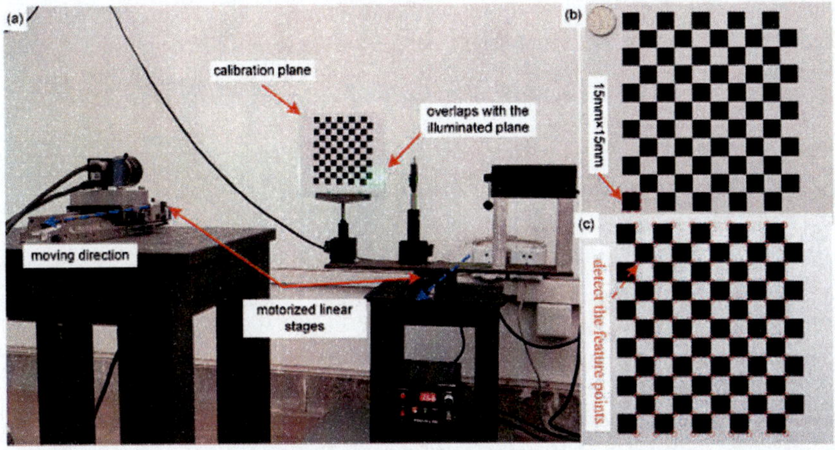

Fig. 10.6 The calibration tests of automatic tomographic scanning measuring device: **a** The calibration device. **b** The calibration plane. **c** Detect the feature points

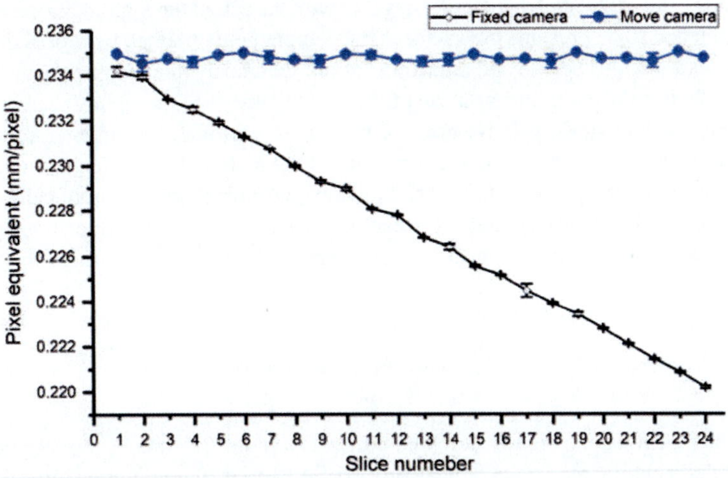

Fig. 10.7 The error lines of pixel equivalent for the two groups of tests

between camera and laser device is a fixed value of about 0.2348. However, in the fixed camera comparison experiment, the pixel equivalent changes with the change of slice position. The closer the slice position is to the camera, the smaller the pixel equivalent is (The position of slice 24 is closest to the camera, and the position of slice 1 is furthest from the camera). This is called the near large far small effect. Therefore, in the tomographic scanning measuring device, the design of camera and laser device moving synchronously can ensure the pixel equivalent in a stable state. In addition, the error line of pixel equivalent reflects a small error, roughly ± 0.3 μm.

This means that the high repeatability accuracy of tomographic scanning measuring device can ensure that the camera and laser device are located in the pre-determined position in each test, which fundamentally guarantees the measurement accuracy.

10.5.2 The Accuracy of the Optimized Image Processing Algorithm

In order to verify the accuracy of the optimized image processing algorithm for large deformation measurement, a translational test of the speckle image was performed. The translation test system is shown in Fig. 10.8a. The speckle image was generated in MATLAB by randomly projecting thousands of white dots onto a black background to subpixel positional resolution (Stanier et al. 2015). The speckle image is shown in Fig. 10.8b with a size of 105 mm $\times$ 105 mm. In the translation test, each time the target speckle image is controlled to move at a constant interval (0.5 mm), the post-processing program is used to analyze and calculate the two speckle images before and after the motion, and the accuracy of the post-processing program is evaluated by comparing the calculated displacement value with the actual translation displacement value. Figure 10.8c presents the vector displacement field obtained by processing the two speckle images before and after translation. Obviously, the displacement vector clearly depicts the overall translation of the speckle image to the right. It can be seen from Fig. 10.8d that the relative error of the 10 translational experiments fluctuates up and down. The average relative errors in 10 translational experiment is 0.04668 mm, which is approximately 0.2 pixel equivalent (the pixel equivalent of calibration experiment is 0.2215 mm/pixel). The results show that the calculated accuracy of the post-processing algorithm meets the requirements of image measurement.

10.6 Modified 3D Reconstruction Method

Three-dimensional reconstruction method is to obtain 3D image model and the spatial deformation by means of post-processing a series of continuous images with computer software [19, 20]. An important challenge in 3D reconstruction is how to build a 3D model from a series of 2D slices information. In order to obtain accurate 2D speckle image information, incremental RG-PIV algorithm was used in the image post-processing. The newly developed 3D reconstruction method is volume rendering method. To clearly describe the steps of modified 3D reconstruction method based on the medical 3D reconstruction, the flowchart of procedures is given in Fig. 10.9. Firstly, the 2D image sequences were obtained by the tomographic scanning measuring device. The deformation data of 24 slice images were calculated by the incremental RG-PIV algorithm, and the coordinates and deformation data of 24 slices were extracted by MATLAB. Then the discrete 3D volume data matrix (x, y,

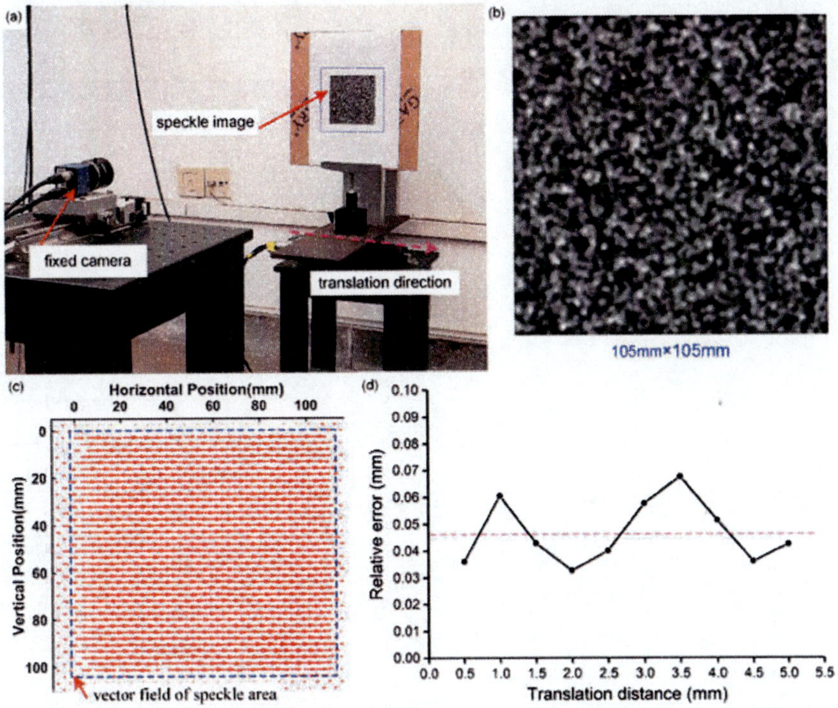

Fig. 10.8 The calibration tests of accuracy of post-processing algorithm: **a** The translation test system. **b** The speckle image. **c** The vector disolacement vector. **d** The relative error value of the 10 times of translational calibration experiments

z, data) was constructed. Subsequently, the cubic spline interpolation function is used to obtain the missing data between the two slices in order, and then the experimental data and the interpolated data are integrated into a complete 3D volume data matrix. The construction principle of 3D data set is shown in Fig. 10.10. After constructing the data set, the image information is smoothed, filtered, the grayscale transformation and equalization processing are conducted by using the toolbox in MATLAB. Then, the volume rendering method is used to project the 3D data field. To accurately reflect the spatial details, the well-developed Ray Casting algorithm is used to map the planar projection [21]. In order to accurately reflect the internal information of the 3D displacement field, the self-written isosurface rendering function is used to present the changes of the three-dimensional displacement field. Subsequently, the three-dimensional image fragments are constructed and the normal direction of the fragments are calculated. Finally, the ideal 3D reconstruction image is obtained.

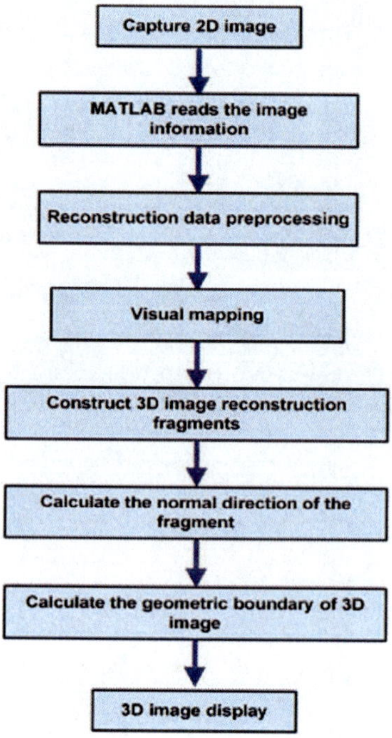

Fig. 10.9 The flowchart of 3D reconstruction procedures

10.7 Application to Jacked-Pile Penetration

A series of 2D speckle images about jacked-pile penetrating model tests obtained through the automatic tomographic scanning measuring device were shown in Fig. 10.11. As can be seen from Fig. 10.11, the first slice is at the axial section of the model pile, and the remaining slices extend outward along the Y-axis with fixed spacing.

10.7.1 Comparison of the Displacement Pattern Between Flat-Ended Pile and Cone-Ended Pile

Figure 10.12 presents the penetration depth-time curves and the penetration resistance-time curves for flat-ended pile and cone-ended pile. As the display of the penetration depth-time curve indicated, the pile penetration test adopts displacement control method. For each penetration state, the penetration depth is 0.5R, and then the slices at all positions are scanned during image acquisition. Comparing the penetration resistance-time curves of flat-ended pile and cone-ended pile, the penetration resistance of flat-ended pile is always greater than that of cone-ended pile.

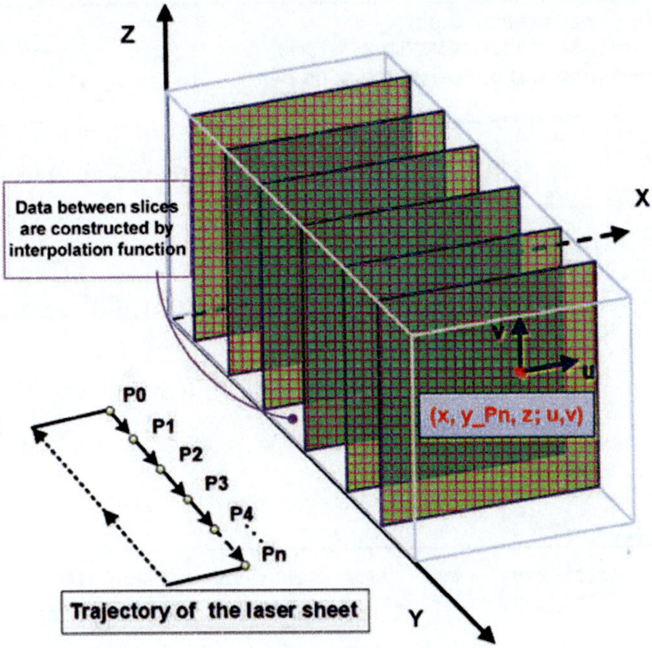

Fig. 10.10 Construction of three-dimensional data set

Fig. 10.11 The actual rendering of model tests

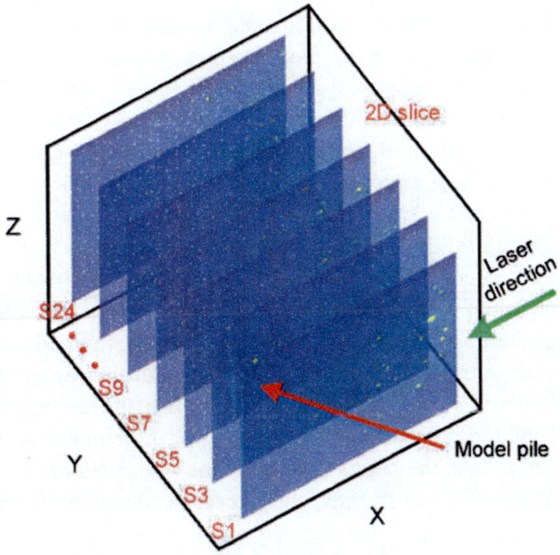

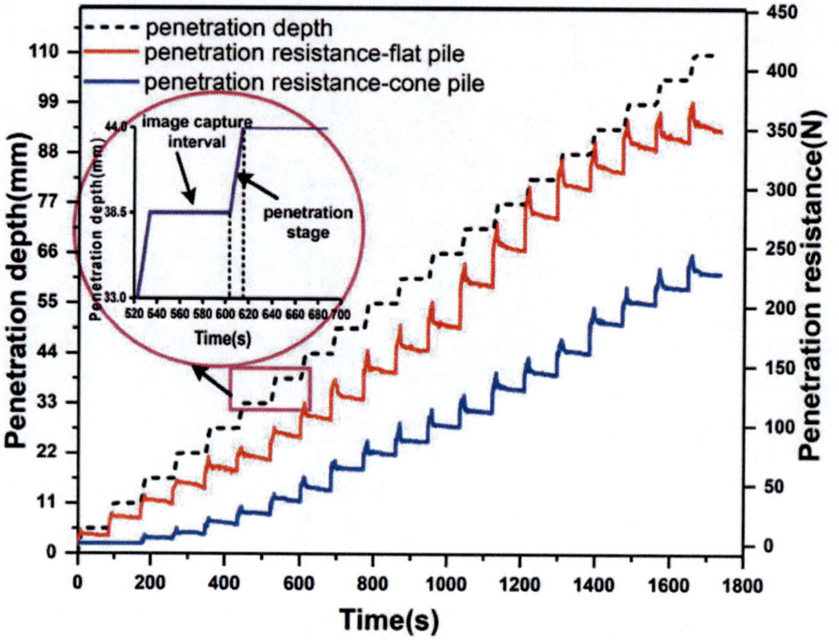

Fig. 10.12 The penetration depth-time and the penetration resistance-time curves

For each penetration cycle: the penetration resistance gradually increases during the penetration phase; the penetration resistance decreases instantaneously during the image acquisition phase, and then slowly decreases. This is because during the penetration phase, the pile tip is subjected to continuous penetration resistance, and the pile has the same speed as the soil under the pile tip. When the pile stops penetrating, the pressing effect of the pile on the soil is no longer enhanced. At the same time, due to the inertia effect, the soil under the pile tip still moves further downward. Therefore, the penetration resistance decreases instantaneously. However, because the soil mass at the bottom of the pile tip is in a state of high stress, the compression effect on the surrounding soil mass due to the penetration of the pile will not dissipate quickly. At the same time, because the interval of image acquisition is short, the change trend of penetration resistance decreases suddenly and then becomes slow. Regarding the relationship between penetration resistance and penetration depth, a similar trend was observed by [9]. For the penetration resistance-time curve of cone-ended pile, the penetration resistance was zero at the initial stage of the penetration process. When the pile penetrates to the depth of 1R, the penetration resistance gradually increases. This is possibly due to that the sandy soil has no cohesive force, the surface soil cannot withstand the pressure. Although the surface of the model pile is rough, the contact area between the pile tip and the soil is small for the cone-ended pile at the initial stage of penetration. In addition, the penetration form of cone-ended pile is punching into the soil, which is different from the form of squeezing soil of flat-ended pile. In the initial stage of penetration, the confining pressure on the upper soil is small and the restriction effect on the soil around the pile tip is small, so the penetration resistance of the cone-ended pile is almost zero. However, for the penetration resistance-time curve of flat-ended pile, due to the large contact area between pile tip and soil, the pile tip is subject to penetration resistance at the initial stage of the penetration process.

In order to improve the understanding of the different penetration mechanism of the two types of piles, the spatial displacement fields for flat-ended pile and cone-ended pile are compared in Figs. 10.13, 10.14 and 10.15. To quantify and display the differences of soil movement, the isosurface of displacement normalized by pile radius R. Outward and downward movements are labelled as positive, and inward and upward movements as negative. Figure 10.13a exhibited that the front view of spatial vertical displacement isosurface for flat-ended pile penetration when the pile penetrates to the depth of 10R. It can be seen that the spatial vertical displacement isosurface can be roughly divided into two parts: compression region at the bottom of pile tip and uplifted region around pile body. For the uplifted region, the vertical displacement decreases radially in the radial direction. Regarding the compression region, it is shaped like a light bulb with the maximum vertical displacement at the center of the bulb [22–25]. The height of the bulb is approximately 5R, and the width of the bulb is about 4R. Thus, this means that the high pressure stressing zone during the continuous penetration extends roughly outward within the range of 4R. Figure 10.13b shows the side view of spatial vertical displacement isosurface for flat-ended pile. The vertical displacement value of 0.02R is taken as the standard to quantify the maximum influence range of soil disturbance, and the influence range of flat-ended pile penetration on the surrounding soil disturbance is 13.6R.

Figure 10.14a, b show the front view and the side view of spatial vertical displacement isosurface for the cone-ended pile penetration when the pile penetrates to the depth of 10R, respectively. The vertical displacement of flat-ended pile penetration is greater than that of cone-ended pile penetration. It was observed that there

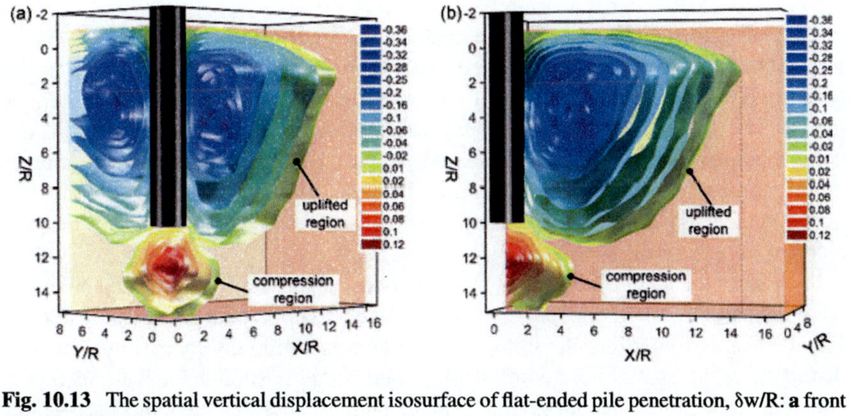

Fig. 10.13 The spatial vertical displacement isosurface of flat-ended pile penetration, δw/R: **a** front view; **b** side view

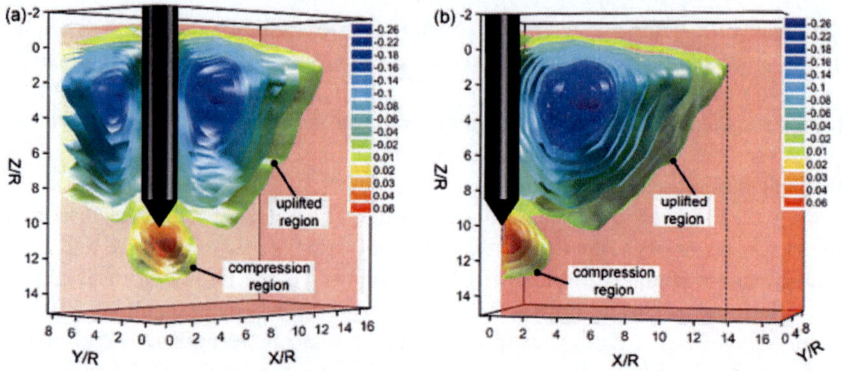

Fig. 10.14 The spatial vertical displacement isosurface of cone-ended pile penetration, δw/R: **a** front view; **b** side view

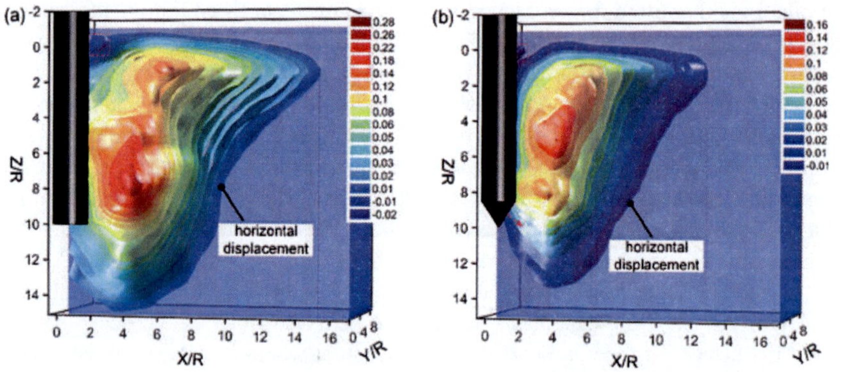

Fig. 10.15 The spatial horizontal displacement isosurface, δu/R: **a** flat-ended pile; **b** cone-ended pile

is general agreement between flat-ended pile and cone-ended pile for the spatial vertical displacement contour. Nevertheless, there is a significant difference in the compression region: the direction of the isosurface of the vertical displacement at the bottom of cone-ended pile is approximately along the normal direction of the conical plane. This is attributed to the reason that the cone-ended pile penetration is mainly through the pile tip punching into the soil, and continuous shearing process occurs between the pile and soil along the cone surface, pushing the soil to the two sides.

Note that the spatial horizontal displacement isosurface for flat-ended pile and cone-ended pile when the pile penetrates to the depth of 10R as indicated in Fig. 10.15a, b. Regarding the influence of pile penetration on the surrounding soil disturbance, the spatial horizontal displacement trend is similar for flat-ended pile and cone-ended pile. The horizontal displacement is caused by the lateral expansion in the process of pile penetration. The lateral expansion can be roughly divided into two stages: when the pile tip contacted the soil, the soil disturbance around the pile tip will produce a compressive expansion effect; when the pile tip has passed, it has the effect of secondary accumulation expansion on the upper soil. Due to the friction force between the surface of pile shaft and soil, a small part of negative horizontal displacement area appears near the upper surface of pile in Fig. 10.15. The horizontal displacement of flat-ended pile penetration is greater than that of cone-ended pile penetration. The horizontal displacement value of 0.01R is taken as the standard to quantify the maximum influence range of soil disturbance, the influence range of flat-ended pile and cone-ended pile penetration on surrounding soil disturbance is 14.3R, 11.8R, respectively.

Figure 10.16a, b displays the spatial displacement vector field for the flat-ended pile and the cone-ended pile when the pile penetrates to the depth of 10R. In the case of a shallow penetration (the penetration depth of the pile is less than 10 times the pile diameter), it can be observed that the direction of the vector arrow on the outside of the pile is inclined upward, and there are a few downward arrows at the bottom of the pile, which further verifies that the deformation of soil around the pile is mainly inclined upward for the so-called shallow penetration [26]. Meanwhile, the angle between the motion direction of the vector arrow and the X-axis for slice far from the pile is greater than that of slice near the pile. Comparing Fig. 10.16a with 10.16b, it can be concluded that the size and influence range of displacement vector for flat-ended pile penetration are larger than that of the cone-ended pile penetration. Considering the vector field pattern surrounding the pile tip, the vector direction at the bottom of the pile tip is initially vertical downward, then the vector direction of the soil in the transition zone gradually rotates obliquely upward, resulting in obliquely upward movement of displacement vectors on one side of the penetrating pile [24]. This is because the process of pile penetration is a cavity expansion process, and soil particle will undergo the resistance caused by passive earth pressure in the process of movement. Since the upper confining pressure is less than the lateral confining pressure, the soil particle will move upward.

10.7.2 Deformation Behaviour During Continuous Penetration

Figure 10.17 shows the spatial vertical displacement isosurface of flat-ended pile for penetration depth of 3.5R, 7R. Regarding the different stages of vertical displacement for flat-ended pile penetration, a systematic comparison for penetration depth of 3.5R, 7R, 10R is shown in Figs. 10.13 and 10.17. It is clear that the variation trend of vertical displacement at different stages is similar. As the penetration depth increases gradually, the vertical displacement increases gradually, and the influence range on the surrounding soil disturbance increases gradually. For the compression region, when the penetration depth increases from 3.5R to 7R, the vertical displacement

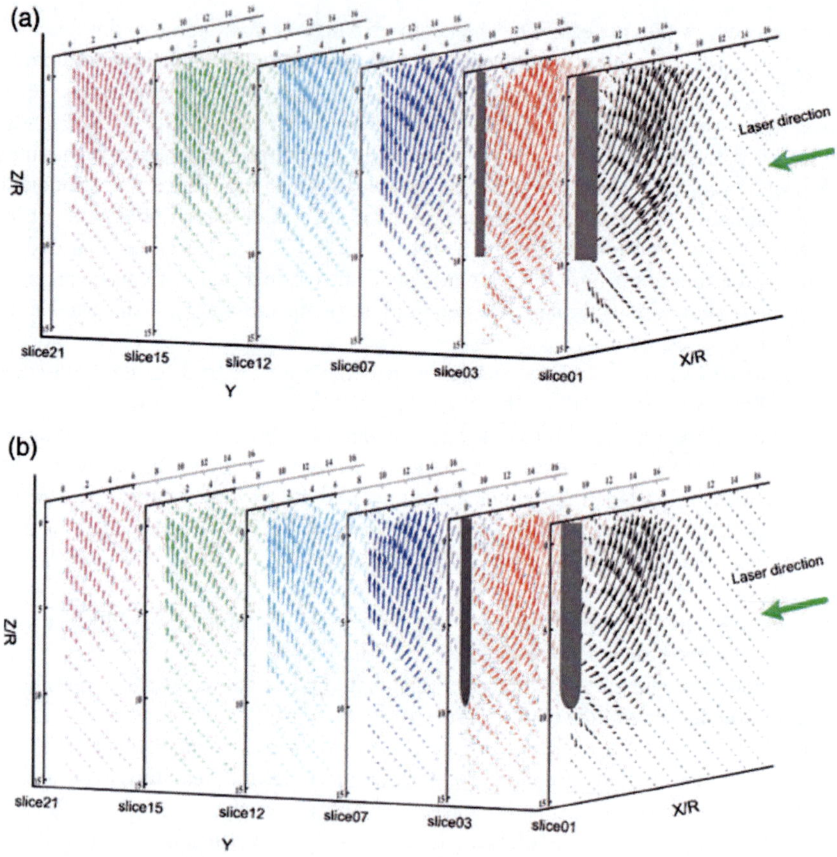

Fig. 10.16 The spatial displacement vector field: **a** flat-ended pile; **b** cone-ended pile

and influence range of the compression region become larger. However, when the penetration depth increases from 7 to 10R, the vertical displacement is generally in a stable trend, but the influence range on the surrounding soil disturbance gradually increases.

Examples of the surface heave of different stages for flat-ended pile penetration is shown in Fig. 10.18 obtained from the measured displacement data. It can be observed that the variation trend of ground heave is roughly similar at different penetration stages, which increases first and then decreases. With the increase of penetration depth, the position of the maximum ground heave gradually moves away from the pile shaft. At the same time, with the increase of penetration depth, the ground heave and influence range gradually increase. Notice that adjacent to the driven pile, it is common to find a gap or little heave between the pile shaft and the surrounding soil. The effect is probably due to friction between the pile shaft and the surrounding soil,

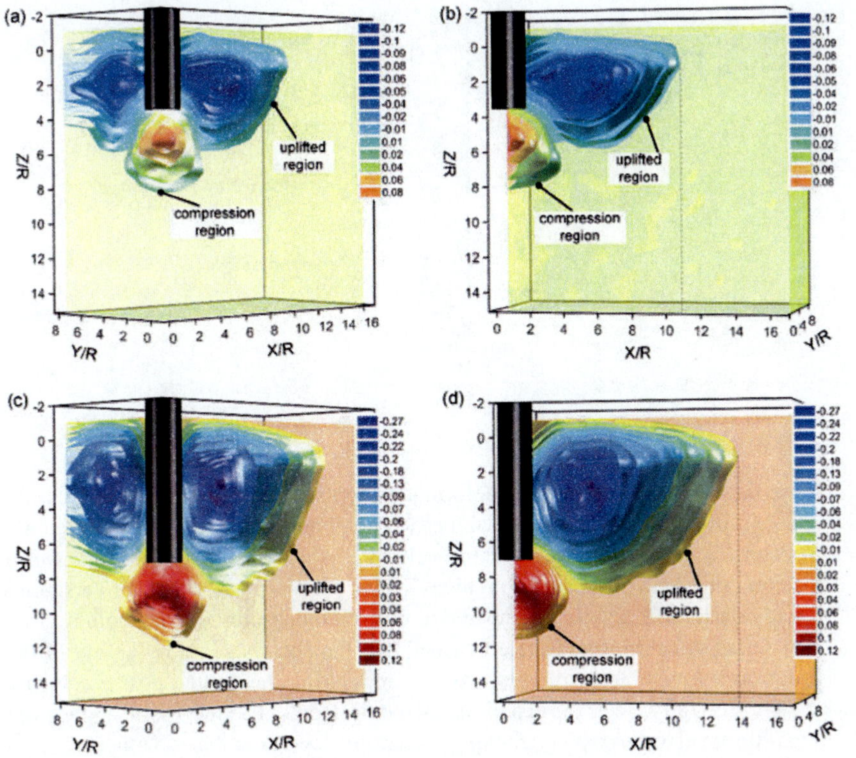

Fig. 10.17 The different stages of vertical displacement isosurface for flat-ended pile penetration, δw/R: **a** front view, Z/R = 3.5; **b** side view, Z/R = 3.5; **c** front view, Z/R = 7; **d** side view, Z/R = 7

which dragged down the soil close to the shaft during pile progressive penetration [24, 25].

10.8 Summary and Conclusions

A novel automatic tomographic scanning measuring device was developed and a modified 3D reconstruction procedure with the transparent soil model and incremental RG-PIV technique was proposed. They are applied to investigate the penetration mechanisms and the disturbance effect for jacked-pile penetration quantitatively. The main conclusions obtained in this study are as follows:

(1) The layout of the novel automatic tomographic scanning measuring device is composed of a laser sheet apparatus and a CCD camera which moves synchronously. By first processing the images with RG-PIV method, 2D deformation information for a series of images can be obtained. These 2D deformation

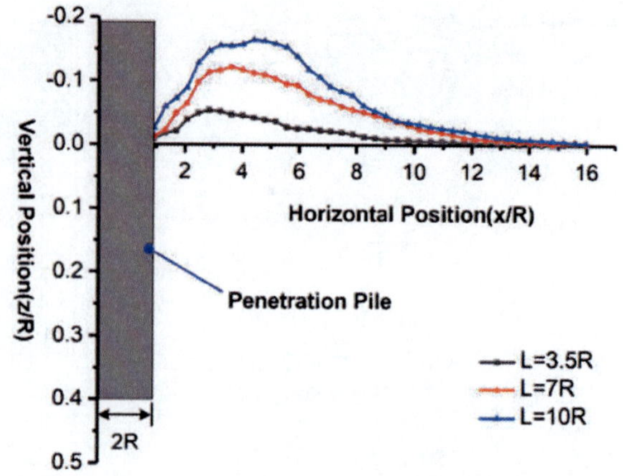

Fig. 10.18 The ground heave of different stages for flat-ended pile, δw/R

are used to construct the 3D deformation by the cubic spline interpolation to obtain the missing data between the two neighboring slices. The modified 3D reconstruction procedure can intuitively plot 3D displacement field. It is worth noting that, this novel tomographic scanning measuring device is of versatility and relative affordability compared to other 3D measurement technology.

(2) A series of calibration tests confirmed that the accuracy and reliability of the developed novel tomographic scanning measuring device and post-processing algorithm fully meets the experimental requirements. The successful application of this novel tomographic scanning measuring device combined with a modified 3D reconstruction method will provide a new perspective for understanding geotechnical engineering problems.

(3) One key mechanism was found to occur during jacked-pile penetration, namely, the displacement pattern is mainly divided into compression region and uplift region. In the case of the compression region, the pressurized bulb moves progressively downward as the pile penetrates into the ground; for the uplift region, the deformation of soil around the pile is mainly inclined upward for the so-called shallow penetration.

(4) The stages of lateral expansion can be roughly divided into two phases: when the pile tip is in contact with the soil, the soil disturbance around the pile tip will produce a compressive expansion effect; when the pile tip has passed, it has the effect of secondary accumulation expansion on the upper soil.

(5) Due to friction between the pile shaft and the surrounding soil, which dragged down the soil close to the shaft during pile progressive penetration. Accordingly, the variation trend of ground heave is roughly similar in different penetration stages, which increases first and then decreases.

References

1. Garcia D, Orteu JJ, Penazzi L (2002) A combined temporal tracking and stereo-correlation technique for accurate measurement of 3D displacements: application to sheet metal forming. J Mater Process Technol 125:736–742. https://doi.org/10.1016/S0924-0136(02)00380-1
2. Scarano F (2012) Tomographic PIV: principles and practice. Meas Sci Technol 24(1):012001. https://doi.org/10.1088/0957-0233/24/1/012001
3. Yuan B, Liu J, Chen W, Xia K (2012) Development of a robust Stereo-PIV system for 3-D soil deformation measurement. J Test Eval 40(2):256–264. https://doi.org/10.1520/JTE103856
4. Yuan B, Chen R, Teng J, Peng T, Feng Z (2015) Investigation on 3D ground deformation and response of active and passive piles in loose sand. Environ Earth Sci 73(11):7641–7649. https://doi.org/10.1007/s12665-014-3935-9
5. Yu L, Pan B (2017) Color stereo-digital image correlation method using a single 3CCD color camera. Exper Mech 57(4):649–657. https://doi.org/10.1007/s11340-017-0253-7
6. Hall SA, Bornert M, Desrues J, Pannier Y, Lenoir N, Viggiani G, Bésuelle P (2010) Discrete and continuum analysis of localised deformation in sand using X-ray μCT and volumetric digital image correlation. Géotechnique 60(5):315–322. https://doi.org/10.1680/geot.2010.60.5.315
7. Higo Y, Oka F, Sato T, Matsushima Y, Kimoto S (2013) Investigation of localized deformation in partially saturated sand under triaxial compression using microfocus X-ray CT with digital image correlation. Soils Found 53(2):181–198. https://doi.org/10.1016/j.sandf.2013.02.001
8. Viggiani G, Andò E, Takano D, Santamarina JC (2015) Laboratory X-ray tomography: a valuable experimental tool for revealing processes in soils. Geotech Test J 38(1):61–71. https://doi.org/10.1520/GTJ20140060
9. Iskander M, Liu J (2010) Spatial deformation measurement using transparent soil. Geotech Test J 33(4):314–321. https://doi.org/10.1520/GTJ-102745
10. Sang Y, Wang Z, Yu S, Zhao H (2019) The loading test on the singe pile with pile cap in transparent soil model. Geotech Test J 42(2):385–406. https://doi.org/10.1520/GTJ20170153
11. Ezzein FM, Bathurst RJ (2011) A transparent sand for geotechnical laboratory modeling. Geotech Test J 34(6):590–601. https://doi.org/10.1520/GTJ103808
12. Liu J, Iskander MG (2010) Modelling capacity of transparent soil. Can Geotech J 47(4):451–460. https://doi.org/10.1139/T09-116
13. White DJ, Bolton MD (2004) Displacement and strain paths during plane-strain model pile installation in sand. Géotechnique 54(6):375–397. https://doi.org/10.1680/geot.2004.54.6.375
14. Stanier SA, Blaber J, Take WA, White DJ (2016) Improved image-based deformation measurement for geotechnical applications. Can Geotech J 53(5):727–739. https://doi.org/10.1139/cgj-2015-0253
15. Pan B, Dafang W, Yong X (2012) Incremental calculation for large deformation measurement using reliability-guided digital image correlation. Opt Lasers Eng 50(4):586–592. https://doi.org/10.1016/j.optlaseng.2011.05.005
16. Vanne J, Aho E, Hamalainen TD, Kuusilinna K (2006) A high-performance sum of absolute difference implementation for motion estimation. IEEE Trans Circ Syst Video Technol 16(7):876–883. https://doi.org/10.1109/TCSVT.2006.877150
17. Wang P, Sang Y, Shao L, Guo X (2019) Measurement of the deformation of sand in a plane strain compression experiment using incremental digital image correlation. Acta Geotech 14(2):547–557. https://doi.org/10.1007/s11440-018-0676-z
18. Zhang Z (2000) A flexible new technique for camera calibration. IEEE Trans Pattern Anal Mach Intell 22(11):1330–1334. https://doi.org/10.1109/34.888718
19. Cnudde V, Boone MN (2013) High-resolution X-ray computed tomography in geosciences: a review of the current technology and applications. Earth Sci Rev 123:1–17. https://doi.org/10.1016/j.earscirev.2013.04.003
20. Kashuk S, Mercurio SR, Iskander M (2015) Methodology for optical imaging of NAPL 3D distribution in transparent porous media. Geotech Test J 38(5):603–619. https://doi.org/10.1520/GTJ20140153

21. Levoy M (1988) Display of surfaces from volume data. IEEE Comput Graphics Appl 8(3):29–37. https://doi.org/10.1109/38.511
22. Chong MK (2013) Soil movements due to displacement pile driving. In: Proceedings of 7th international conference on case histories in geotechnical engineering, paper no. 2.59. https://scholarsmine.mst.edu/icchge/7icchge/session02/30
23. Kobayashi T, Fukagawa R (2003) Characterization of deformation process of CPT using X-ray TV imaging technique. In: Third international conference on deformation characteristics of geomaterials, pp 43–47
24. Massarsch KR, Wersäll C (2013) Cumulative lateral soil displacement due to pile driving in soft clay. In: Sound geotechnical research to practice: honoring Robert D. Holtz II, pp 462–479. https://doi.org/10.1061/9780784412770.031
25. Ni QCCI, Hird CC, Guymer I (2010) Physical modelling of pile penetration in clay using transparent soil and particle image velocimetry. Géotechnique 60(2):121–132. https://doi.org/10.1680/geot.8.P.052
26. Zhao H, Liu C, Tand Xi, Wei H, Zhu F (2020) Study of visualization measurement system of spatial deformation based on transparent soil and three-dimensional reconstruction technology[J]. Rock Soil Mech 41(9):3170–3180

Part III
Application of Transparent Soil Modelling in Geotechnical Engineering

Chapter 11
Application of Transparent Soil Modeling Technique to Investigate Pile Foundation

Abstract Pile foundation is an important type of deep foundation. It is widely used in various building and structures. The interaction mechanism between soil and pile, as well as soil deformation during pile driving and loading are critical for the design and analysis of pile foundation. Transparent soil modelling technique provides an innovative tool to conduct the research on pile foundation. Applications of transparent soil modeling technique to investigate pile foundation problems are reported in this chapter.

11.1 Visualization Model Test on Construction Process of Tapered Pile Driving and Pile Base Grouting in Transparent Soil

11.1.1 Introduction

Conventional tapered piles are known to offer improved side friction under compressive loads due to changes in cross-section area along the length of the pile. Additionally, tapered piles can reduce negative skin friction caused by ground settlement relative to the pile. Conventional belled piles can improve tip resistance due to the enlarged pile base. One new pile construction method has been developed to install a tapered pile with belled bottom to combine the benefits of both tapered piles and belled piles [1].

Model tests and numerical simulations of tapered piles under compressive, lateral, uplift load, or negative skin friction were presented [2, 3, 4]. These results showed that the side resistance of tapered piles can be larger than that of a pile with uniform cross section, taper angle is one of effective ways to reduce the negative skin friction effects. However, the tip resistance of a tapered pile is often smaller than that of a uniform cross section pile. Field tests on comparative analysis between tapered piles and uniform cross section piles were carried out [5–7]. These field test results affirmed that the bearing capacity of tapered piles can be nearly 30–50% higher than that of a uniform section pile under the same conditions. One of bearing capacity

H. Zhao et al., *Transparent Soil Modelling Technique and Its Application*, Springer Tracts in Civil Engineering, https://doi.org/10.1007/978-981-19-6825-9_11

prediction methods using CPT test data was proposed for tapered piles in medium to dense sands [8].

Field tests for belled piles embedded in soft soils were reported [9, 10], in which test data were used to provide insights on the negative skin friction development, compressive bearing capacity, and uplift capacity of belled piles. For layered ground, simplified approach for predicting bearing capacity of belled piles, using limit equilibrium method with an assumption of a power function for the slip surface was presented.

In summary, it is evident that visualization studies on tapered pile installation or enlarged pile base grouting are relatively little. In this study one visualization model test system that contains transparent soil, a pile driving load system and an optical measurement system (including an optical platform, a CCD (Charge-coupled Device) camera and a linear laser, etc.) was designed. The soil displacement distribution during tapered pile installation was measured non-intrusively and the squeezing effect during enlarged pile base grouting was visually observed.

11.1.2 Construction Process In-Situ

Construction methods

Pieces of prestressed tapered piles are fabricated in the factory. The construction procedures for belled wedge piles can be described as follows: (1) The first piece of prestressed tapered pile section is driven into soils together with pile shoe by a pile driving machine; (2) The second piece of prestressed tapered pile section is connected with the first piece of prestressed tapered pile section by electric welding, then the jointed tapered pile section is driven into soils; (3) The inside soils of prestressed tapered pipe piles are excavated by an excavator; (4) Pile shoe is removed and the soils in the bottom of tapered pile are excavated; The bottom hole is nearly 2–3 times of the diameter of the tapered pile tip in width and depth; (5) The prefabricated steel cage is lowered into the bottom hole; (6) The concrete is poured into the enlarged pile tip, and the enlarged pile tip is formed.

In order to improve the quality of the constructed piles and reduce the soil disturbance caused by pile driving, the pile driving rate is usually controlled within 1.2–1.5 m/min under normal circumstances. However, a slower rate of 1.0–1.2 m/min is used for a loose to medium-loose sand layer.

Quality assurance and quality control

The accepted quality check practice of the constructed drilled shafts or driven piles can be extended to belled wedge piles. Since excavation of the enlarged pile base is small in size, therefore, the caving and necking can be minimized or prevented. Expect traditional static pile load testing and low-strain integrity testing for tapered pile and equal section pile; the quality of the belled wedge piles after their formation in the ground can also be checked by the following two methods: (1) excavate the inside soils of the tapered pile for visual inspection, and (2) monitoring of the amount of concrete poured during construction of the enlarged pile base.

11.1.3 Model Description

Transparent clay made of amorphous silica powder [11] has been used by many researchers to study pile penetration problems [12–15]. And transparent sand made of silica gel [16] has been used to study shallow foundation [17, 18] and tunneling [19]. Transparent soil made of aquabeads [20] has been used to study 3-D flow [21]. The transparent soil surrogate, 0.1–1 mm, used in this study was fused quartz (described by Ezzein and Bathurst [22]) and the fluid was a blend of N-Paraffin C 12 and Technical White Oil 15 [23]. After mixing, the transparent soil was de-aired under vacuum. The physical properties of the used fused quartz are shown in Table 11.1. Although the viscosity of the used oil mixture is higher than that of water, Ezzein and Bathurst pointed out that the materials behave similarly to natural sands with angular particle shapes and have similar friction angles. Moreover, for sand-structure interaction problems carried out at slow rates of loading, the relatively high viscosity is unlikely to modify model results.

In this research, a Plexiglas model box measured 120 mm × 120 mm in plan, 200 mm high was used. In addition to the model box, the set-up also included a black and white charge-coupled device (CCD) camera, a 2 W line laser light source, a loading frame, an optical table, a plastic syringe for grouting and a PC for image processing. The set-up is shown in Fig. 11.1. A distinctive speckle pattern was generated by the interaction between the transparent sand and the light sheet when it spread into the transparent sand. The CCD camera with a resolution of 1280 × 960 was set approximately 300 mm away from the laser sheet with its optical axis perpendicular to the laser sheet plane. A macro zoom lens was used to adjust the size of the captured area. During installation, the top of the model pile was connected to the loading frame through a drive shaft which provided a stable vertical speed of 1 mm/s in this study. About 80 pictures were captured at an average rate of 1 frames/s and stored in the computer through the software "MV-Capture" provided by the camera manufacturer. Hot chocolate modeling grouting slurry were firstly stored in the plastic syringe. After the model pile installation, hot chocolate was injected into the transparent soil mass artificially through a plastics hose with a diameter of 5 mm. During this stage, about 50 pictures were taken to record the enlarged pile base grouting process.

Table 11.1 Properties of fused quartz (reported by Ezzein and Bathurst [22])

Density (Mg/m^3)	Specific gravity	Mohs hardness	Modulus of elasticity (GPa)	Poisson's ratio	Refractive index
2.2	2.24	4.9	72	0.16	1.4586

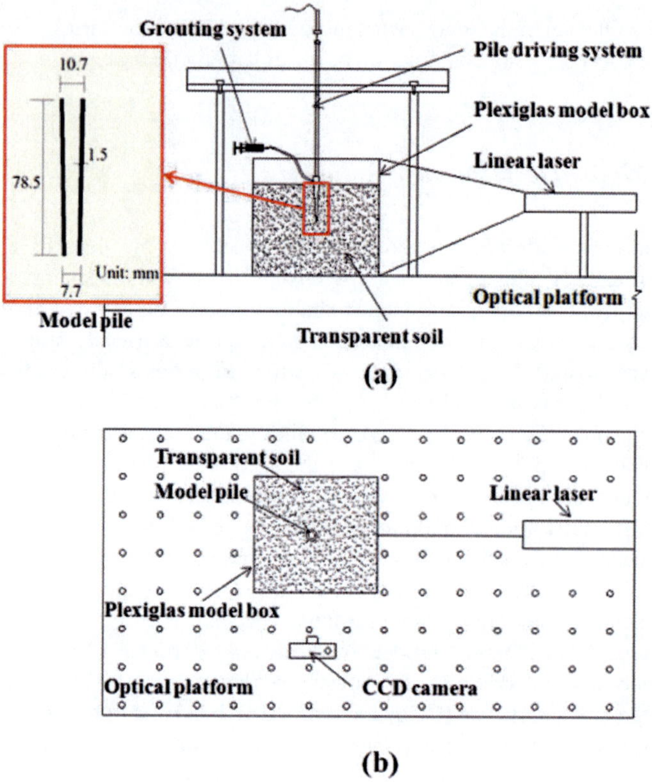

Fig. 11.1 Model test set-up: **a** elevation; and **b** plan view

11.1.4 Results Analysis and Discussion

Displacement patterns

Two same model tests were conducted under the same conditions and one was chosen for analysis. Typical soil displacement fields around the pile during installation can be displayed in the form of vector quiver plots. Figure 11.2a shows the incremental displacement field as the pile tip depth L increased from 6 to 7D where D is the pile average diameter. The horizontal axis is the radial distance r from the centerline of the pile, normalized by D. The vertical axis is the normalized depth below ground level, z/D. The cumulative displacement field after a penetration of 7D is shown in Fig. 11.2b. From Fig. 11.3, it is clear that the displacement patterns around the model pile are quite different from those around the solid circular pile depicted by Ni et al. [15]. To display the components of the displacement fields, corresponding incremental and cumulative displacement contours were produced and shown in Fig. 11.3 where the horizontal and vertical displacements, δ_r and δ_z, are also normalized by D.

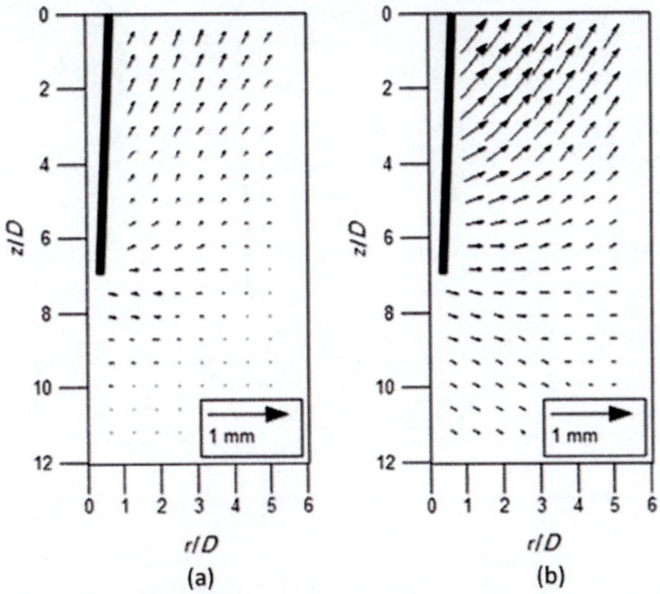

Fig. 11.2 Displacement vectors: **a** from $L = 6D$ to $7D$; **b** from $L = 0D$ to $7D$

In Fig. 11.3a and b it is seen that, when the tip moved from $L = 6D$ to $7D$, both the horizontal and vertical movements cover a region not just around the pile tip but also along the pile shaft, while for the solid circular pile, the incremental soil movements are located in a round-shaped region found by Ni et al. [15]. The 'nose cone' with a stationary core (relative to the pile) observed by both White and Bolton [24] and Ni et al. [15] was not seen in this model test. This is attributable to the typical variable diameter along the pile shaft. Cumulative displacement contours are shown in Fig. 11.3c and d for a penetration range from $L = 0D$ to $7D$. It is clear that the zone of maximum movement is located along the pile shaft and close to the ground level. This is probably due to the larger diameter of the pile top. It is also noticeable that both incremental and cumulative displacements around the model pile in this study are much smaller than those around the solid circular pile. This mainly because that the used model pile is a tapered pipe pile.

Grouting process

During the grouting process, no vertical effective stress was exerted on the specimen. Hot chocolate with a volume of 4000 mm^3 which is insoluble in the pore fluid was injected into the transparent soil at a rate of 50 mm^3/s. In this study the horizontal distance from the centreline of the pile to the boundary was 55 mm and the vertical distances from the pile bottom to the smaple surface was 63 mm. The whole process of development of a grout bulb was visually observed. During injection about forty pictures were captured and four of them were chosen as four different step to depict

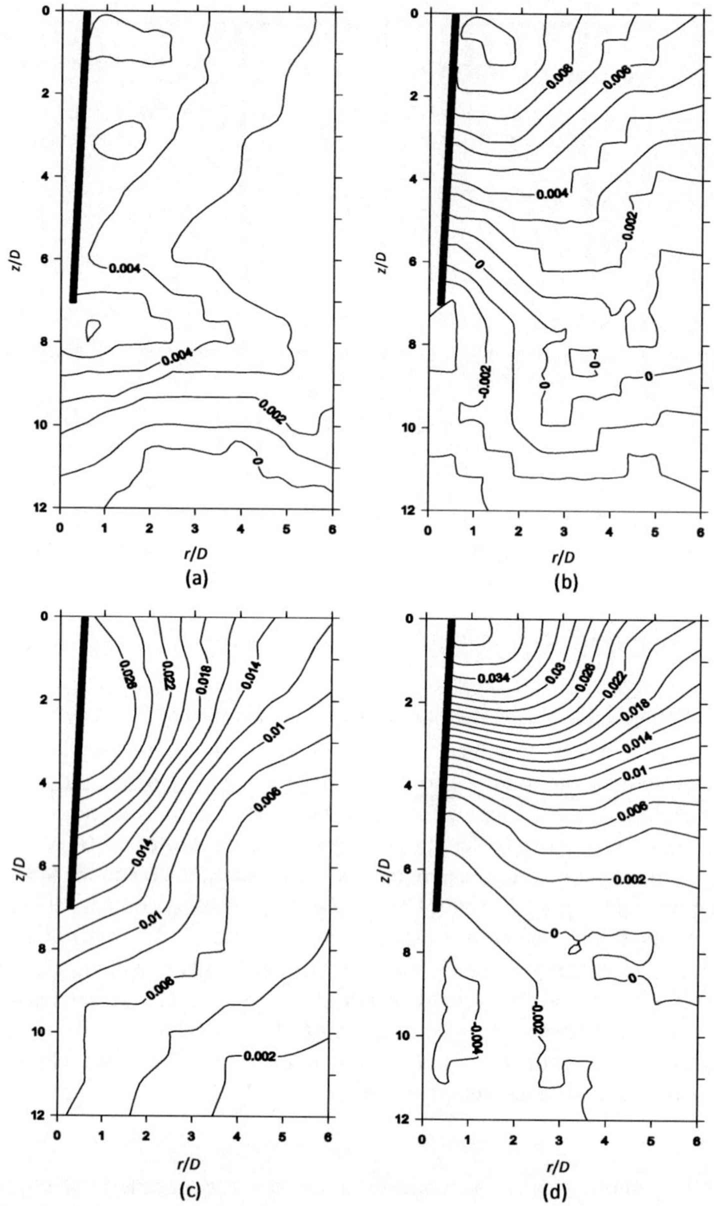

Fig. 11.3 Displacement contours normalized with pile diameter D: **a** from $L = 6D$ to $7D$, δ_r/D; **b** from $L = 6D$ to $7D$, δ_z/D; **c** from $L = 0D$ to $7D$, δ_r/D; **d** from $L = 0D$ to $7D$, δ_z/D

the grouting process. In Fig. 11.4 it is clear that with the increase in injection volume the grout bulb grows bigger. Finally an elliptical cavity with a major axis in the vertical direction was formed. This is probably due to the degree of lateral confinement and the state of stress of the ground. Results of cavity aspect ratio (CAR) defined by Au et al. [25] which is basically the ratio of the maximum horizontal dimension to the maximum vertical dimension of the cavity with different injection volume during grouting process are presented in Fig. 11.5. For both the two tests, values of CAR decrease initially and then increase. Values of CAR are generally smaller than unity indicating that formation of elliptical cavities with a major axis in the vertical direction. After injection of 4000 mm^3 hot chocolate, the final size of the formed grout bulb is shown in Fig. 11.6 and then compared with the numerical results simulated by Au et al. [25]. From Fig. 11.6 it is clear that the numerically simulated cavity has the maximum horizontal dimension of 19.2 mm and the maximum vertical dimension of 20.8 mm with a final CAR of 0.92. While in this study the excepted experimental cavity should be a thorough cavity with the maximum horizontal dimension of 17.4 mm and the maximum vertical dimension of 20.9 mm and have a final CAR of 0.83. In fact, because of the existence of the pile which may block the flow of the grouting, the experimental results was not a exact cavity which has some flaw in the upper left of the cavity seen in both Figs. 11.3 and 11.6.

11.1.5 Modeling Limitations

The interim results of a modeling test have been presented and there were some technical problems that may limit the accuracy of the results. Firstly, while the used material has been studied to model natural sand, some of its properties differ from those of a natural soil. And it has not yet been possible to reproduce transparent non-saturated soils due to the difference between the refractive indices of water and air [21]. What's more, boundary effects imposed by rigid sidewalls which may have influence on the shape of the formed grout bulb need further study.

Despite these limitations, it is believed that the techniques similar to those described in this paper may have a wider application in geotechnical engineering.

11.1.6 Conclusions

Model tests of visualizing construction process of belled wedge pile have been performed. The developed technology is a reliable tool in the prediction of soil behavior during penetration. It can be utilized in the application of grout injection process and to develop a better understanding of the fundamental behavior of grout injection. For the belled wedge pile, it was found that soil movement pattern around the pile during installation is quite different from that around a solid circular pile

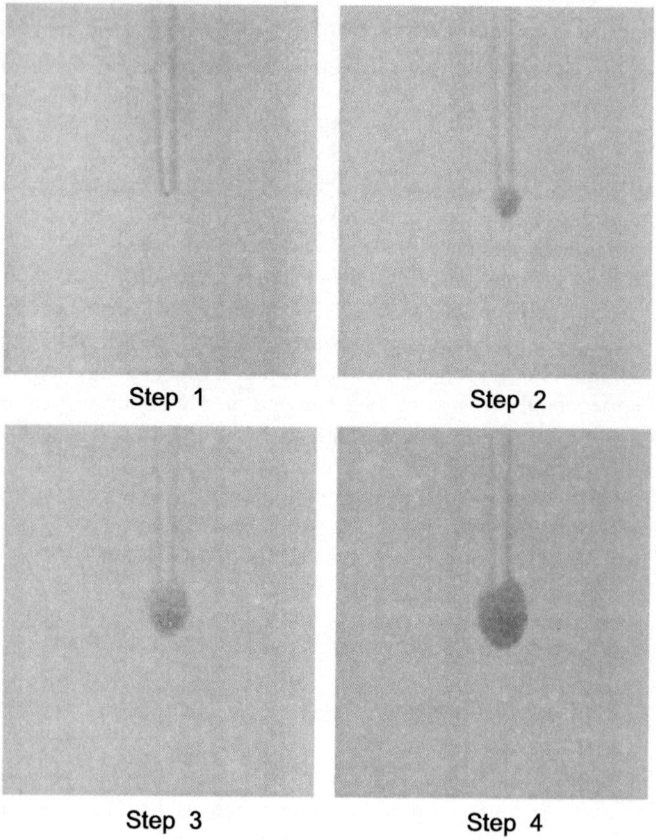

Fig. 11.4 Four different steps during the pile base grouting process

Fig. 11.5 CAR during grouting process

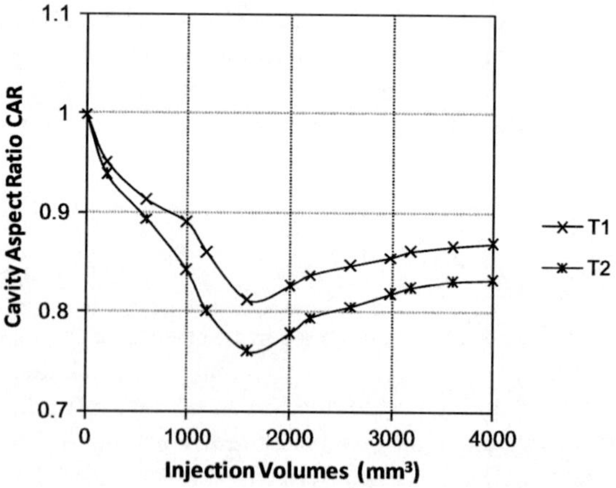

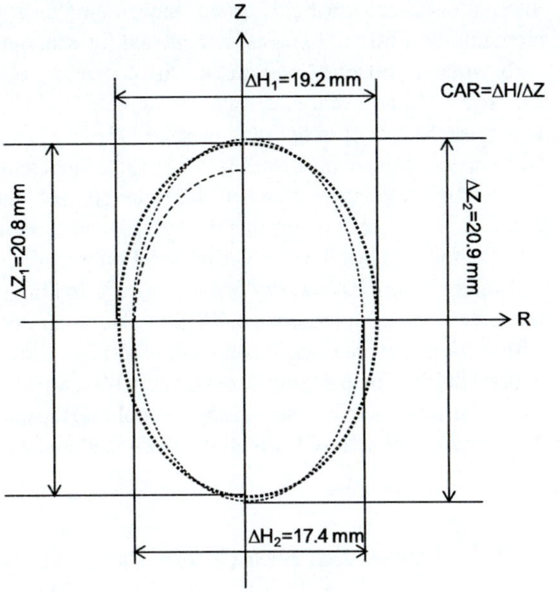

Fig. 11.6 Final size of the formed grout bulb

which has been depicted by previous research. And after grout injection, a grout bulb with elliptical shape was formed and visualized non-intrusively.

Further testing is on going to study how the parameters, such as conditions of the outer boundary, injection volume and injection pressure, affect the shape of the grout bulbs.

11.2 Visualization Model Test on Bearing Capacity of Pipe Pile Under Oblique Pulling Load

11.2.1 Introduction

Pipe pile foundation is one of economical piles under vertical load because of its internal side friction. Majority, pile foundation will be worked in vertical and lateral combined load condition [26]. However, the researches focused on bearing capacity of pipe pile under oblique pulling load was relatively little. Series of studies on ultimate axial uplift capacity or ultimate lateral resistance of single pile and pile group was carried out [27–30]. Based on traditional model tests and numerical analysis method, bearing capacity of pile under oblique load and negative skin friction were carried out [31]. It is imperative to understand the pile-soil-loading interaction phenomenon and the ultimate resistance of piles under oblique loads. However, the internal displacement fields or soil movements during pile-soil-loading interaction are naturally difficult to measurement and little studied.

In order to observe internal soil displacement, transparent soil which is manufactured by transparent materials with a pore fluid having the same refractive index was made to model either clay or sand [11, 20, 22, 32]. Combined with nonintrusive particle image velocimetry (PIV) technique, transparent soils have been used in model tests to study flow and deformation problems [12, 15, 33 21, 34, 35, 36]. These research results show that transparent soil is one of suitable methods to analysis internal soil displacement under pile-soil interaction.

Hence, an optical system consisting of a loading system, model piles, a linear laser, a charge-coupled device (CCD) camera, an optical platform and a computer with a frame grabber was developed. Two pipe piles with different diameter under oblique pulling loads at angles $\alpha = 0°, 45°, 60°$ and $90°$ were conducted in transparent sand. The load–displacement response, oblique ultimate pulling resistances, and soil deformation fields have been qualitatively and quantitatively studied.

11.2.2 *Model Test Description*

11.2.2.1 Model Test Set-Up

The model test set-up, as shown in Fig. 11.7, consists of a loading system, a linear laser, a CCD camera, an optical platform, and a computer for image processing. The loading system is composed by a sliding bracket, two movable pulleys and several loads. The linear laser, which is manufactured by Nanjing Laichuang Laser Technology Co., Ltd., China, could produce a vertical light sheet. The CCD camera with a resolution of 1280 × 960 was set approximately 500 mm away from the laser sheet with its optical axis perpendicular to the laser sheet plane. A macro zoom lens was used to adjust the size of the captured area. The optical platform provides a secure and stable platform for experiment, and all devices can be fixed in this experimental platform by the bolts. The images were captured and stored in the computer through the software "MV-Capture" provided by the camera manufacture. A Plexiglas square box with a thickness of 5 mm and inner dimensions of 120 mm × 120 mm × 250 mm (L × W × H) was used for model tank, and it was fixed on the optical platform by the bolts. Two model piles with 100 mm pile length, 15 mm (Pile A) and 20 mm (Pile B) pile out-diameter were used in this paper. Model piles are made by Plexiglas material and its friction is coarse. The model piles were subjected to oblique pulling loads at angles, α, 0°, 45°, 60° and 90° with the vertical axis of the pile.

11.2.2.2 Transparent Sand

In this study fused quartz, which is manufactured by Xuzhou Xinyi Wanhe Minerals Co. LTD in China, was chosen for transparent material. The size of fused quartz

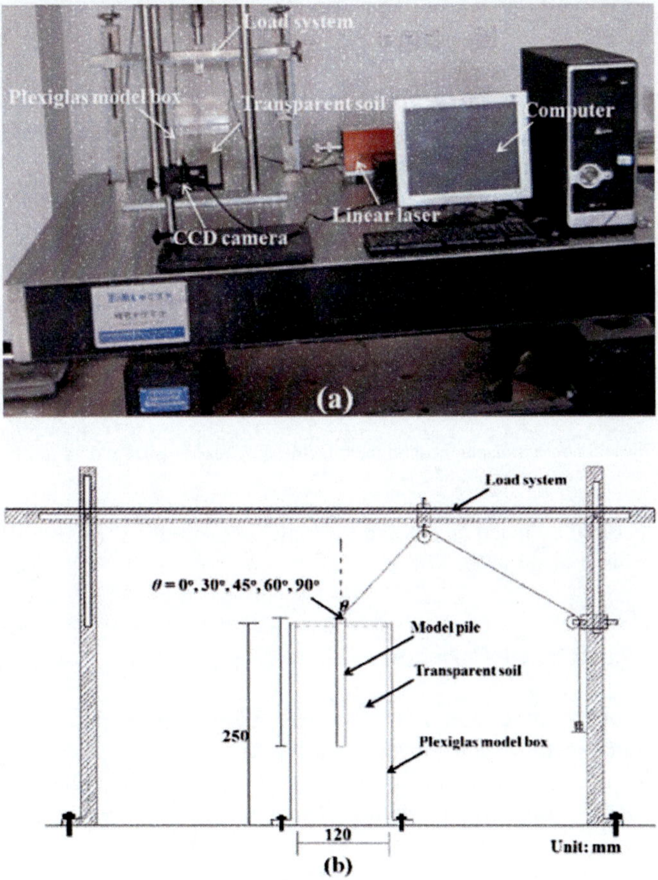

Fig. 11.7 Physical photo and schematic diagram of model test setup

ranges from 0.5 mm to 1.0 mm, and its physical photo and microstructure with 80 times magnification are shown in Fig. 11.8a and b, respectively.

The specific gravity of fused d quartz is 2.186, the uniformity coefficient, Cu, and gradation coefficient, Cc, were found to be 1.83 and 1.00, respectively. The minimum dry density, ρ_{min}, and maximum dry density, ρ_{max}, were 0.970 g/cm3 and 1.274 g/cm3, respectively.

The refractive index of fused quartz was 1.4585. The matching pore fluid was 25:75 blend by weight of colorless 15 mineral oil and a normal paraffinic solvent Norpar 12 fluid. The mixed oil has a refractive index of 1.4585 at 24 °C. Nearly 50% relative density of soil sample was artificial manufactured. The friction angles of transparent soil from triaxial tests with 30% and 70% relative density were 37.1° and 39.4°, respectively. The distinctive speckle pattern of transparent sand which

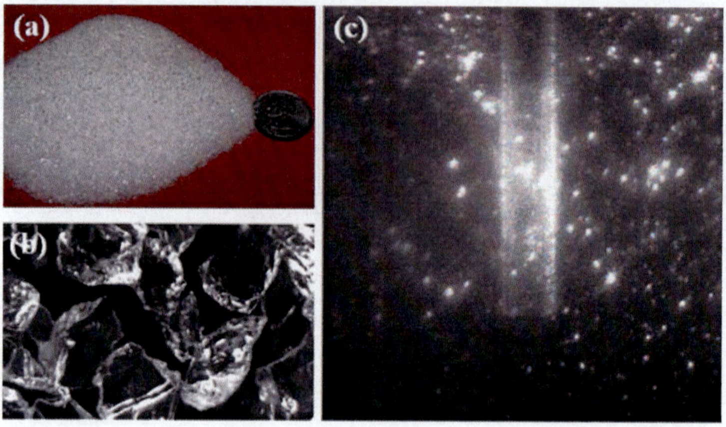

Fig. 11.8 Physical photo transparent sand manufactured by fused quartz and its speckle pattern

was manufactured by fused quartz and the matching pore fluid (mixed oil) with the model pile was shown in Fig. 11.8c.

11.2.2.3 Digital Image Correlation

The interaction between laser light and transparent soils produces a distinctive speckle pattern. Small particle movement will result in change in the speckle distribution in the plane of measurement. Images captured before and after deformation were analyzed by using the particle image velocimetry (PIV), which is a classical pattern recognition technique based on using the correlation function to locate the best matching position of two images and predicting movements. For each of these interrogation windows the image before and after each decrement of load pressure are correlated to produce an average particle displacement vector. Doing this for all interrogation regions could produce a vector map of average particle displacements. The discrete form of standard cross-correlation function is as follows:

$$C(\Delta x, \Delta y) = \frac{1}{MN} \sum_{m=0}^{M-1} \sum_{n=0}^{N-1} f(m, n) g(m + \Delta x, n + \Delta y) \quad (11.1)$$

where M and N are the dimensions of the interrogated images, f and g are the grayscale intensities of two images being interrogated. For displacements measurements with the PIV, the whole image area is divided into many smaller interrogated windows, and the spatial displacement field can be obtained by obtaining the displacement in each interrogated window.

11.2.3 Results and Discussions

11.2.3.1 Load–displacement Response

Typical diagrams of oblique pulling load versus vertical or lateral displacement for pipe pile with two pile out-diameters (D = 15 mm and 20 mm) are shown through Figs. 11.9 and 11.10. Figure 11.9a and c show the oblique pulling load versus vertical displacement of two different pile types (D = 15 mm and 20 mm), respectively. Figure 11.9b and d show the oblique pulling load versus lateral displacement of two different pile types, respectively. Four types of load–displacement curves were observed for the piles under oblique pulling load (α = 90o, α = 60o, α = 45o and α = 0o). In general, all the load–displacement curves are non-linear. Under the same load number, the vertical or lateral displacement increased with the oblique pulling load angles decreasing. Figure 11.10a, c, e, g and i show the oblique pulling load versus vertical displacement of five difference pulling load angles (α = 90°, α = 60°, α = 45°, α = 30° and α = 0°). Figure 11.10b, d, f and h show the oblique pulling load versus lateral displacement of four difference pulling load angles (α = 90°, α = 60°, α = 45°, α = 30°). The oblique pulling load versus (vertical or lateral) displacement curves obtained by Patra and Pise [37], and Krishna and Patra [38] are also shown in Fig. 11.10 for comparative analysis. The load–displacement curves obtained in this paper is steeper than those obtained in previous reference results. It may be caused by the scale of model piles.

11.2.3.2 Ultimate Bearing Capacity

The mode of failure is determined from load–displacement diagram. In case of piles under pure lateral loading ($\alpha = 90°$), the failure mode is lateral or normal failure; piles under vertical pulling ($\alpha = 0°$), the failure mode is axial failure. Similarly, piles under oblique pulling, the minimum failure load obtained from displacement diagram is taken as the ultimate load. The failure mode is taken as the displacement corresponding to minimum failure load. For example, at $\alpha = 60°$, the displacement corresponding to minimum failure load is lateral displacement; hence, the failure mode for this case is lateral failure. In this study, the ultimate failure load has been found out where a large displacement occurs for a small increment of applied load [37, 38]. It is taken as the point at which the curve exhibits a peak or maintains a continuous displacement increase with no further increase in pulling load. The oblique ultimate resistance of Pile A (with 15 mm pile diameter) is a little smaller than those of Pile B (with 20 mm pile diameter). The ultimate bearing capacity of piles under oblique pulling loads ($\alpha = 90°$, $\alpha = 60°$, $\alpha = 45°$, $\alpha = 30°$ and $\alpha = 0°$) are shown in Table 11.1. For example, at $\alpha = 60°$, the oblique ultimate resistance of Pile A and Pile B equal 8.5 N, and 16.0 N, respectively. The ultimate resistance of piles under $\alpha = 90°$ is nearly 4 times of that of piles under $\alpha = 0°$. 1.2.3.3 Soil displacement field under failure load.

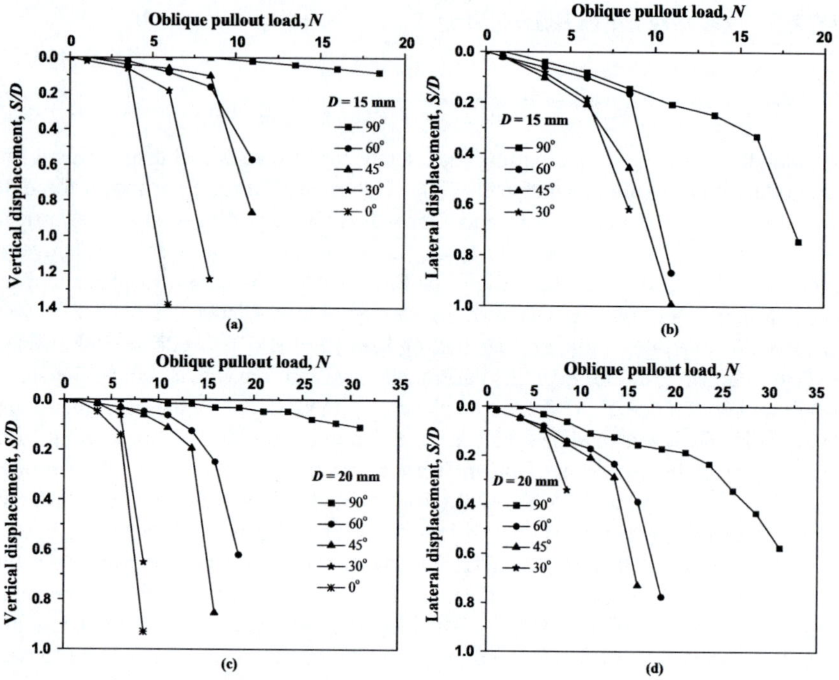

Fig. 11.9 Curves on oblique pulling load versus axial (vertical or lateral) displacement for two pile diameter with different load angles

The ultimate bearing capacity of piles can be seen from Table 11.2. Take piles under $\alpha = 90°$ for example, the ultimate resistance (failure load) of Pile A and Pile B equal 16.0 N and 23.5 N, respectively. The accumulated displacement vectors and contours are calculated from 0 N to failure loads. The displacement vectors of Pile A and Pile B are shown in Fig. 11.11a and b, respectively. It shows that the displacement vectors patterns of piles with different diameters are similar. The (lateral or vertical) displacement contours of Pile A are shown in Fig. 11.12a and b, respectively. The (lateral or vertical) displacement contours of Pile B are shown in Fig. 11.12c and d, respectively. It shows that the displacement contours of piles with different diameters are similar. It also shows that the values of displacement of Pile A are a little larger than those of Pile B. It is because that the pile body displacement of Pile A is larger than that of Pile B under each failure load condition.

11.2.3.3 Soil Displacement Field Under Working Load

Take piles under $\alpha = 60°$ and 8.5 N oblique load for example, the displacement vectors of Pile A and Pile B are shown in Fig. 11.13a and b, respectively. The accumulated displacement vectors and contours are calculated from 0 N to 8.5 N. It shows that

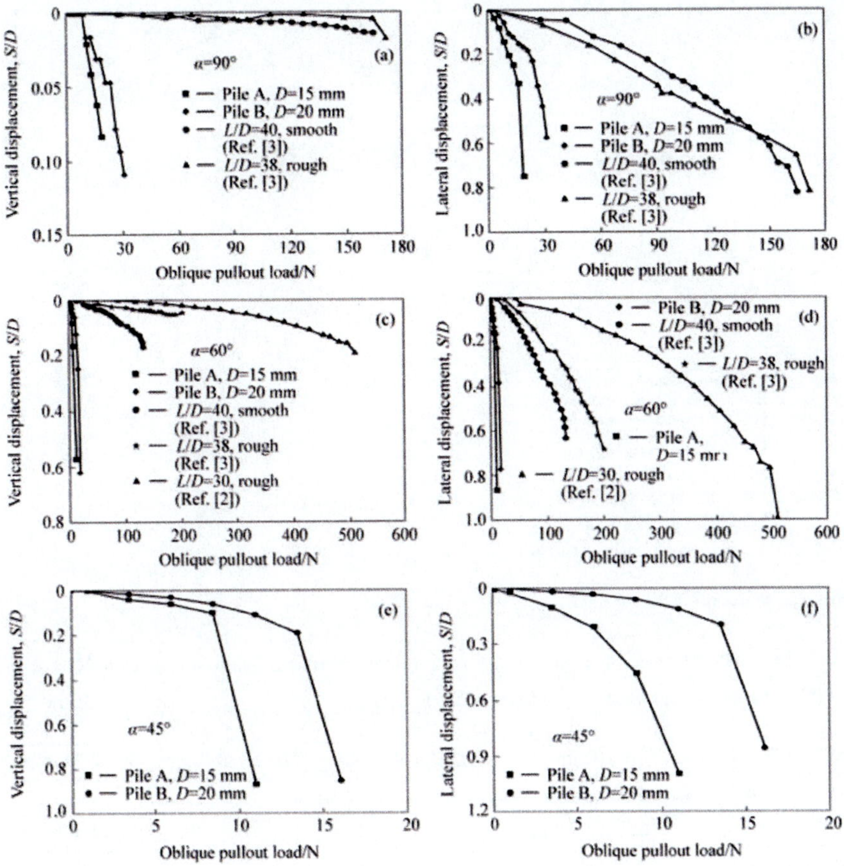

Fig. 11.10 Curves on oblique pulling load versus axial (vertical or lateral) displacement with different pile type

the displacement vectors patterns of piles with different diameters are similar. The (lateral or vertical) displacement contours of Pile A are shown in Fig. 11.14a and b, respectively. The (lateral or vertical) displacement contours of Pile B are shown in Fig. 11.14c and d, respectively. It shows that the displacement contours of piles with different diameters are similar. It also shows that the values of displacement of Pile A are a little larger than those of Pile B in the same oblique load values because of its large diameter.

Take pile B under 6.0 N oblique pulling load for example, the displacement fields (displacement vectors and displacement (radial or vertical) contours) of soil around the pile B under different oblique pulling load angles $\alpha = 0°$, $\alpha = 30°$, $\alpha = 45°$, $\alpha = 60°$ and $\alpha = 90°$ were displayed in Figs. 11.15a–e, respectively. The accumulated displacement vectors and contours are calculated from 0 N to 6.0 N loads. Take 6.0 N oblique pulling load and Pile B for example, the radial or vertical

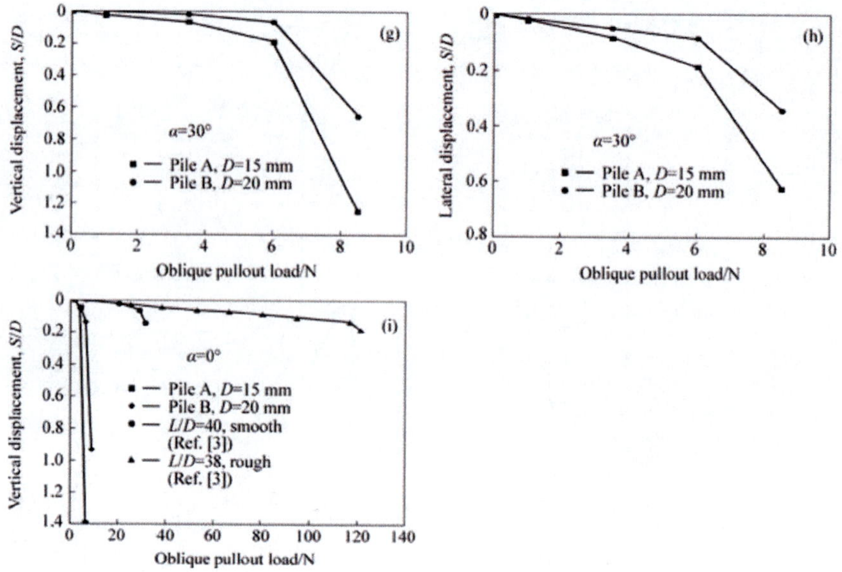

Fig. 11.10 (continued)

Table 11.2 Ultimate bearing capacity (Q_{ult}) of piles under oblique pulling loads (Unit: N)

α (°)	Pile A	Pile B	Patra and Pise [37]		Krishna and Patra [38]	
	$D = 15$ mm	$D = 20$ mm	Smooth $L/D = 40$	Rough $L/D = 38$	$L/D = 30$	$L/D = 20$
0	3.5	6.0	20.2	116.4	204.2	118.5
30	6.0	6.0	27.9	204.6	364.4	156.9
45	6.0	13.5	–	–	–	–
60	8.5	16.0	130.7	186.9	330.0	161.8
90	16.0	23.5	149.6	164.3	394.4	232.2

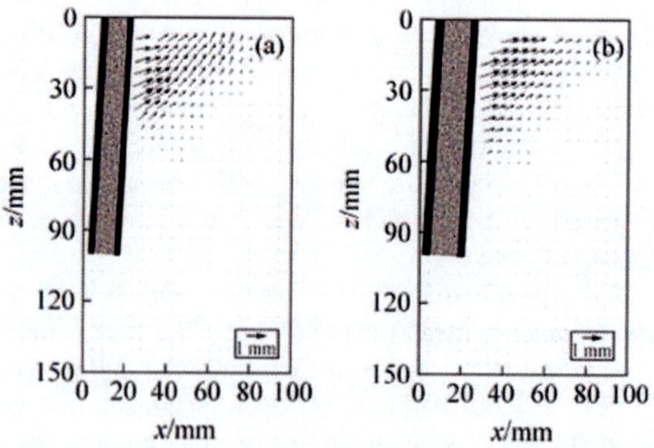

Fig. 11.11 Displacement vectors of piles with different diameter under failure load

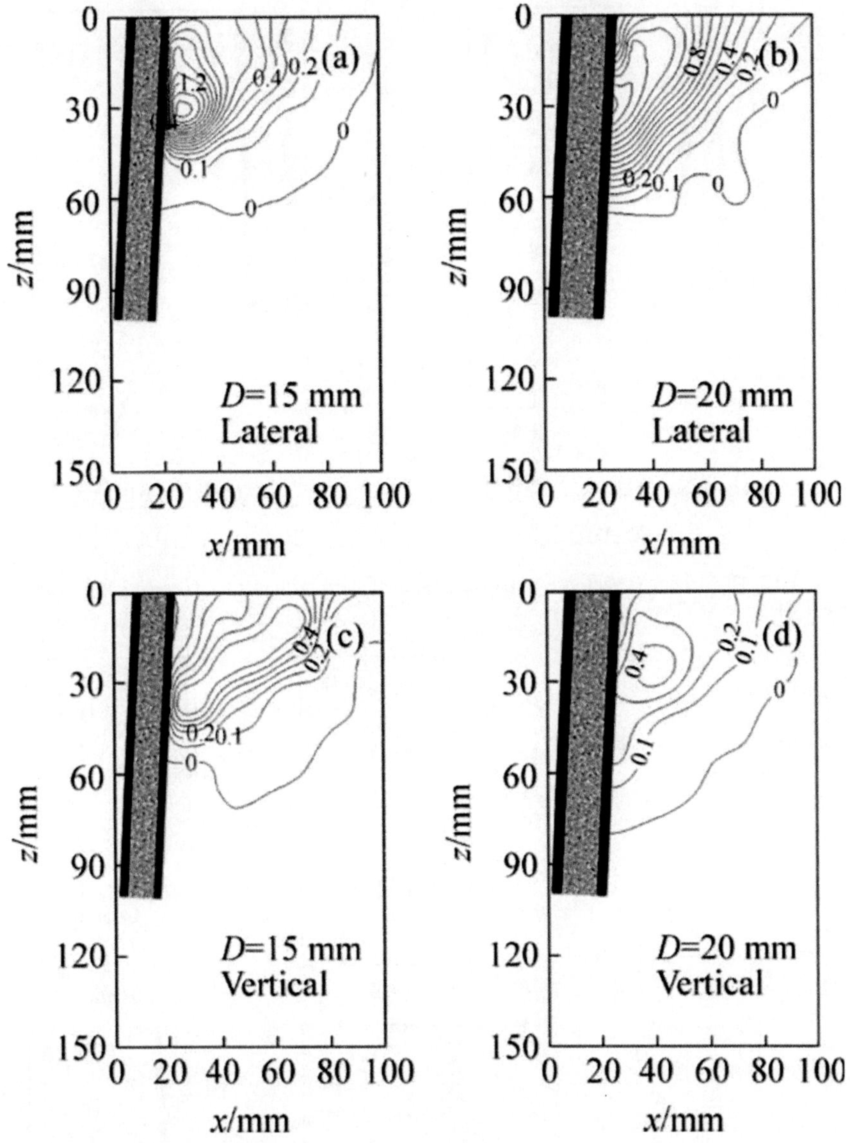

Fig. 11.12 Displacement contours (radial δ_r or vertical δ_z) of piles with different diameter under failure load

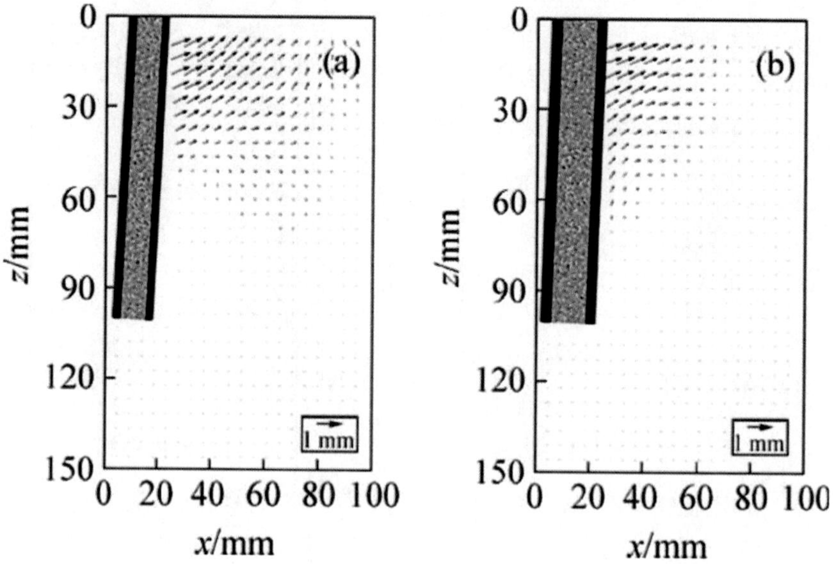

Fig. 11.13 Displacement vectors of piles with different diameter under 8.5 N and $\alpha = 60°$ oblique load

Fig. 11.14 Displacement contours (radial δ_r or vertical δ_z) of piles with different diameter under 8.5 N and $\alpha = 60°$ oblique load

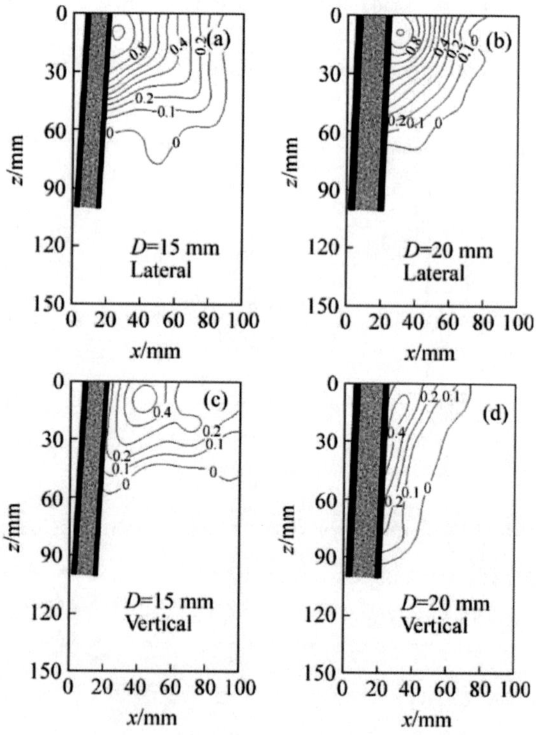

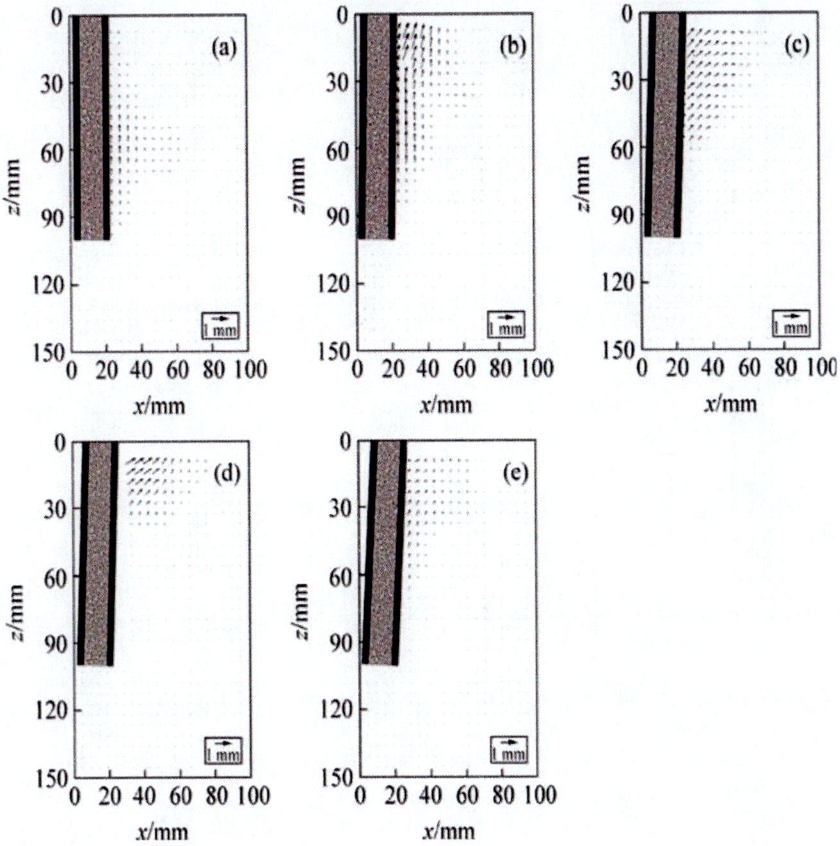

Fig. 11.15 Displacement vectors of piles with different oblique load angles and 6.0 N loads

displacement contours of pile B with different oblique load angles are shown in Fig. 11.16a–i, respectively. It can be seen that the displacement fields of pile B under different oblique pulling load angles have some noticeable difference. The displacement vectors and (radial or vertical) displacement contours of pile B under oblique pulling load ($\alpha = 30°$, $\alpha = 45°$ and $\alpha = 60°$) are close to that of pile B under lateral load ($\alpha = 90°$), but different with that of pile B under axial uplift load ($\alpha = 0°$).

11.2.4 Conclusions

This study demonstrated the combined use of transparent soil and PIV technique in the model tests to study internal soil deformation caused by pile moving under

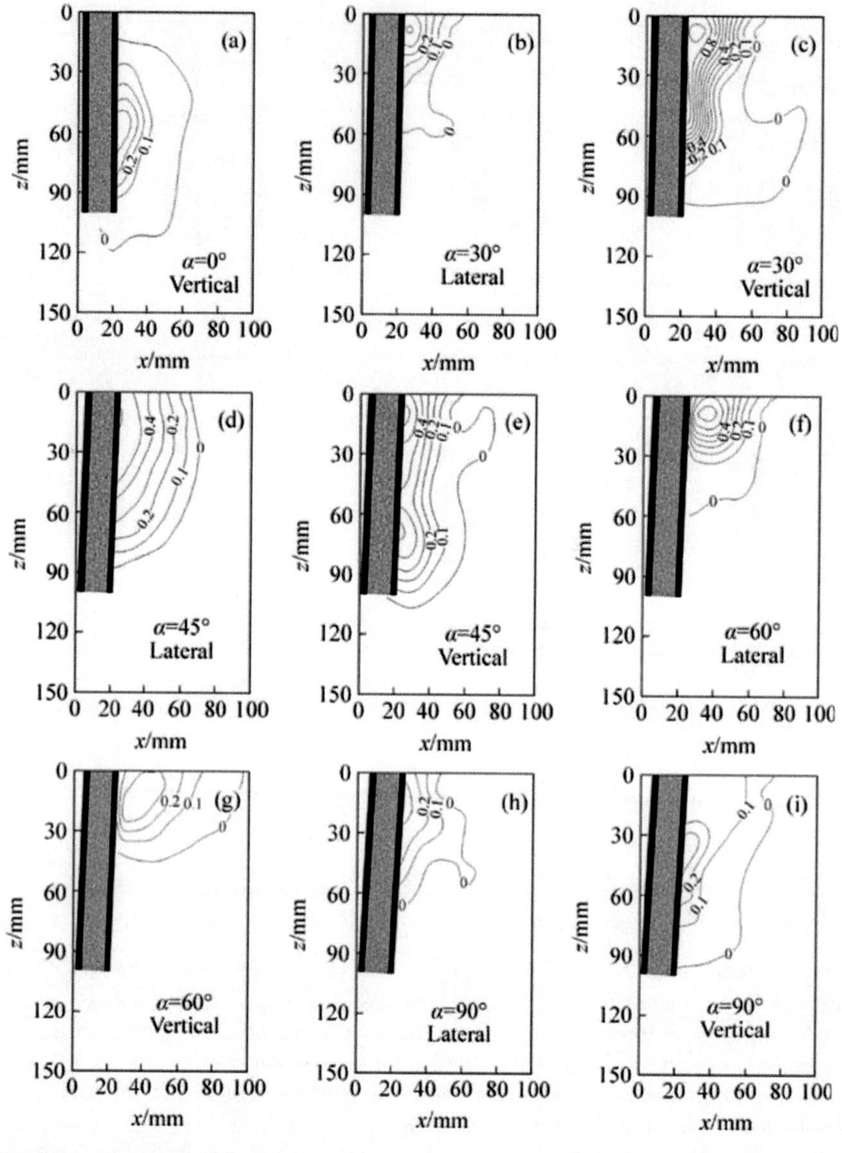

Fig. 11.16 Displacement contours (radial δ_r or vertical δ_z) of piles with different oblique load angles and 6.0 N loads

oblique pulling loads. The visualization of spatial deformation inside soil masses will improve the understanding on the capacity of piles under oblique pulling loads. Some conclusions can be obtained as follow:

1. Test results showed that the developed optical system and transparent sand are suitable for studying pile-soil interaction under oblique pulling loads problems. The visualization of spatial deformation inside soil masses will improve the understanding on the capacity of pipe piles under oblique pulling loads.
2. The displacement vectors and (radial or vertical) displacement contours of pile under oblique pulling load ($\alpha = 30°$, $\alpha = 45°$ and $\alpha = 60°$) are close to that of pile under lateral load ($\alpha = 90°$), but different with that of pile under axial uplift load ($\alpha = 0°$).

However, there are some limitations in this study. The same problem with other physical models is that the tests are performed on a scaled model at gravity condition, the stress-dependent behavior of natural soils cannot be properly modeled in this study; hence, transparent soil used in centrifuge physical model test will be one of important research directions.

11.3 Soil Plugging Effects in Pipe Pile

11.3.1 Introduction

Open-ended pipe piles are widely used in urban construction, marine engineering and offshore port wharf engineering because of the easy penetration characteristics of the pipe pile and the bearing capacity of the pipe pile which is close to the bearing capacity of the closed-ended pipe pile. When the open-ended pipe pile penetrates into the ground, the soil will enter the interior of the pipe pile from the bottom in the initial stage of penetration. When the pipe pile continues to penetrate deeper, the soil column inside the pipe pile can be referred to as a 'plug' when it inhibits additional soil intrusion into the interior of the pipe pile [149, 65]. This is known as soil plugging effect. The existence of plugging effect leads to the distinctive characteristics of open-ended piles compared to the equivalent closed-ended piles.

Many scholars have conducted a series of studies on the plugging effect of open-ended piles. O'Neill and Raines [150], Randolph et al. [39] and Liu [151] established static equilibrium formulas based on the principle of equilibrium to evaluate the axial bearing capacity of open-ended pipe piles. Paik and Lee [40] proposed to quantitatively evaluate the degree of soil plugging by defining IFR or PLR. On the basis of his work, many scholars pointed out that IFR is more suitable as an index for evaluating soil plugging effect than PLR [41, 42, 43]. Since the IFR cannot be easily obtained in actual projects or field tests, it is very important to develop a formula for accurately predicting the IFR value based on the PLR value for calculating the plug capacity. Meanwhile, a large amount of valuable work has been carried out to investigate the responses of soil plugging effect through field tests [41, 44, 45, 46, 152] and physical model tests [47, 48, 65]. The test results show that the degree of soil plugging is highly dependent on the pile diameter, relative density, penetration depth and installation [49–52]. A key component of the experimental design was to adopt the double-walled pipe pile system because the instrumented measurement scheme cannot be applied to the inside of pipe pile [46, 53, 65]. Based on the results of field tests as well as model test data obtained from the double-wall pipe pile system, the mobilization mechanisms of the three components (external frictional resistance, annulus resistance and plug resistance) of the pipe pile's bearing capacity were discussed [53]. Given an increasing demand for open-ended pipe piles, considerable effort has been made to develop the calculation method of the bearing capacity of open-ended pipe piles [53, 54, 152, 153]. However, these methods are empirical formulations based on CPT data. The difference in the mobilization mechanism of the frictional resistance between the internal and external pile-soil interface leads to a problem involving greater uncertainty in accurately calculating the bearing capacity of pipe piles. Although some literatures have studied the dynamic formation mechanism of soil plug during the penetration process, it is still necessary to further reveal the complexity of plugging characteristics to serve the optimization for the design method of pipe piles.

In order to realize the visualization research of the soil plugging phenomenon, some scholars have developed different test methods to study the soil plugging effect of pipe piles. The advanced non-destructive techniques, such as Industrial X-ray CT scanners, have been applied in small-scale geotechnical model tests [154, 155]. In addition, half-model tests with image acquisition system have been used to measure the soil movement [55]. Although the behavior of the surrounding soil mass under the pipe pile was discussed based on the observation of the movement of soil particles, there was still considerable uncertainty about the change of soil plugging state inside the pipe pile during the penetration process compared to its initial condition. Consequently, the visualization study of the formation process of soil plug is of great significance to further guide numerical simulation or theoretical calculation.

This investigation provides a visualization method based on transparent soil modelling to investigate the dynamic formation mechanism of soil plug during the penetration process of open-ended pile. The specific objectives of this study are: (i) to better visualize the formation process of soil plug and obtain the displacement vector field; (ii) to analyse the influencing factors on the plugging effect such as pile diameter, penetration speeds, wall thickness and relative density; and (iii) to elucidate the plugging mechanism of open-ended pile. Considering the complex influence of plugging effect on the bearing capacity of pipe piles, the test results of this study can provide a new perspective for further improving the theoretical calculation of the bearing capacity of pipe piles, and can be used as a verification case of numerical simulation of pipe piles.

11.3.2 Laboratory Tests

11.3.2.1 Test Apparatus

The experimental setup in the investigation was designed to enable visualization of the plugging phenomenon inside the pipe pile under static load test. The experimental system consisted of nine main components: (1) micro-force testing machine, (2) model pile, (3) electrified electromagnet, (4) digital camera, (5) transparent soil model box, (6) linear laser device, (7) loading control system, (8) image capturing and processing system, (9) optical shockproof platform. The micro-force testing machine is the Instron3345 single column micro-force testing machine produced by Instron Company in America. It has a 5kN force sensor. The displacement measurement accuracy can be controlled within the range of 0.02 mm, and the measurement acquisition frequency can reach up to 100 Hz. The upper and lower parts of the electrified electromagnet are provided with two connectors (connector A and connector B). The electrified electromagnet is vertically connected to the universal testing machine through connector A, and the electrified electromagnet is vertically connected to the lower model pile through connector B. The internal dimensions of transparent soil model box are 320 mm long by 180 mm wide by 290 mm high. The linear laser device, with a maximum output power of 2000mW and wavelength of 532 nm, was placed on the right side of the transparent soil model box. Since the transparent soil model tests need to be carried out in a closed dark environment, taking into account that the photos taken by the low-resolution CCD industrial camera in the low-sensitivity state will have high-noise images [56, 57]. In order to obtain a clear soil plug phenomenon inside the pipe pile, the image acquisition device is a digital camera, and the resolution of the digital camera is set to 6000 × 4000 pixels. The photograph of the test apparatus is depicted in Fig. 11.17.

11.3.2.2 Test Materials

The transparent soil material used in this study was made from fused quartz (Fig. 11.18) and refractive index matching pore fluid [11, 18, 56]. Ezzein and Bathurst [22], Guzman et al. [32] and Zhao and Ge [58] have proved that the mechanical properties of transparent soil can appropriately simulate the mechanical properties of natural sand soil. The purity of the fused quartz is 99.99%, which was obtained from high specification quartz processing plant. Figure 11.2 illustrated the grain size distribution curve of the fused quartz. In order to better simulate the characteristics of natural sand, the grain size distribution curve of fused quartz is approximately consistent with that of Fujian standard sand. The grain size range of the fused quartz was 0.1–2 mm. A summary of properties of fused quartz was given in Table 11.3. The mixed pore fluid comprised of N-dodecane and White Oil 15 in a volume ratio of 1:6.175 (due to the slight difference in the refractive index of the White Oil 15

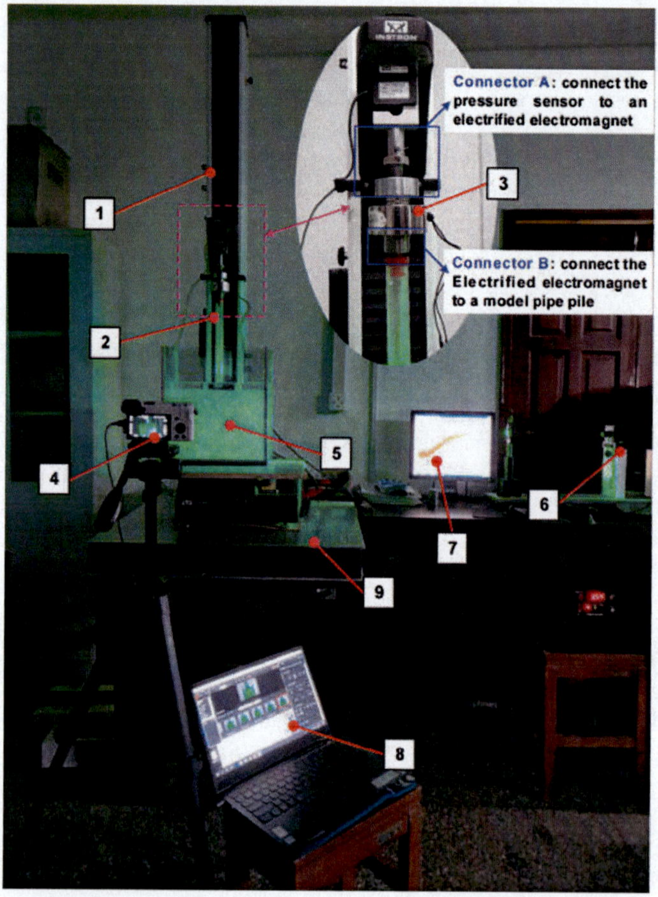

Fig. 11.17 Components of the experimental setup (1: micro-force testing machine, 2: model pile, 3: electrified electromagnet, 4: digital camera, 5: transparent soil model box, 6: linear laser device, 7: loading control system, 8: image capturing and processing system, 9: optical shockproof platform)

from different batches, the volume ratio will change slightly). The refractive index of the pore fluid is 1.4590 at room temperature of 20° [23, 56].

Figure 11.19 shows a series of model pipe piles used for the model tests. The length of model pipe piles is 270 mm. The inner and outer sides of all model pipe piles are polished and roughened by machine tools in the same way, which can ensure the consistency of the roughness of the interior and exterior of the pipe piles. The upper part of the model pipe pile was embedded with a special casing screw, and the model pile was vertically connected with connector B through the casing screw. In order to ensure the balance of air pressure between inside and outside of the pipe pile during the test, two holes were opened in the symmetrical position of the upper part of the model pipe pile. For the model piles, both the ratio of the container size

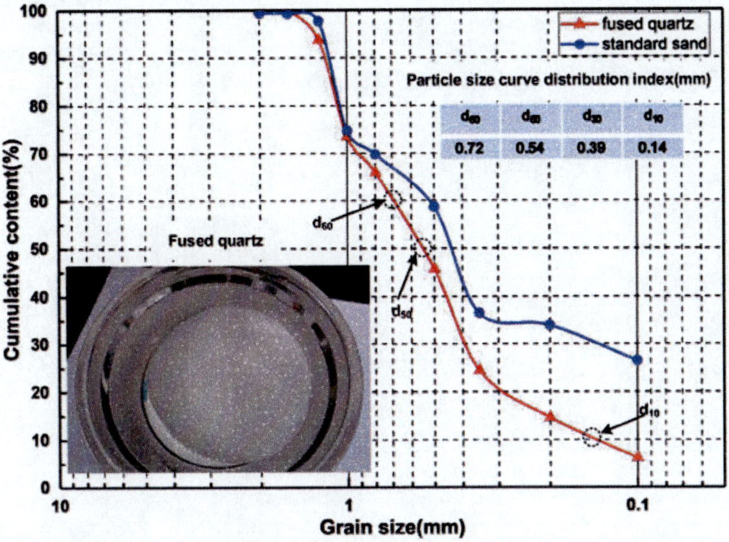

Fig. 11.18 The grain size distribution curve of the fused quartz

Table 11.3 Properties of fused quartz in the model tests

Property	Description or value
Specific gravity, G_s	2.33
Mean particle size, D_{50}	0.54
Friction angle, φ, measured in direct shear test	38.4°
Uniformity coefficient, C_u	5.45
Gradation coefficient, C_c	2.04
Maximum dry density, ρ_{dmax} (kg/m^3)	1.56
Minimum dry density, ρ_{dmin} (kg/m^3)	1.18

Note The fast direct shear tests were conducted at vertical stress of 50, 100, 150, 200 kPa, respectively

to the pipe pile diameter ($L/D_{max} = 11.4$; $D_{max} = 28$ mm) and the ratio of the model pipe pile diameter to the mean grain size of the fused quartz ($D/d50 = 46$) meet that suggested by Mo et al. [59] in order to reduce the boundary and grain size effects [56].

11.3.2.3 Test Preparation and Procedures

Four test groups with a total 10 experimental conditions were performed in this study. Table 11.4 summarizes all tests conducted. The test preparation started by pouring

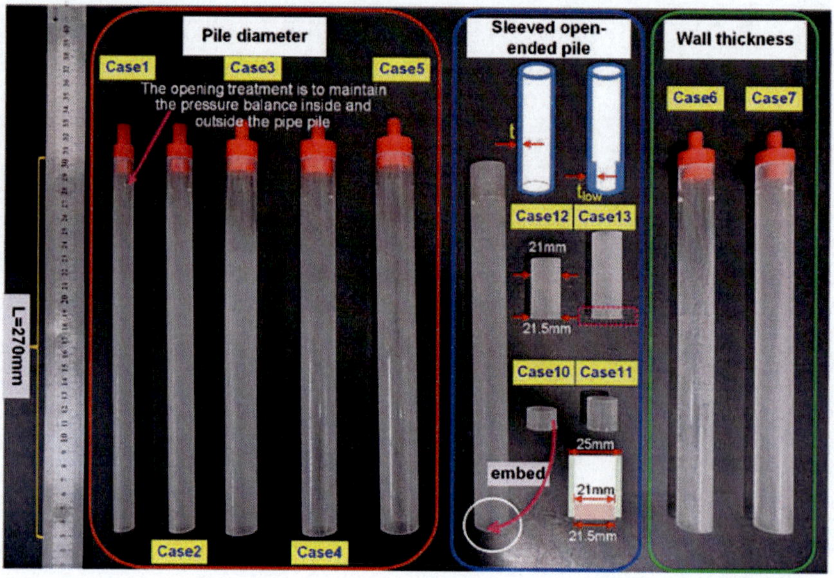

Fig. 11.19 Model pipe piles

Table 11.4 Summary of the experimental program in this study

Test groups ID	Exp.ID	Pile diameter/D (mm)	Wall thickness/t (mm)	Penetration speed/v (mm/min)	Relative density (%)
Test group 1 (pile diameter)	Case 1	14	2	4	MD/58
	Case 2	18	2	4	MD/58
	Case 3	22	2	4	MD/58
	Case 4	25	2	4	MD/58
	Case 5	28	2	4	MD/58
Test group 2 (wall thickness)	Case 4	25	2	4	MD/58
	Case 6	25	3	4	MD/58
	Case 7	25	5	4	MD/58
Test group 3 (penetration speed)	Case 8	25	2	2	MD/58
	Case 4	25	2	4	MD/58
	Case 9	25	2	6	MD/58
Test group 4 (relative density)	Case 4	25	2	4	MD/58
	Case 10	25	2	4	D/76

the fused quartz and the mixed pore fluid into the box. In the process of preparing transparent soil samples by 'layered vibrating compaction' method, the difference of compactness between different soil layers is prone to occur. And it is not conducive to the standardization control of parallel tests conditions. Therefore, this test adopts the 'vacuum preloading' method to prepare transparent soil samples. During this stage, the mass M1 of the mixed sample of fused quartz and pore fluid and the mass M2 of the glass model box are measured. The total mass M1 of the transparent soil sample is strictly controlled to be consistent in each test. The mixed sample of fused quartz and pore fluid was put into a large-volume vacuum box and vacuumed for 4 h. After vacuuming, the air inlet speed is controlled by the air inlet value on the vacuum box to compact the transparent soil sample. It should be noted that the transparent soil (Fig. 11.20) was densified to the target relative density according to the control of vacuuming time, inlet velocity and the vibration of the model box. Thereafter, the position of the experimental setup should be adjusted to meet the requirements. The position of the pipe pile for each test is located in the middle of the model box, which meets the distance requirement of avoiding construction effect [48]. By adjusting the linear laser device and the location of the transparent soil model box, the laser sheet should be aligned with the central axis of model pipe pile. The digital camera was positioned at a proper distance to the model box so that the whole speckle image is clearly visible.

Each parallel experiment was carried out in two separate stages: the first stage is penetration process; and the second stage is static load test. During the penetration test, in order to ensure the integrity of the pipe pile and the pressure sensor, and at the same time to ensure the verticality of the pipe pile penetration, the electrified electromagnet is in working condition. And the universal testing machine presses the pipe pile to the specified depth of 150 mm at a constant rate. After the penetration test, the electrified electromagnet is disconnected from the power supply and is in a state of no magnetic force. The electrified electromagnet is disconnected from the connector B. Based on Chinese Technical code for Building Pile Foundations (JGJ106-2014) and Zhang [156] introduced the static load test method, the loading method of the static load test in this investigation is the constant rate penetration method. The static load test was carried out 12 h after the penetration test. The test is terminated when the load no longer increases, and the ultimate load is determined according to the load-penetration curve. All tests reported in this investigation were conducted in a darkroom to obtain the clear speckle images.

11.3.2.4 Parameter Definition

Image measurement technique was implemented to accurately obtain the change of soil plug at different time of pipe pile penetration test. High resolution images were obtained about every 1 mm of pipe pile's penetration by the digital camera in this test. As shown in Fig. 11.21, the degree of soil plugging discussed in this study can be defined by the following parameters. L_0 is the initial height of the transparent soil model. Ln represents the height from the bottom of the pipe pile to the bottom of

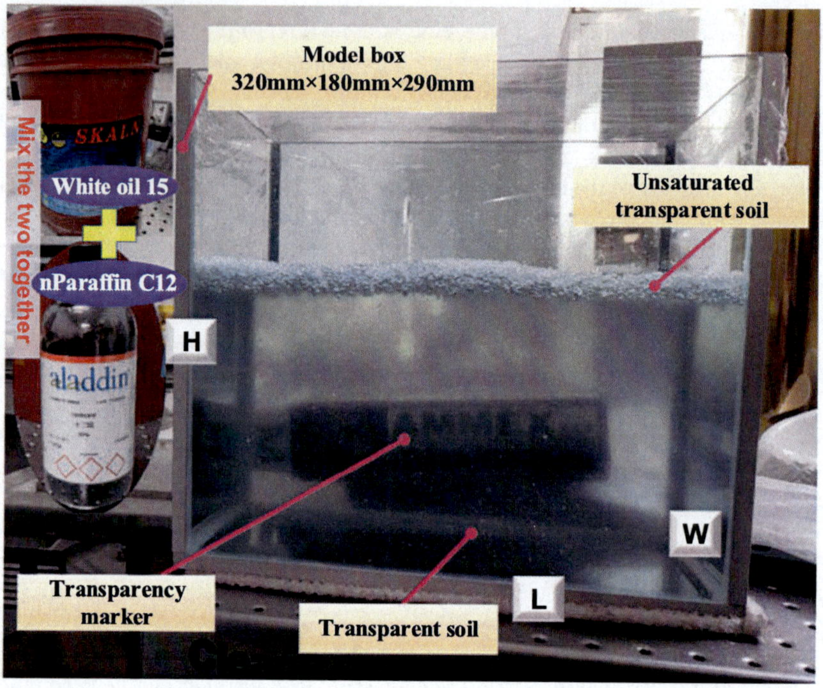

Fig. 11.20 Transparent soil model

the transparent soil sample under different penetration stages. H_n is defined as the height of the soil plug, which is the distance from the highest point of the soil plug on the central axis of the pipe pile to the bottom of the pipe pile determined from the images taken at each penetration depth. The degree of soil plugging for open-ended pipe piles can be quantified using PLR and IFR in this study, defined as

$$PLR = H\_n/((L\_0 - L\_n) \tag{11.2}$$

$$IFR = dH/dL \times 100(\%) \tag{11.3}$$

As given in Eq. (11.2), PLR is the ratio of the soil plug height to the penetration depth at different penetration stages. In Eq. (11.3), IFR represents the ratio between the increment of the soil plug and the increment of the penetration depth in each penetration increment step. According to the degree of soil plug, the soil plugging effect can be divided into three modes: unplugged, partially plugged, and fully plugged [42, 43, 46]. The total penetration resistance and the load bearing capacity of open-ended pile can be understood as consisting of external frictional resistance, annulus resistance and plug resistance [46, 49–52, 48, 53].

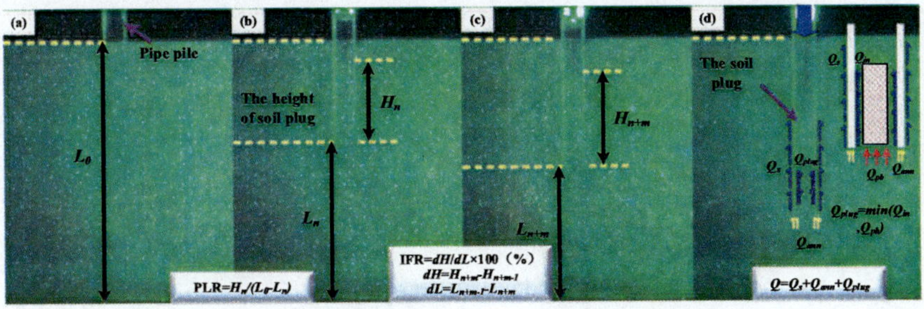

Fig. 11.21 The evaluation index of the degree of soil plugging

$$Q = Q\_s + Q\_ann + Q\_plug \tag{11.4}$$

$$Q\_plug = \min(Q\_in, Q\_b) \tag{11.5}$$

For an open-ended pipe pile, the overall axial penetration resistance Q applied to the top of the pipe pile is composed of three resistance parts: the external shaft resistance, Q_s, provided by the friction between the external wall of the pipe pile and the soil around the pile. The annulus resistance of pile base, Q_ann, provided by the resistance produced by the annulus contact surface of the pipe pile base and the soil. The soil plug resistance, Q_plug is the smaller value of either the inner frictional resistance, Q_in, or the bearing capacity of the soil beneath the plug base, Q_b.

11.3.3 Test Results and Analysis

11.3.3.1 Displacement Vector Field of Open-Ended Pipe Pile

To explore the disturbance of the soil around the pile in the process of pipe pile penetration, the model test of a pipe pile with a diameter of 28 mm was taken as an example, and the displacement vector field of different penetration stages was analyzed as follows. As shown in Fig. 11.22, the length of the vector arrow represents the magnitude of the displacement, and the angle of the vector arrow represents the direction of movement. The size of the example vector in the right corner of subplot is equal to the displacement of 5 mm. The soil plugging was formed progressively during the pipe pile penetration [60]. In the early penetration stage (penetration depth = 10 mm), because the soil plug had not formed inside the pipe pile, the soil intrusive inside the pipe pile moved upward, and the soil outside the pipe pile showed a small oblique displacement (Fig. 11.22a). When the penetration depth reached 60 mm, the displacement vector arrows inside the pipe pile showed a vertical downward pattern, and the length of the upper vector arrow was the largest. Compared with the initial stage of penetration, the lateral disturbance effect on the soil outside of the pipe

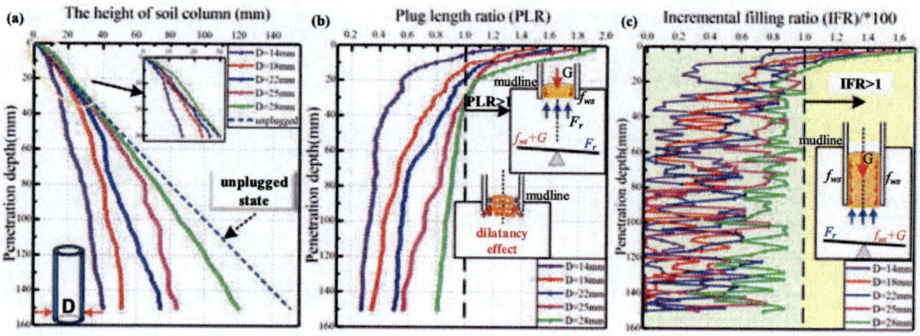

Fig. 11.22 Evaluation index of soil plugging under different pile diameter condition during installation: **a** the height of soil plug, **b** PLR, **c** IFR

pile increases. In addition, there were downward arrows at the bottom of the pipe pile, which also proves that the soil at the bottom of the pile tip was compressed vertically by the soil plug effect of the pipe pile. When the final penetration depth 150 mm was reached (Fig. 11.22c), the downward vertical displacement inside the pipe pile became more significant, and the maximum vertical displacement at the top of the soil column was approximately 30 mm. The lateral disturbance of the soil on both sides around the pile gradually increased, and the lateral disturbance range was about 4R. Moreover, as the degree of soil plugging increased, the soil at the bottom of the pile end was compressed more obviously. It can be intuitively seen that the displacement variation inside the pipe pile was much larger than that outside the pipe pile. It indicates that the disturbance effect of pipe pile penetration on the soil around the pile is small, and the plugging effect plays a crucial role in the mechanism of pipe pile.

11.3.3.2 Parametric Studies

To examine the influencing factor on the plugging effect for open-ended pipe piles, a series of static penetration tests of model pipe piles were performed based on the major influencing parameters, such as pile diameter, penetration speeds, relative density and wall thickness.

A penetration depth of 2 mm (the value of dL in Eq. 11.3) is taken as an image acquisition interval to obtain the changing height of soil plug, and a series of images are taken continuously at the interval. Each set of tests has 75 data values, according to which the evaluation index of the change of soil plug state during the penetration process is calculated.

Effects of pile diameter

Figure 11.23 shows the evaluation index of plugging effect under different pile diameter condition during installation. The height of soil plug, PLR and IFR were measured on five sets of pipe pile model tests (Test group1: case1, case2, case3,

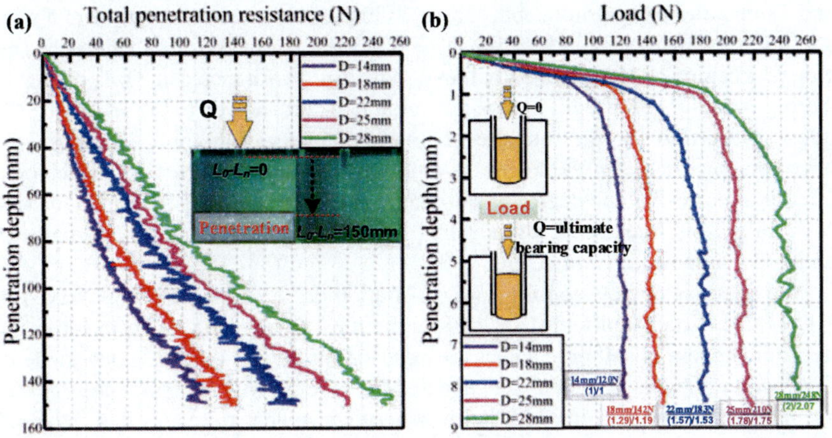

Fig. 11.23 **a** The total penetration resistance versus penetration depth with different pile diameter condition during installation, **b** Load-penetration depth curves under different pile diameter

case4, case5) with different pile diameters to study the influence of pile diameter effect on the degree of plugging. Figure 11.23a shows the height of soil plug versus penetration depth in the vertical penetration model test of pipe piles with different pile diameters. In the Fig. 11.23a, the dashed line represents the unplugged mode for the height of soil column is equal to the penetration depth of pipe pile. In the initial penetration stage, the height of soil column in five groups of pipe piles with different pile diameters is all greater than the height of soil column in the case of unplugged indicated by the dotted line. This feature was also found by Jeong et al. [41] in a field test of pipe piles. According to the force equilibrium analysis, the soil column inside the pipe pile is subjected to the self-weight of soil column (G), the internal friction resistance (f_{ws}) and the upward pushing force (F_r). At the initial penetration, $G + f_{ws} < F_r$, so it causes the soil to gradually moving upward. Beyond the initial stage of penetration, the internal soil column may develop sufficient frictional resistance to start to limit the continued intrusion of the soil. The height of the soil column under five conditions starts to be gradually smaller than the depth of penetration. The soil plugging length increased with the pile penetration depth, and with the increasing pile diameters at the same penetration depth. Meanwhile, the increase rate of the soil plug height is slowing down with the increase of penetration depth. For the case of smaller pile diameters, in the later stage of penetration, the increase rate of the corresponding soil plug height is slowing down. Clearly, as the pile diameter increases, the height of the soil plug gradually increases.

The results are shown in Fig. 11.23b in terms of PLR versus penetration depth. Firstly, in the initial stage of penetration, PLR is greater than 1. Similar characteristics has been reported by Pail (1993, 2003) and Doherty and Gavin [61]. It is because the transparent soil used in the model test is in a dense state, so there is a dilatancy appeared. For pipe piles with pile diameters of 14 mm, 18 mm, 22 mm, 25 mm and 28 mm, the penetration depths at which PLR starts to be less than 1 are 5 mm, 13 mm, 19 mm, 30 mm and 32 mm. This indicates that the position where the soil plugging begins to occur gradually moves downward with increasing pile diameter. Then, according to Poiseuille's law, the volume of soil allowed to intrude inside the pile is proportional to the square of the inner diameter of the pile. Therefore, the larger the pile diameter, the lower the ability to restrict soil intrusion into the pipe pile. This means that the degree of soil plugging is inversely proportional to the size of the pile. Therefore, the PLR gradually increases with the increase of pile diameter under the same penetration depth.

The relationship between IFR and penetration depth can be clearly observed in Fig. 11.23c. Compared with PLR, IFR is the most sensitive index to quantify the dynamic process of soil plug and as the criterion for dividing the plugging mode. It is difficult to obtain the parameters updated in real time with each incremental step by invasive observation method, but it is very convenient to record this 'moving' parameters with the help of transparent soil visualization measurement technology. A key point noted in Fig. 11.23c is that the relationship between IFR and penetration depth is fluctuating, which means that the formation of soil plug is not a continuous stationary process. Similar characteristics of the dynamic change of the soil plug has been reported by Paikowsky [62] and Seo (2018). The reason for the fluctuation of IFR is that the arch effect of the soil plug not only increases the unit internal friction but also enhances the inhibition of soil intrusion, and the squeezing effect of the pipe pile also promotes the increase of upward pushing force. As a result, the equilibrium state of the weight of soil column, internal friction resistance and the upward pushing force is in the process of dynamic change. As observed from Fig. 11.23c, IFR is unsteadily reducing with the penetration depth. IFR in the initial stage is greater than 1, which is similar to the change of PLR. This was also observed from large-deformation numerical analyses of open-ended jacked pipe piles by Doherty and Gavin [61] and Wang et al. [60].

When the pipe pile penetrates into the range of 60 –80 mm, IFR of the small-diameter pipe pile starts to reach 0 (fully plugged mode). After that, the IFR value increased to a smaller value at next increment, this means that the soil plugging mode changes from fully plugged to partially plugged. And then, reached 0 again at the subsequent penetration. This is consistent with some previous findings by other researchers [42, 43, 53, 63]. In addition, as the pile diameter increases, IFR also increases correspondingly. It can also be observed that for pipe pile with a diameter of 28 mm, IFR is basically greater than 0.5.

The total penetration resistance (Q) versus penetration depth (L) curves obtained from five model tests with different pile diameter are shown in Fig. 11.24a. It can be seen that the total penetration resistance increases with the penetration depth. At the same time, the maximum penetration resistance of the five pipe piles (D from 14 mm to 28 mm) to the penetration depth of 150 mm increases correspondingly, which are 118 N, 139 N, 183 N, 220 N, and 252 N, respectively. During penetration, the growth rate of penetration resistance in the initial stage is less than that in the later stage. The Q–L curve have a small fluctuation. Some strong similarities also exist for the results obtained by Paikowsky [62]. This was because in the process

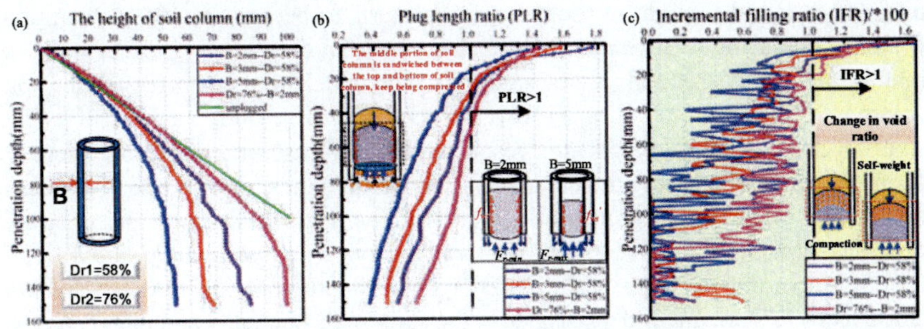

Fig. 11.24 Development of the height of soil column, PLR and IFR under different wall thickness and relative density during installation: **a** soil column height, **b** PLR, **c** IFR

of statically pressing the pipe pile, the compaction effect of pile tip on soil around the pipe pile is nonlinear, and the compaction action to overcome the resistance of soil around the pile end is discontinuous. Secondly, the force variations match the expected mechanism of intermittent formation of soil plug, showing a decrease in the penetration resistance when the soil plug destroyed, and build up as the fully plugged state reached again [62].

Figure 11.24b shows the curve of load and penetration depth. As mentioned above, the static load test method selected is the constant rate penetration method. The ultimate load is determined according to the load and the penetration depth curve under the relatively stable state when the load is no longer increasing. The ultimate bearing capacity of pipe piles with diameters of 14 mm, 18 mm, 22 mm, 25 mm and 28 mm measured in static load test is 120 N, 142 N, 183 N, 210 N and 248 N, respectively. The ultimate bearing capacity of pipe piles with a diameter of 14 mm is designated to be one, and the ultimate bearing capacities of the piles with different diameters are 1.19, 1.53, 1.75 and 2.07 respectively. The contact area of the outer wall of the pipe pile and the pile end with the soil is proportional to the size of the pile diameter. With the increase of the pile diameter, the degree of soil plug is gradually reduced, but the bearing capacity of the pipe pile has not been suppressed. The greater degree of soil plug indicates that the coefficient of lateral earth pressure inside the pipe pile increases, which in turn increases the unit internal friction. However, the actual influence on the internal friction resistance of pipe pile is not only related to the degree of soil plug, but more importantly, it depends on the effective height of soil plug. For the problem of marine pile foundation, it is proved that large diameter pipe piles can be developed based on the test results [64].

Effects of wall thickness and relative density

In order to study the influence of wall thickness and relative density on the soil plugging during the penetration process of pipe pile, four different conditions (Test 2: case4, case6, case7 and Test 5: case 14) were considered in this study. Figure 11.25a presents the development of the height of soil plug under different wall thickness and relative density during installation. Evidently, the larger the wall thickness of pipe pile, the earlier the height of the soil column starts to be less than the penetration depth for the same relative density, which means that the plugging effect starts to inhibit the intrusion of soil particles into the pipe pile. The final heights of the soil columns formed by pipe piles with wall thicknesses of 2 mm, 3 mm, and 5 mm were 84 mm, 69 mm, and 54 mm, respectively. Obviously, the plugging effect of open-ended pile increases with the increase of wall thickness of pipe pile in a certain range. The previous research results of several scholars have given the contrary conclusions about the influences of relative density on soil plugging [Nicola 1997, 40, 65]. According to our test results, the height of soil plug is 84.6 mm under medium dense and 97.7 mm under dense state for the same wall thickness and pile diameter. This is consistent with previous studies [50, 55, 65], which demonstrated that the loose soil is conducive to the formation of soil plugs. A brief explanation for this phenomenon is proposed as follows. The continuous compression and deformation of the soil column itself causes the relative density to change before and after the formation of the soil column. A larger change in relative density also means a higher degree of plugging. The allowable change of relative density of loose soil is much larger than the dense soil. Therefore, the degree of soil plugging in the medium dense state is greater than that in the dense state.

In order to reveal the influence of wall thickness and relative density on the plugging behavior of pipe piles, Fig. 11.25b and c show the development of PLR and IFR under different wall thickness and relative density during penetration, respectively. IFR and PLR exhibit similar changing trends with the penetration depth, and both show a gradual decrease with penetration depth. The obvious difference is that for a pipe pile with a wall thickness of 5 mm, the IFR reaches 0 when the penetration depth is 65 mm, while for a pipe pile with a wall thickness of 2 mm, the IFR is 0 for the first

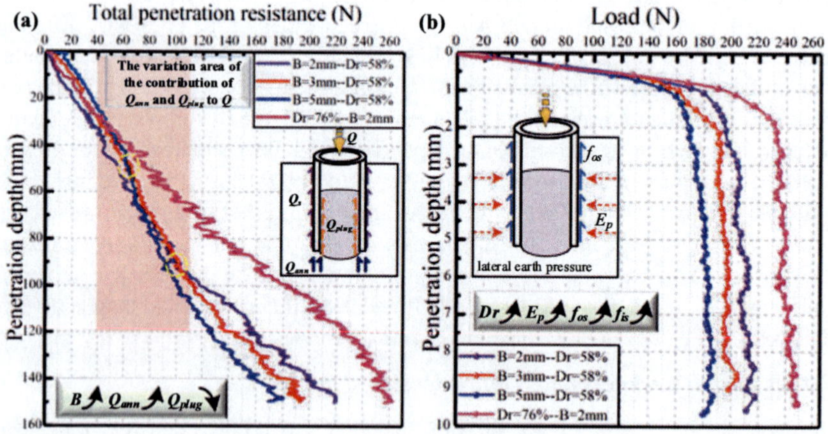

Fig. 11.25 **a** The total penetration resistance versus penetration depth with different wall thickness and relative density during installation, **b** Load-penetration depth curves under different wall thickness and relative density

time when the penetration depth is 100 mm. To some extent these differences are attributable to the reduction of space inside the pipe pile as all the other parameters are set to be same in the tests.

The measurement results are presented in Fig. 11.26a in terms of the total penetration resistance versus depth. The final total penetration resistance of pipe pile with different wall thickness (case4, case6 and case7) are 220 N, 196 N, 181 N, respectively. Evidently, at the initial stage of penetration, the greater the wall thickness of the pipe pile, the greater the corresponding total penetration resistance. The reason why the total penetration resistance of the pipe pile with thicker wall in the initial stage of penetration is the largest is because the annular contact area of pile is large and the annulus resistance of pile contributes to the majority of the total penetration resistance. Moreover, the degree of plugging is relatively low in the initial stage of penetration. On the contrary, when the pipe pile continues to penetrate to a depth of 60 mm, the total penetration resistance of the pipe pile with a wall thickness of 5 mm begins to be less than that of the pipe piles with a wall thickness of 2 mm and 3 mm. The main difference in this stage is that the height of the soil column formed during further penetration increases sequentially for 5 mm, 3 mm and 2 mm wall thicknesses of the pipe piles. This means that the contribution of the annulus resistance to the total penetration resistance decreases gradually, while the contribution of the internal friction mobilized by the formed soil column to the total penetration resistance increases gradually. This confirms that the mobilization of internal friction is directly proportional to the height of soil column. The degree of soil plugging can only represent the magnitude of unit internal friction.

Figure 11.26b exhibited that the overall load-penetration depth response of pipe pile with different wall thickness and relative density. The general variation of three sets of tests with different wall thickness displays a similar tendency and matches fairly well with the total penetration resistance during penetration. The ultimate bearing capacity of three sets of model tests with different wall thickness are 186 N, 205 N, and 218 N, respectively. With the increase of wall thickness, the ultimate bearing capacity of pipe pile decreases gradually. Although increasing the wall thickness of the pipe pile strengthens the annulus resistance of pile tip, it weakens the

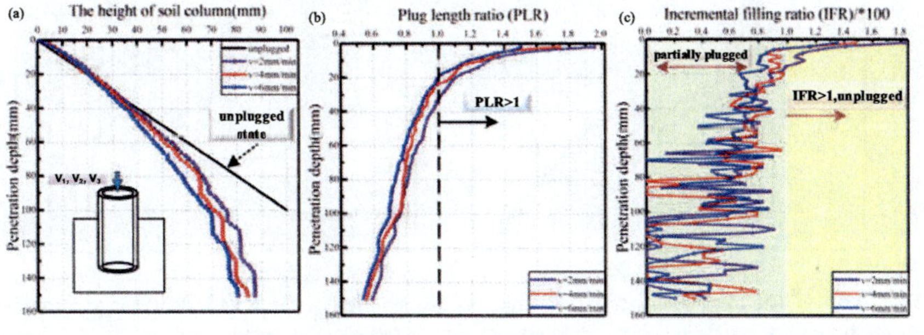

Fig. 11.26 Evaluation index of soil plugging under different penetration speeds condition during installation: **a** the height of soil column, **b** PLR, **c** IFR

internal friction resistance, so increasing the wall thickness suppresses the bearing capacity of the pile instead. With respect to the ultimate bearing capacity of pipe piles under different relative density, the ultimate bearing capacity of pipe piles under medium dense condition and dense condition are 186 N and 242 N respectively. This means that the ultimate bearing capacity increases obviously with increasing the relative density.

Effects of penetration speed

From the obtained images, the height of the soil column inside the pipe pile can be clearly identified. This is the advantage of using the transparent soil model test. The height of soil column, PLR and IFR were measured on three sets of model test (case 4, case8 and case9) to investigate the plugging effect for different penetration speeds (Fig. 11.27). From Fig. 11.27a, it can be seen that the soil plug height under the three model test are almost identical in magnitude at the initial shallow penetration phase (approximately 0–30 mm). As the penetration progressed further, the plugging effect gradually became active. For the penetration speed of 2 mm/min, the final height of the soil column is 87 mm, which is the largest among the three sets of tests; The state of fully plugged (IFR = 0) first appeared at penetrate depth of 104 mm. In comparison, the final height of the soil column with a penetration speed of 4 mm/min is 80 mm; The degree of soil plugging increases accordingly, and the fully plugged state was first found at the depth of 82 mm. For the penetration speed of 6 mm/min, it is the fastest to reach the state of fully plugged (L_0-L_n = 70 mm), and the degree of soil plugging is the largest among the three sets. Based on the results, the degree of soil plugging increases with the increase of penetration speed. This result was consistent with the previous findings [48].

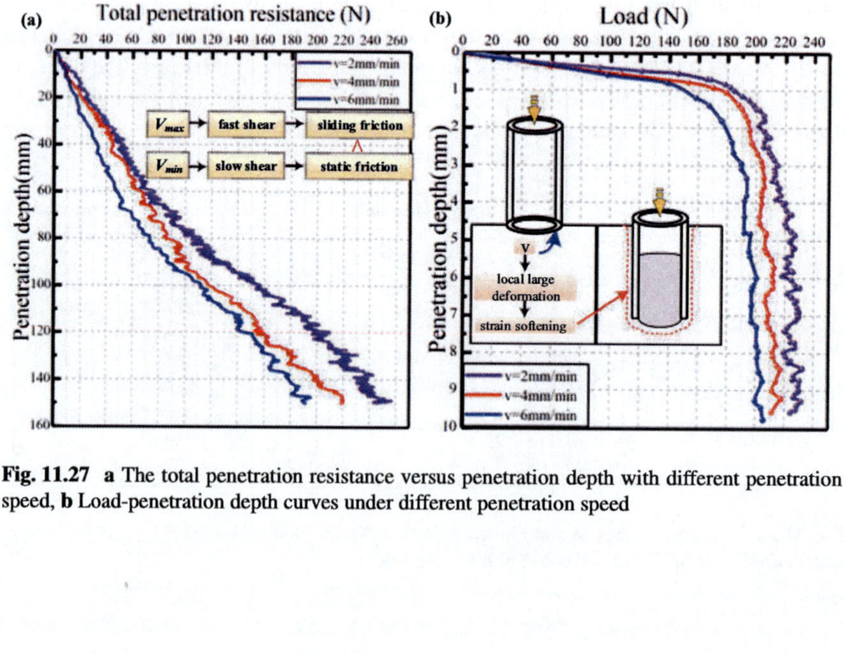

Fig. 11.27 **a** The total penetration resistance versus penetration depth with different penetration speed, **b** Load-penetration depth curves under different penetration speed

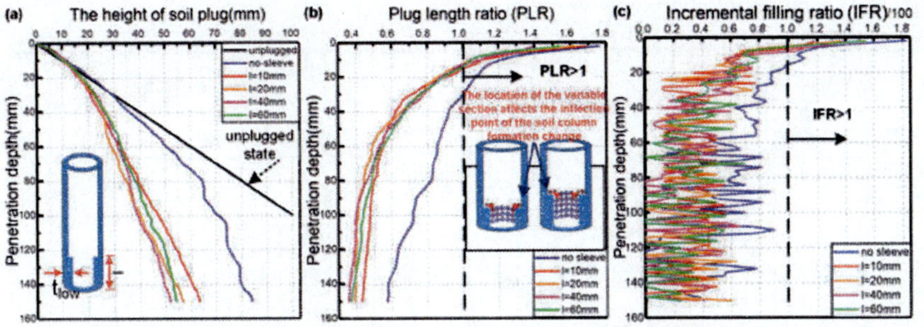

Fig. 11.28 Effect of sleeved heights on plugging behavior: **a** the height of soil plug, **b** PLR, **c** IFR

Both the total penetration resistance and the ultimate bearing capacity were monitored in these model tests to investigate the effect of penetration speeds on the plugging behavior. The results are shown in Fig. 11.28 in terms of the penetration depth versus the total penetration resistance and load. From the results, it can be concluded that for the same penetration depth, the greater the penetration speed, the lower the penetration resistance. Moreover, in the static load test, the smaller the static load rate, the greater the ultimate bearing capacity monitored. The local large deformation of the pile-soil interface is often accompanied by strain softening, which leads to a reduction of the external frictional resistance. Secondly, the height of the soil column formed under the faster penetration speed is smaller, so the internal friction resistance that can be mobilized is also smaller. Regarding the annulus resistance, Liu [151, 157] reported that the unit annulus resistance is independent on pile scale and plugging mode, so it can be considered that the annular resistance is not affected by the penetration speed. Therefore, the bearing capacity of the pipe pile is inversely proportional to the penetration speed. This further demonstrated that a large degree of soil plugging does not mean that the bearing capacity of the pipe pile is large, and the bearing capacity of the pipe pile is directly proportional to the internal frictional resistance mobilized.

11.3.3.3 The Dynamic Evolution Process of Soil Plug

With the aid of transparent soil visualization test technology, and based on the observed real plugging phenomenon, Fig. 11.29 summarizes the key evolution processes of plugging formation.

(a) When pipe pile begins to penetration into the soil, the soil beneath the pipe pile squeezes into the inside of open-ended pile to form a soil column. The shear strength of the pile-soil contact surface is less than the shear strength of the soil itself, so the circumferential shear process of the pipe pile on the soil is mainly manifested as the overall slip damage of the soil along the contact surface. The circumferential shear causes the soil at the bottom of pile tip to

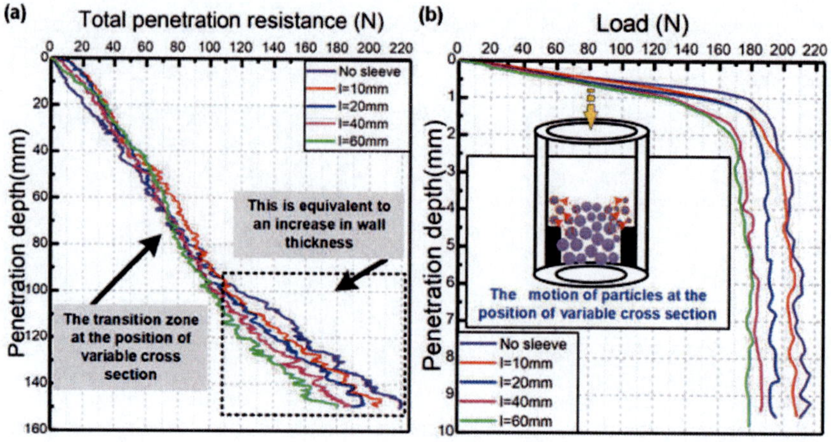

Fig. 11.29 **a** Effect of sleeved heights on the total penetration resistance, **b** Effect of sleeve heights on the ultimate bearing capacity

expand upward, which releases the stress in the vertical direction. The vertical downward penetration of pipe pile produces a downward drag force on the soil particles, causing the initial top surface of soil column to take on an upward convex form (Fig. 11.29a). The height of the soil column at the initial stage is greater than the penetration depth. As a result, the PLR and IFR obtained during the initial stage of penetration are greater than 1. This initial penetration phase of the pipe pile is called the fully unplugged.

(b) As penetration continues, more soil particles intrudes into the inside of pipe pile to form a soil column. As illustrated in Fig. 11.29b, there is a certain amount of settlement on the top surface of soil column, which is different from the state of fully unplugged. The height of soil column starts to be smaller than the penetration depth at this stage, which also indicates that the plugging effect starts to inhibit the soil intruding into the pipe pile. The PLR and IFR are in the range of 0 to 1. This stage is called developing plugged. Compared with the fully unplugged stage, the soil column is formed by the overall sliding of the contact surface and also produces its own compression deformation at this stage. This is because the middle portion of the soil column is sandwiched between the top and bottom portion of the soil column and therefore keeps being compressed as penetration continues. The result of continuous compression of the middle part of the soil column is not only to increase the unit internal friction resistance at different depth, but also to increase the upward pushing force. However, there is a difference between the increments of the two in terms of the change in the equilibrium state of the forces. Therefore, the IFR value is in an intermittently fluctuating decreasing trend in the depth direction, rather than a gradual decrease.

(c) As shown in Fig. 11.29c, with further penetration, a large settlement appears on the surface of soil column in this stage. The height of soil column keeps constant, and the soil column does not move relative to the inner wall of the pipe pile, but the soil below the pipe pile can still continue to further intrusion due to the increased upward pushing force In this state, the PLR is gradually reduced, and the IFR value is 0. This state of soil plug is called fully plugged.

(d) When the pipe pile continues to penetrate, a smaller settlement occurs at the top of the soil column compared to the stage of the fully plugged (Fig. 11.29d). The height of soil plug is slightly increased, and the IFR also became greater than 0. After reaching the state of fully plugged, the increase of the upward pushing force mobilized by the plugging effect is greater than the increase of the combined force of the mobilized internal friction and self-weight, resulting in the destruction of the fully plugged. Based on the interaction of the increment in penetration resistance of the various parts mobilized under different soil plug states, this results in an unstable state alternating from fully plugged to partially plugged. This stage is termed as destroying and reproducing plugged. The soil column is gradually transformed from an active arch to passive arch by alternating soil plug state [57].

Discussions

The IFR is the most accurate parameter to describe the degree of soil plugging. In order to consider the plugging effect on the bearing capacity of the pipe pile, various design methods such as Purdue method and University of Western Australia method introduce IFR to quantify the degree of soil plugging. However, it is not practical to measure IFR in real time for practical engineering. Paik (2003) and Murthy et al. [47] proposed corresponding empirical formulas between IFR and PLR based on the results obtained from the field tests and model tests. But the data selected for fitting the empirical equation are more based on the data points of different penetration stages in the whole penetration process, and the sample data for different pipe pile variables are less. In view of this, this study summarizes the data provided in the published literature for PLR and the corresponding IFR [41, 47, 65, 158] and proposes a new empirical formula in combination with a series of data points obtained from this model test.

The empirical formula between PLR and IFR can be expressed as follows:

$$\mathrm{IFR} = 132.64 \times \mathrm{PLR} - 38.53 \tag{11.6}$$

Relationship given by Paik (2003) and Murthy et al. [47] are also plotted in Fig. 11.30 to compare the results of the present experiments with the published data. It can be seen that the current model test data are basically consistent with these published data of small-scale model test. Evidently, the empirical formula proposed by Murthy underestimates the IFR. Compared with the empirical equation proposed by Paik (2003), Eq. 11.5 is more accurate for predicting the IFR values of larger

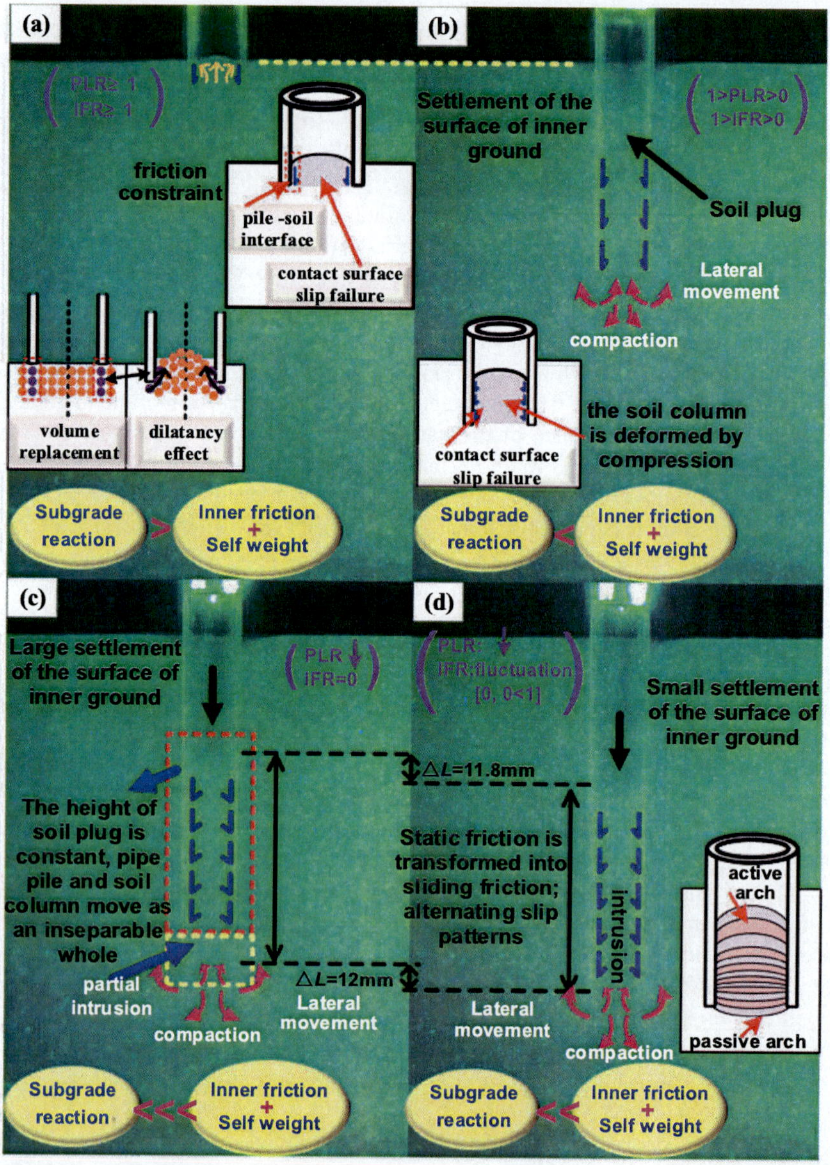

Fig. 11.30 The evolution of soil plug for open-ended pipe pile: **a** fully unplugged, **b** developing plugged, **c** fully plugged, **d** destroying and reproducing plugged

diameter piles. Considering that most of the practical applications involve large-sized pipe piles, it is thus reasonable to believe that this empirical formula has a certain range of applicability.

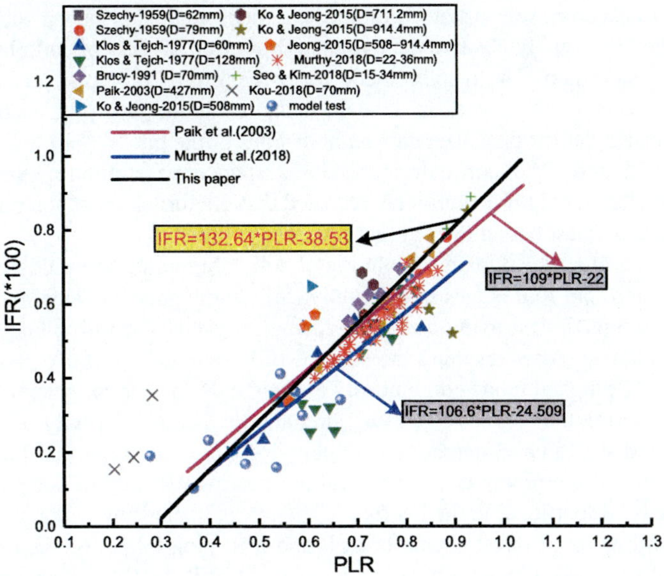

Fig. 11.31 Relationship between IFR and PLR

The ratios of the pile diameter to soil particle size are around the lowest value due to the limitation of the model box size. Although the designed test program satisfies the size effect conditions required by the model test, the results of the small-scale model tests can only be used as a reference for guiding engineering practice, and cannot be directly applied to actual projects without calibration and verification. However, the plugging behaviors revealed by the current tests are in good agreement with the previous laboratory tests and field tests. Therefore, the 1 g visual model test can also provide a new perspective for qualitatively revealing the mechanism of the plugging effect. In future, it can be considered to carry out a visualized transparent soil test under the centrifuge conditions to more realistically and in-depth explore the problem of plugging effect (Fig. 11.31).

11.3.4 Conclusion

In this study, a series of tests to visualize the dynamic formation of soil plugging were carried out to investigate the influencing factors such as pile diameter, wall thickness, relative density, and penetration speed. The plugging behaviors and load-bearing characteristics of the pipe piles were explored by analyzing the evaluation indexes of the soil plugging and the monitored forces. The primary conclusions can be summarized as follows:

(1) The transparent soil visualization measurement method has been successfully applied to visually observe the dynamic formation process of soil plug during pile penetration. The displacement vector pattern of pipe pile penetration is dominated by larger vertical displacements inside the pipe pile, accompanied by smaller lateral displacements on both sides of the pile.
(2) The PLR and IFR obtained during the initial stage of penetration are greater than 1. The fluctuating IFR intuitively revealed that the formation of soil plugging is not continuous, but an intermittent process.
(3) By integrating the dynamic evolution of soil column, the soil plugging can be classified into four stages in this study: fully unplugged, developing plugged, fully plugged, destroying and reproducing plugged. The fully plugged is not an end state. After reaching the state of fully plugged, the fully plugged and partially plugged mode continuously exchange during further penetration.
(4) Considering the degree of soil plugging and the bearing capacity of pipe pile, it found that a large degree of soil plugging does not mean that the bearing capacity of the pipe pile is large, and the bearing capacity of the pipe pile is directly proportional to the internal frictional resistance mobilized.
(5) According to the development of PLR and IFR, plug formation seems to more sensitive to pile diameter, relative density and wall thickness than penetration speed. For pipe piles with larger diameter and higher relative density, the plugging effect are weak, but the total penetration resistance and the ultimate bearing capacity are larger.
(6) The pipe piles with larger diameter and higher relative density produce the largest total penetration resistance, while piles with larger wall thickness and penetration speed produce the lowest resistance. For pipe piles with larger wall thickness, although the percentage of annular resistance increases, the percentage of internal shaft resistance decreases, and the total penetration resistance generated decreases instead.

11.4 Pile-Soil-Cap Interaction Investigation

11.4.1 Introduction

Pile foundation is an important deep foundation widely used in civil engineering. It supports the upper structure by transferring the load to the deep soil. Pile foundations can greatly reduce the settlement of the structure and the risk of non-uniform settlement. The settlement of pile foundations has been studied by numerous researchers. Early at 1943, Terzaghi proposed a theoretical settlement calculation method for pile foundation and a bearing capacity failure theory [66]. Later, Meyerhof [67] proposed limit equilibrium theory for the bearing capacity of pile foundation and 68 proposed a cavity expansion theory to describe the failure of the pile foundations. Poulos and Davis [69] presented settlement calculation method for pile foundations based on elastic theory. Numerical methods such as finite element and boundary element method are used to analyze the settlement of pile foundations [70, 71]. Recently, Loukidis and Salgado [72] and Han et al. [73] have conducted many studies on pile foundation. However, only a few studies carried out on the pile cap effects. Poulos and Aust [74] have computed the settlement of an axially-loaded pile with rigid cap using the interaction influence factor method. Chow et al. (2001) presented a numerical analysis on pile-cap-pile-group interaction in nonhomogeneous soils. They found that the effect of the cap in contact with the ground has a small influence on the stiffness of the group compared to the case without a ground-contacting cap. Chen et al. [75] conducted a comparing experimental test on single pile and single pile with pile cap in Shanghai soft clay. By measurement of axial force of pile, pile cap resistance, pile head settlement and pile bottom settlement, they analyzed the interaction mechanism of short piles and pile cap. Lou et al. [76] placed a transducer below the pile cap and the pile body and analyzed the load transfer mechanisms of single pile and single pile with cap. They found out that the pile cap made the pile axial force at the top change more slowly and reduced the shaft friction resistance. Lou et al. [77] compared three friction piles with different cap sizes through loading test. They concluded that the increase of the upper load will increase the portion of load carried by pile cap, it can modify the average foundation stiffness by adjusting the pile cap; the shaft friction transfer function is close to elastic–plastic with very small displacement limit. Wang et al [78] performed laboratory model test on single piles with cap in medium coarse sand. Li et al. [79] conducted a series of small scale model tests on single pile, single sheet pile and single pile with cap in clayey silt.

In the above described test, the pile body deformation is gained through strain gages. However, soil deformation information is usually difficult to obtain. Traditional method to obtain the soil deformation information is through placing discrete transducers into the soil mass. Because the soil is opaque, the embedded transducers are only providing scattered information. Therefore, the soil deformation obtained through numerical analysis and theoretical assumptions are not fully proved. Soil deformation data such as soil compression and soil sliding is important to understand the mechanism of load transfer because the soil, pile and pile cap are working together to take the upper load. The soil deformation reflected the load transfer mechanisms of pile foundations.

In this study, the full field soil deformation was investigated for a series of loading test on the small-scale pile models with different pile caps using transparent soil modeling technique. Transparent soils have evolved in the past twenty years as a useful tool for the physical modeling of soil-structure interaction mechanisms. Transparent soils are two-phase media made by refractive index (RI) matching of solids representing the soil skeleton and the saturating fluids [80]. Transparent soil was first created in 1993 by consolidating a transparent slurry [81]. Amorphous silica powder, silica gel, aquabeads, and fused quartz sand are the four types of primary materials used in making transparent soils [11, 22,23, 58, 82]. Ezzein and Bathurst [22] were the first to demonstrate that fused quartz can be used as a transparent soil surrogate by matching it with a pore fluid made of a blend of two mineral oils. The static and dynamic properties of the fused quartz with matching pore fluid are similar to natural angular sand [32, 83], which is more useful for the researchers to study the geotechnical engineering problems. Hence, fused quartz is used in this investigation to make the small-scale models. A laser sheet goes into the transparent soil model and generates a laser speckle effect. The digital camera placed in front of the model will capture the digital image of the laser speckles. After further processing the digital images with the image pattern matching technique, such as adaptive digital image cross correlation [84], the soil deformation can be obtained.

11.4.2 Experimental Program

A small-scale loading testing system based on transparent soil was set up in the investigation shown in Fig. 11.32. This testing system is including a small loading frame and loading control system, a sheet laser system, an image capturing and processing system, and a digital camera. The sheet laser system is generated with a line laser through a cylindrical mirror. This was developed in our laboratory.

A series of small-scale model piles were designed to investigate the interactions between the soil and pile foundations with different pile and cap configurations. The small-scale pile models are shown in Fig. 11.33. The detailed information for the model piles are listed in Table 11.5. The pile model is made from ABS plastic materials with Young's modulus of 2 GPa. The dimension of the pile model is selected according to the transparent soil model. At present, the transparent soil model is limited in its dimension. Therefore, a large size soil model is not allowed. Based on the similitude theory proposed by [85], the scaling factor λ for length is 100, the corresponding the prototype piles are listed in Table 11.5. Assuming a reasonable

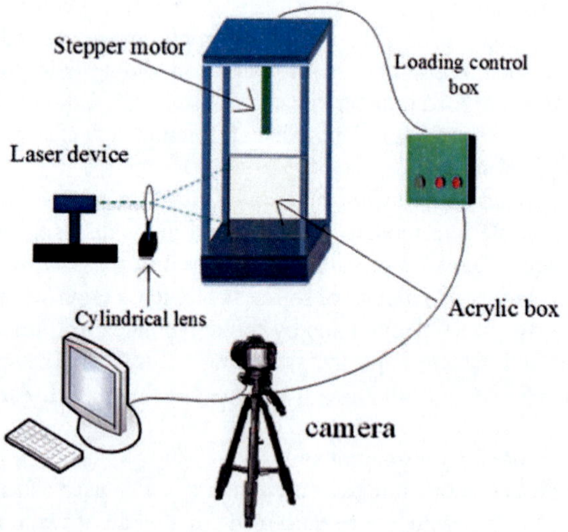

Fig. 11.32 The experimental set up for the model pile loading test

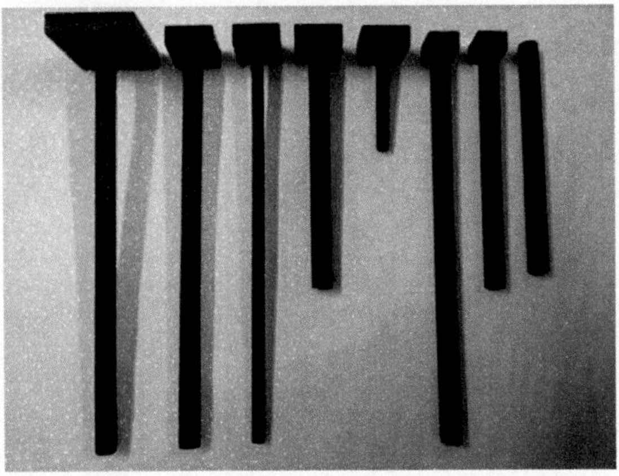

Fig. 11.33 The small-scale pile models

Table 11.5 Size of the test pile model and the prototype pile

Pile models	Pile model size			Prototype pile size		
	Pile cap size	Diameter	Length	Pile cap size	Diameter	Length
	(mm × mm)	(mm)	Mm	(m × m)	(m)	m
0MM	0	8	100	0	0.8	10
SMM	13 × 13	8	100	1.3 × 1.3	0.8	10
MMM	20 × 20	8	100	2.0 × 2.0	0.8	10
LMM	34 × 34	8	100	3.4 × 3.4	0.8	10
MSS	20 × 20	5	60	2.0 × 2.0	0.5	6
MSM	20 × 20	5	100	2.0 × 2.0	0.5	10
MSL	20 × 20	5	160	2.0 × 2.0	0.5	16
MLM	20 × 20	12	100	2.4 × 2.4	1.44	12

value of elastic modulus for the concrete piles of 30GPa, the scale factor of longitudinal rigidity of is 10000, which implies that $(EA)_p = (EA)_m*10{,}000$, this is a very large value. To simplify the analysis, a rigid pile is assumed since the corresponding stiffness of the prototype pile is very large. Since this study is mainly focusing on the mechanism of pile-soil-cap interaction, the analyses were also carried out on the pile models rather than the prototype.

Three sets of pile models were designed, one to consider the effects of pile cap, the second is to consider the effects of pile length, and the third to consider the effects of pile diameter. In the first set, a single pile without cap with a diameter of 8 mm and a length of 100 mm were made as a basic case denoted as 0MM. To examine the influence of pile cap size, three different pile cap sizes of 13 mm × 13 mm, 20 mm × 20 mm, 34 mm × 34 mm were selected for the pile model with pile length of 100 mm, denoted as 0MM, SMM, MMM respectively in this study. To examine the influence of pile length, the second group of pile models was designed with different

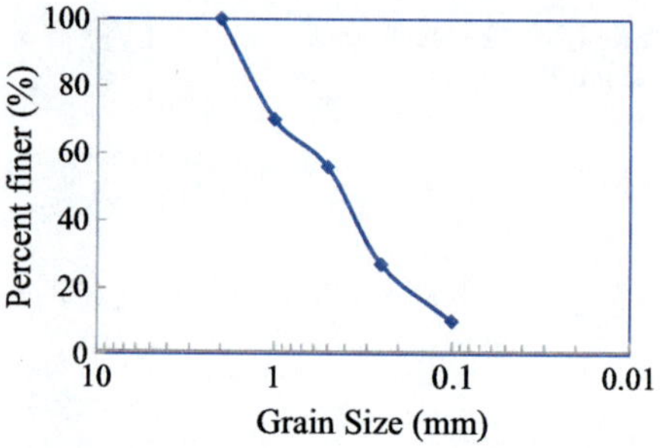

Fig. 11.34 Gradation curve of the fused quartz sand

length of 60 mm, 100 mm, 160 mm with the same pile cap size of 20 mm × 20 mm, pile diameter of 5 mm, denoted as MSS, MSM, MSL. The third group model piles was designed with different pile diameter of 5 mm, 8 mm, 12 mm, the same pile cap size of 20 mm × 20 mm and the same pile length 100 mm, denoted as MMM, MSM, MLM respectively.

Transparent soil used in the experimental study is made of 0.1–3 mm fused quartz and matching pore fluid, which is a mixture of n-dodecane and white oil. The grain size distribution curve for the fused quartz is shown in Fig. 11.34, $d_{60} = 0.55$ mm, $d_{30} = 0.275$ mm, $d_{10} = 0.1$ mm, $C_u = 5.5$, $C_c = 1.375$. It is classified as well graded sand. The maximum dry density $\rho_{dmax} = 1.479$ g/cm^3, the minimum dry density $\rho_{dmin} = 1.238$ g/cm^3. The relative density $D_r = 0.85$ for the fused quartz used for the model test which is a type of dense sand.

The transparent soil is placed in a model box made of glass with a dimension of 300 mm × 70 mm × 280 mm. The pile model is first pushed into the soil. Then the soil model is stirred to let the pore pressure generated disappear. After resting for 24 h, the load is applied slowly through a stepper motor (TP-S6PD) to the pile head. The load applied and the pile head displacement can be controlled and measured. The laser sheet goes into the transparent soil model and induces laser speckle effects due to the interaction between the laser and fused quartz particles. The digital camera (2048 × 1536 pixels) is placed in the front of the transparent soil model and captures the images during the loading test. The captured images are stored onto the computer for image processing. By adjusting the laser sheet, transparent soil model box, and the location of the camera, the laser sheet is able to pass through the pile central axis, making sure that the image plane is on the laser sheet.

Based on Chinese Technical code for Building Pile Foundations (JGJ106-2014), the load is applied to the pile foundation by equal step loading method. Each step load is 1/10 of the maximum load or estimated bearing capacity. After applying each step load, the pile head settlement is recorded with time until the pile stops settling. Then next step load is applied to the pile. The basis for judging the completion of the settlement at each step is that the pile head settlement is less than 0.1 mm per hour and occurs continuously twice according to JGJ106-2014. When sharp settlement or settlement is greater than the allowable value at certain loading step, the loading experiment is finished. The tests were carefully prepared and conducted to ensure

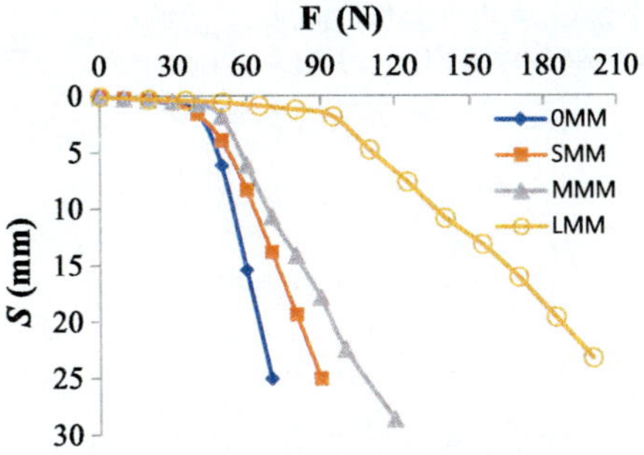

Fig. 11.35 Load-settlement curves for pile model with different cap size

the consistency of the test results. Each model test was repeated three times, similar values were obtained.

11.4.3 Testing Results and Analysis

From the testing data, the influences of pile cap size, pile diameter, and pile length on the soil deformation and loading capacity are analyzed.

11.4.3.1 Piles with Different Pile Cap Size

The loading settlement results for pile models with different cap size are plotted in Fig. 11.35. From the load settlement curves shown in Fig. 11.35, all curves have a linear development before fast settlement. LMM has a much greater linear portion than the other piles. The single pile without cap has the least linear portion. While SMM, MMM has a slightly larger linear portion than 0MM. The linear load limits for these four piles are 36 N, 40 N, 46 N and 92 N. The cap side length ratios for this group of model piles are 1:1.625:2.5:4.25 (the cap area ratio is 1:2.64:6.25:18.06). The elastic load limit versus pile cap area is plotted in Fig. 11.36 which has a linear relationship. The elastic load limit for the pile is increasing with the pile cap size. Beyond the elastic load limit, the pile head settlement develops quickly due to soil plastic deformation. However, the slope of the quickly settlement portion of piles are different for different pile cap sizes. The larger the pile cap size, the smaller the slope for the settlement curve.

The soil deformation for the single pile with different cap sizes is obtained through the laser speckle images obtained by digital camera. MatPIV is used to process these images to give the full field deformation information. MatPIV is an image processing Matlab code developed by based on the cross-correlation theory for pattern matching [86].

For the pile foundation, it mobilizes the shaft friction resistance and the bearing resistance of the surrounding soil to take the load. The stress transfer induces the soil deformation surrounding the pile and the cap. From the captured images during

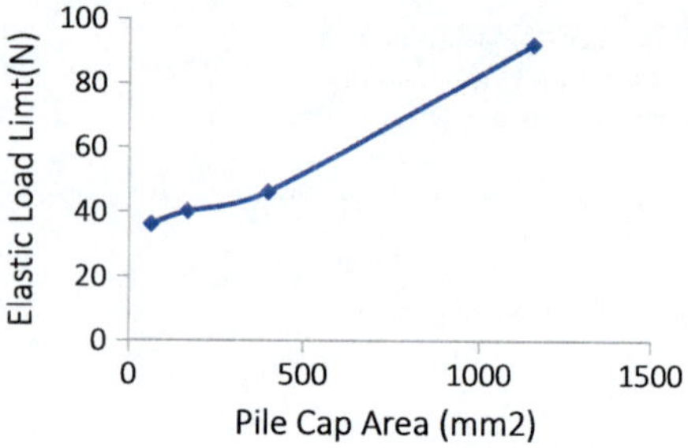

Fig. 11.36 Elastic load limit versus pile cap area

loading the pile, the full field soil deformation is obtained. Figures 11.37, 11.38, 11.39 through Fig. 11.40 show the displacement vector field and the soil deformation field for 0MM, SMM, MMM, and LMM, respectively. Because of axial symmetry and the laser sheet has a better speckle effect on the left of the pile, only left half deformation of soil is shown.

For the single pile without cap-0MM, from Fig. 11.37a, the displacement vector field shows that the soil particles are moving towards up and outward at the pile head. The soil particles at the pile base are moving downward and outward. Along the pile body, the soil particles are moving upwards almost parallel to the pile body. The displacement contour field for 0MM shows a relatively larger displacement happened around the pile head and beneath the pile base. The soil particles taking part in load transfer are few. Maximum displacement for soil particles occurs at the edge corners of the top head and central bottom end of the pile model 0MM. This indicates that the displacement of soil particles is centered at the pile bottom not forming sliding surfaces. The soil deformation zone around the pile base is about 4D away from the pile base center and 6D beneath the pile base. The soil deformation zone is 3D around the pile head and 2D below the pile head.

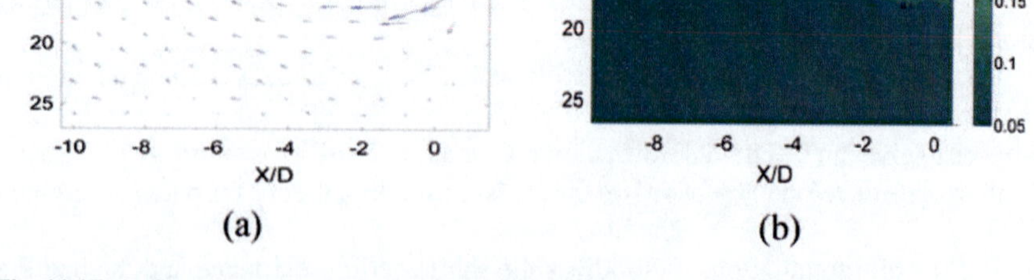

Fig. 11.37 Deformation of soil for pile model 0MM at S = 4 mm, **a** displacement vector filed, **b** displacement contour

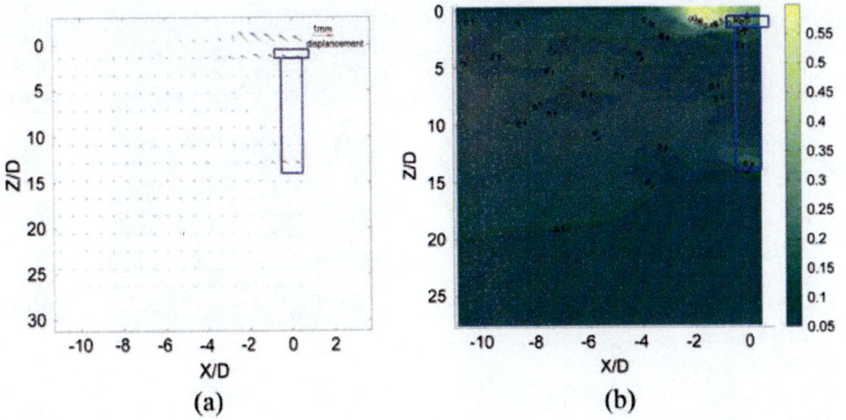

Fig. 11.38 Deformation of soil for pile model SMM at S = 4 mm **a** displacement vector filed, **b** displacement contour

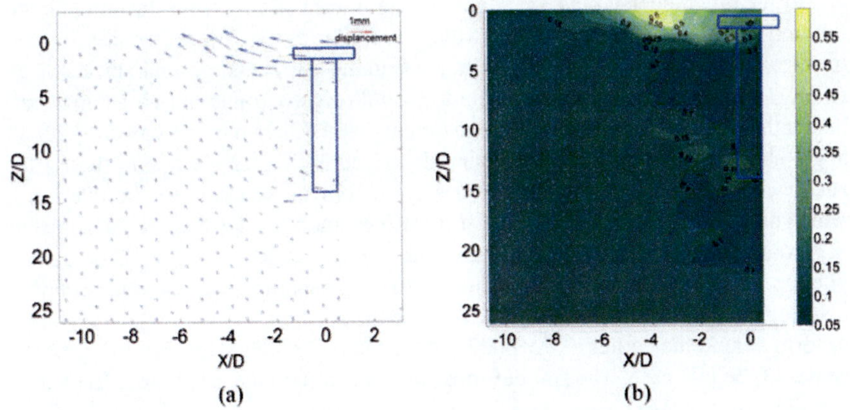

Fig. 11.39 Deformation of soil for pile model MMM at S = 4 mm **a** displacement vector filed, **b** displacement contour

For the pile model SMM, more soil particles are moving surrounding the pile head, even below the pile head for a certain depth, the movement of soils particles can be clearly observed. The soil deformation below the pile base of SMM is smaller than that of pile model 0MM. There are not clear and significant sliding surfaces forming on the left and right of the pile head which is similar to the punch failure pattern for the shallow foundations. The soil particles are pushed to the side of the pile body. The soil is heaving above the ground surface in the deformation zone. The soil deformation zone is about 4D around the pile head and 1.5D below the pile cap. The soil deformation zone around the pile base is about 1D away from the pile base center and 1D beneath the pile base.

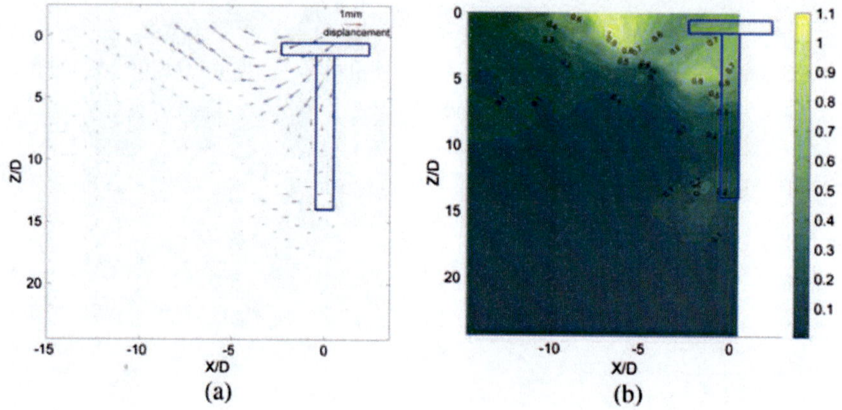

Fig. 11.40 Deformation of soil for pile model LMM at S = 4 mm **a** displacement vector filed, **b** displacement contour

For the pile model MMM, there are two significant sliding zones forming on the left and right side of the pile head, which is similar to the failure surfaces formed under a shallow foundation. The soils are bulging above the ground surface. The sliding surfaces are in a more complete form, with some particles moving downward below the pile head, some moving horizontally, and others moving upward with an angle about 45°.This is similar to what described by Terzaghi's failure theory for shallow foundations. More soil particles are mobilized surrounding the pile head, which helps to take more loads. The soil particles mobilized extended a distance of 8D around the pile head. The influence zone is extended to a depth of 3D below the pile cap. The displacement contour shows that the soil particles are moving very little between the mobilization zone by pile cap and the pile bottom. A sliding zone started to form around the corner of the pile base. The soil particles moved little below the center of the pile base. The soil deformation zone around the pile base is 2D below the pile base and 2.5D away from the pile base center.

For the pile model LMM, the soil particles show a similar deformation pattern as MMM. More soil particles below and surrounding the pile head are carrying the upper load. The soil particles mobilized extended a distance about 13D around the pile head and extended a depth about 9D below the pile cap. Almost all the soil particles surrounding the pile was moving downward for LMM which may induce negative side friction resistance and reduce the shaft resistance. The deformation of soil particles beneath the pile end is similar as MMM except with smaller deformation values. The displacement contour shows that larger values of displacement occurred below the pile cap, on the side of the pile head, and below the pile base. The soil deformation zone around the pile base is about 4.5D away from the pile base center and 3D beneath the pile base.

At settlement S = 4 mm, the pile's bearing resistance is entering a nonlinearly developing stage for these four pile models. With the increase of pile cap size, the

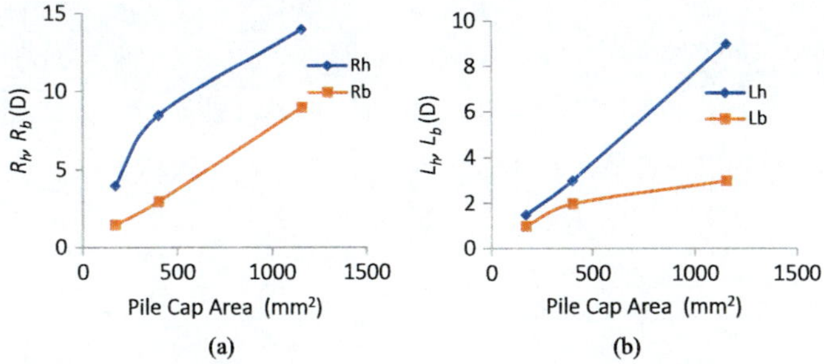

Fig. 11.41 **a** R_h and R_b versus pile cap area, **b** L_h and L_b versus pile cap area

elastic linear load limit is increasing with the pile cap size due to the contribution of the pile cap resistance. Soil particles surrounding the pile cap are going from a punch failure patter to a fully developed bearing capacity failure. However, with the larger pile cap, more soil particles are moving downward which reduces the shaft friction resistance and induces larger pile settlement. Therefore the right selection of the pile cap size should be carefully considered.

The radius of soil deformation zone at pile head, R_h, and the radius of soil deformation zone at pile base, R_b, are plotted in Fig. 11.41a. A nonlinearly increasing trend exists for R_h and pile cap area, R_b first follows a linear trend. The depth of soil deformation zone below pile cap, L_h, and below the pile base, L_b, is plotted in Fig. 11.41b. L_h is increasing linearly with pile cap area almost linearly. However, L_b nonlinearly increases with pile cap area.

11.4.3.2 Soil Deformation with Different Pile Length

For piles with different length, the load settlement curves for MSS, MSM and MSL are shown in Fig. 11.42. These piles have the same pile cap size of 20 mm × 20 mm. It is clear that longer pile settles slower than the short piles. These results are reasonable because the longer pile will mobilize more side friction resistance before developing large settlement. The length ratio for MSS, MSM and MSL is 1: 5/3: 8/3. The elastic limit load for MSS, MSM and MSL is 25 N, 35 N and 50 N, respectively. The elastic load limit versus pile length is shown in Fig. 11.43, which is a linear relation.

The deformation of soil particles was obtained from the captured images by MatPIV. The displacement vector field and the displacement contour field are shown in Fig. 11.44, 11.45 and 11.46 for the pile model MSS, MSM, and MSL respectively.

For the pile model MSS, the deformation of the surrounding soil is fully developed. The soil particles are moving outward and upward on the left and right side of the pile head. There are some soil particles moving horizontally to the left and right, which caused by pushing effects of the pile and pile head. The soil particles along the side of the pile body are moving outward and downward, which might induce the negative side friction. The largest displacement occurred at the ground surface

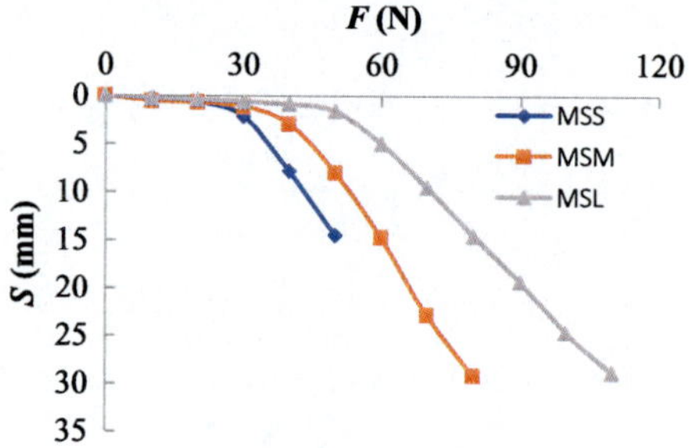

Fig. 11.42 The load settlement curves for pile model MSS, MSM, MSL

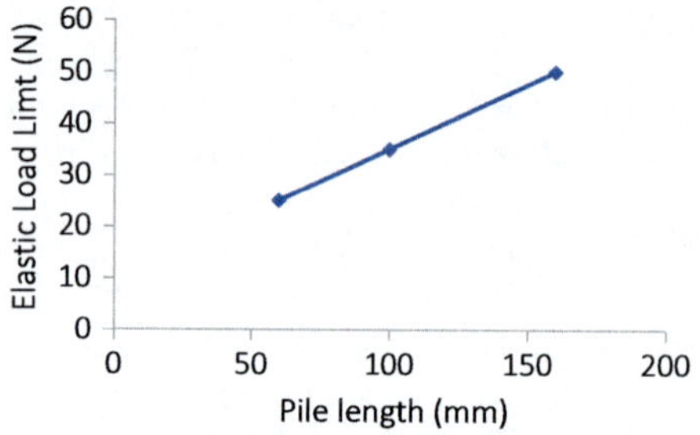

Fig. 11.43 Elastic load limit versus pile length

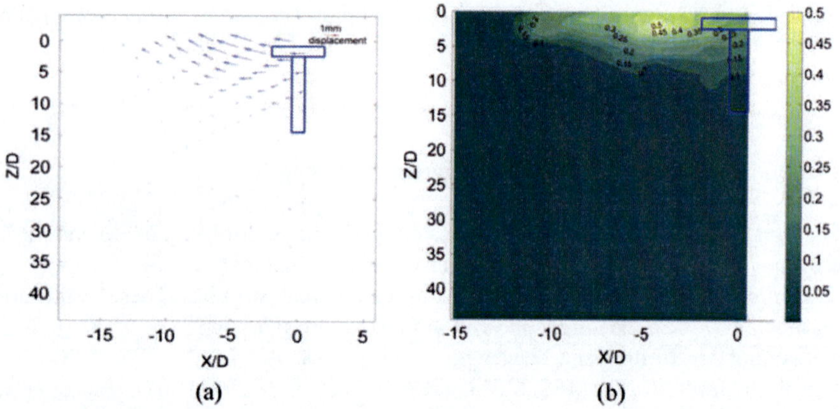

Fig. 11.44 Deformation of soil for pile model MSS at S = 4 mm **a** displacement vector filed, **b** displacement contour

about a distance of 5D from the pile head. The soil fails in shear along a radial shear zone and a passive zone similar to the failure pattern described by Terzaghi's bearing capacity theory. The soil deformation beneath the pile bottom is the smallest and centered at the pile base, indicating that the soil mass is still in an elastic compaction stage. The mobilized zone by the pile cap is 12.5D surrounding the pile head and

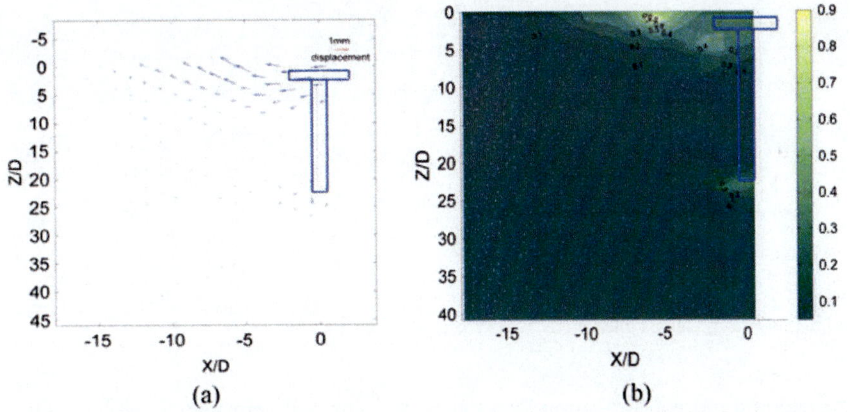

Fig. 11.45 Deformation of soil for pile model MSM at S = 4 mm **a** displacement vector filed, **b** displacement contour

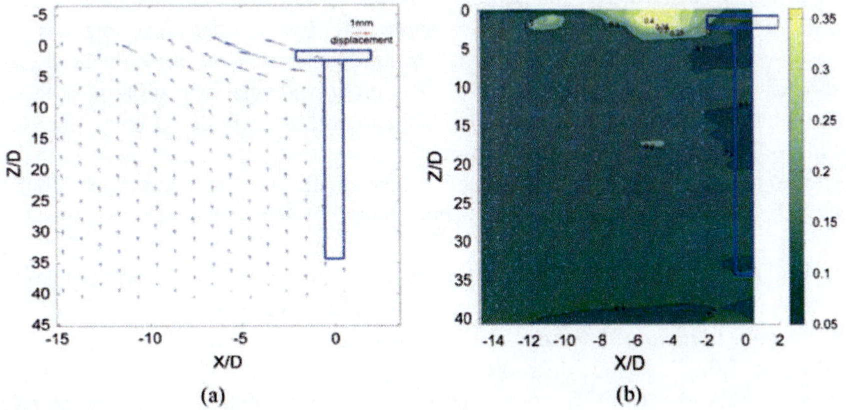

Fig. 11.46 Deformation of soil for pile model MSL at S = 4 mm **a** displacement vector filed, **b** displacement contour

10D below the pile cap. There is no obvious deformation below around the pile bottom. The deformation of the soils surrounding the pile body and cap form a large triangular deformation zone.

For the medium long pile model MSM, soil particles surrounding the pile head are also moving with a similar pattern as the short pile model MSS. But unlike MSS, there are fewer soil particles moving horizontally to the left and right of MSM. The mobilized zone of soil particles has a diameter of 13D surrounding the pile head and with a depth of 8D below the pile cap. The soil deformation value is smaller than that of MSS. The displacement contour shows that the soil beneath the pile base is not sliding and still in an elastic compaction state, but the soil at the pile base corner

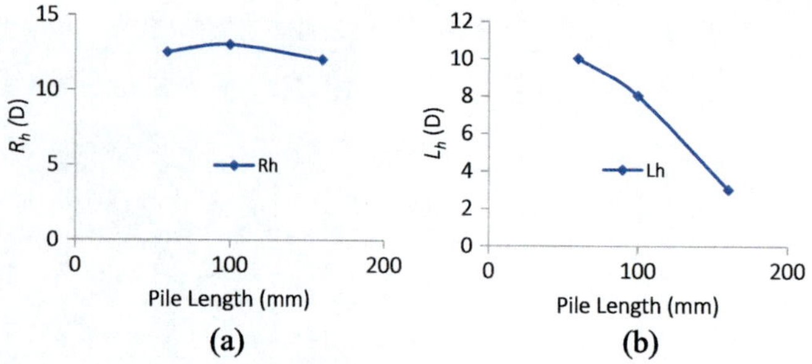

Fig. 11.47 **a** R_h versus pile length, **b** $L_h$versus pile length

developing more deformation than the center. The soil deformation zone is with a radius of 2D away from the pile and a depth of 2.5D below the pile base.

For the pile model MSL, a well develop deformation pattern similar to MSS was observed around the pile head. The soil particles next to the pile head are moving horizontally outward. The soil particles further away from the pile head are moving upward with an angle about 45° to the horizontal. Two soil deformation zones are observed for the model, one is an active deformation zone and one is the radial sliding zone as described by Tergazhi's theory. The mobilized zone of soil particles has a diameter of 12D surrounding the pile head and with a depth of 3D below the pile cap. At S = 4 mm, most of load is taken by soil resistance below the pile cap and the pile tip resistance. The shaft resistance is very small. The soil deformation around the pile base is very small, which indicates that the pile tip resistance developed is also small.

With the same pile cap size and the same pile diameter, the shorter pile, the larger deformation of the soil it induces beneath the pile head. From MSS to MSM, the pile length is increased 40 mm, the influence depth is decreased from 10 to 8D. However, from MSM to MSL, the pile length is increased 60 mm, the influence depth is decreased to 3D. Therefore, when the pile length is short, increasing the pile length can effectively reduce the deep soil deformation. When the pile length is long enough, increasing pile depth no longer reduces the deep soil deformation effectively. These observations can explain why a geosynthetic layer is used to reinforce the pile foundations below embankments [87].

R_h is plotted in Fig. 11.47a versus pile length for MSS, MSM, and MSL. No significant change of R_h with the pile length is observed. L_h is plotted in Fig. 11.47b versus pile length for MSS, MSM, and MSL. L_h decreases quickly with pile length

11.4.3.3 Soil Deformation with Different Pile Diameter

The influences of pile diameter on the load capacity and the soil deformation is investigated through the loading test on pile models of MSM, MMM, and MLM. These pile models have the same pile cap and pile length. The load-settlement curves

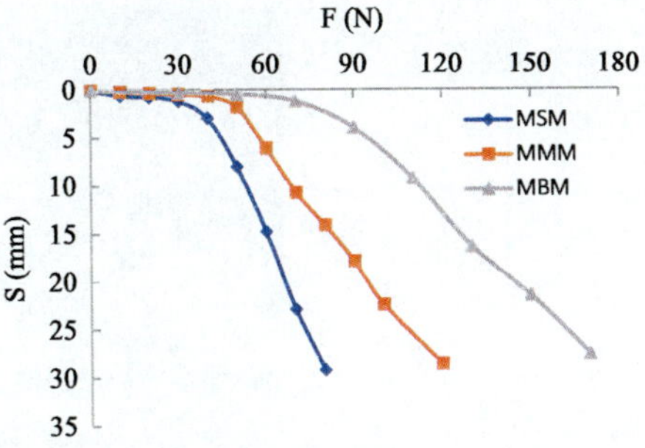

Fig. 11.48 The load-settlement curves for piles with different diameter

Fig. 11.49 Elastic load limit versus pile diameter

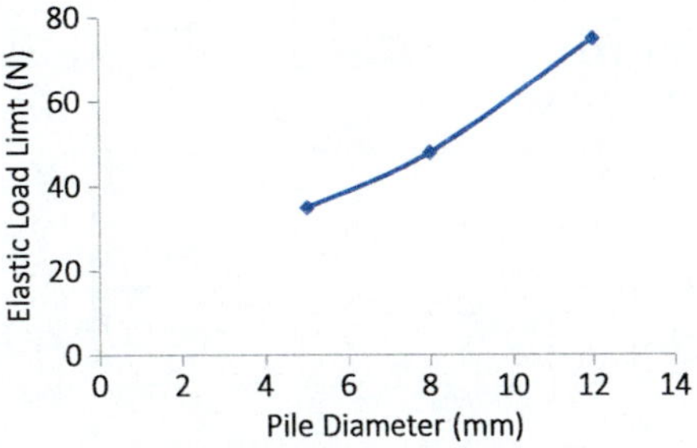

for these three piles are plotted in Fig. 11.48. It shows that the larger diameter piles have higher elastic load limit and settles more slowly at the nonlinear settlement phase. The elastic load limit for the three piles is 35 N, 48 N and 75 N, respectively. Figure 11.49 shows elastic load limit increases linearly with the pile diameter for the three pile models.

The larger piles can reduce the settlement and increase the loading capacity through mobilizing more shaft friction and tip bearing resistance which is recognized in practice. For the displacement contour of three pile models, the soil deformation is centered below MSM, but there are three deformation zones below MMM and MLM. With the increasing pile diameter, it is easier to form two deformation zones around the pile base corner where localized shear failure tends to occur first. The deformation beneath the base of MMM is smallest, and that of MLM is the largest.

Soil deformation was plotted in Figs. 11.39, 11.45 and 11.50 for pile model MMM, MSM, and MLM at a displacement of 4 mm. Pile model MSM and MMM are discussed in the above sections. For MLM, the soil deformation zone has a radius about 4D around the pile head and a depth about 1D below the pile cap. The deformation pattern around the pile cap is much like that a shallow foundation. The soil deformation zone is with a radius of 1D away from the pile base center and a depth of 3.5D below the pile base. For piles model MSM, MMM and MLM, the normalized top influence area is reducing with the pile diameter and the bottom influence zone is also reducing. With the increase of the pile diameter, the deformation zone beneath

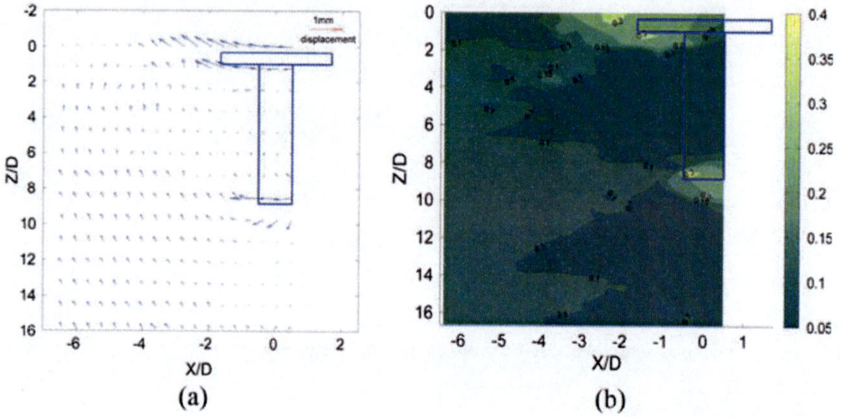

Fig. 11.50 Deformation of soil for pile model MSL at S = 4 mm **a** displacement vector filed, **b** displacement contour

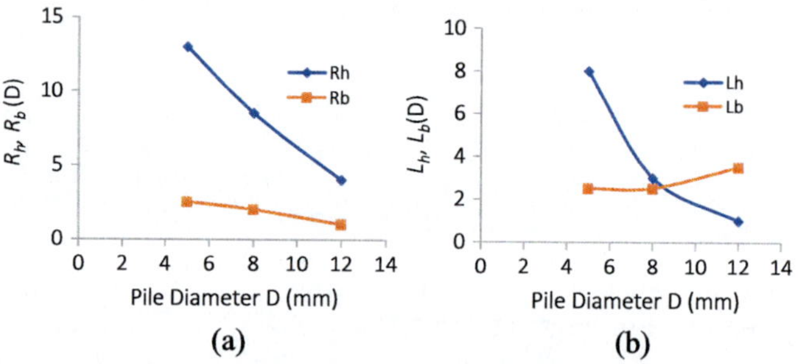

Fig. 11.51 **a** R_h, R_b versus pile diameter; **b** L_h, L_b versus pile diameter

the pile base is more like an oval-shaped zone. The maximum displacement occurs on the left edge corner of the pile base. For MLM, soil particles above the pile base are pushed to move up due to the settlement of the pile.

R_h and R_b are plotted in Fig. 11.51a versus pile diameter for MSM, MMM, and MSL. R_h and R_b are decreasing with the pile diameter, but R_h drops faster. L_h and L_b are plotted in Fig. 11.51b versus pile length for MSM, MMM, and MSL. L_h decreases nonlinearly with pile diameter, but L_b increases slightly with D.

11.4.4 Summary and Conclusions

This study reports a series of small-scale model tests in laboratory carried out on the single piles with pile cap based on the transparent soil model made of fused quartz. Transparent soil model made observations of soil deformation inside soil mass possible. Soil deformation surrounding the pile and the cap is specially discussed at a reasonable value of 4 mm. Pile cap effects, diameter effects and pile length effects are investigated. All the loading tests show that pile cap can effectively increase the bearing capacity of pile foundation. From the testing results, the elastic load limit has a good linear relationship with the pile cap area, pile diameter and pile length.

Soil deformation around the pile cap is similar to that described by Terzaghi's bearing capacity failure theory. The soil particles below the pile cap next to pile body are moving downward, which might reduce the shaft friction or induce the negative shaft friction. The influence zone of pile cap is varying from 4 to 14 times pile diameter at the ground surface and from 1.5D to 9D below the pile cap for three different pile cap size of 1.625D, 2.5D and 4.25D. The radius of soil deformation zone at pile head, R_h, is increasing with pile cap area. The radius of soil deformation zone at pile base, R_b, is linearly increasing with the pile cap area. The depth of soil deformation zone below pile cap, L_h, is increasing with pile cap area almost linearly. However, the depth of soil deformation zone below pile base, L_b, nonlinealry increases with pile cap area.

For shorter pile, the mobilized soil deformation zone is over most of the pile body, but is very small below the pile base. The radius of soil deformation zone R_h is does not change significant with the pile length. L_h of the depth for soil deformation zone below the pile cap decrease with pile length. With the same pile cap size and the same pile diameter, the shorter pile, the larger deformation of the soil will be induced beneath the pile head.

For piles model with larger diameter (D), soil deformation below the pile base becomes smaller at the pile base because the pile tip resistance has not been fully mobilized at a settlement of 4 mm. The larger diameter piles have higher elastic load limit and settles more slowly at the nonlinear settlement phase. R_h and R_b are decreasing with the pile diameter linearly, but R_h drops faster. L_h decrease linearly with pile diameter, but L_b first decrease slightly and then increases.

At settlement value of 4 mm, the radius for soil deformation zone around the pile base varies from zero to 4.5D for all the pile models tested; the depth for soil deformation zone beneath the pile base varies from zero to 6D for all the pile models tested.

11.5 Model Tests of Jacked-Pile Penetration into Sand Using Transparent Soil and Incremental Particle Image Velocimetry

11.5.1 Introduction

Jacked-pile are widespread design solution in pile foundations. It is of great significance to study the soil squeezing effect of jacked-pile penetration and to obtain the ground heave and displacement pattern of surrounding soil for guiding the construction of pile foundation and preventing engineering accidents. To properly interpret and quantify the basic responses of soil movement during jacked-pile penetration, such as vertical displacement and lateral displacement, ground heave and the stress redistributing, it has long been a significant issue and has been discussed by many scholars. Previous studies conducted on the displacements around displacement pile penetration, including theoretical predictions [88, 89, 90, 91, 160], field tests [92–94], numerical simulation [31, 95, 96], and model tests [24, 97, 98, Huang et al. 2019]. Since the cavity expansion theory was proposed by Randolph and Wroth [99], some theoretical analysis methods have been exploited to predict the soil movement around jacked-pile.

The strain path method (SPM) [88] based on the cavity expansion theory, adopted the source theory of fluid mechanics can be used to predict soil movement for deep penetration. For the problem of ground surface ignored by SPM theory, Sagaseta et al. [89] proposed the shallow strain path method (SSPM) for predicting the ground deformations. However, due to the friction between piles and soil not considered, and these theories are only suitable to predict the displacement in undrained clay, the theoretical calculation results cannot be used to study the case of displacement pile penetrating into sandy soil.

Field tests have also been successfully conducted to investigate the issue of jacked-pile penetration [100, 101], but these studies are limited by non-visual of natural soil and influence of embedded transducers on deformation monitoring. In recent years, some scholars have predicted the soil deformation during piles penetration using laboratory tests. Typically, half-model tests with image acquisition system has been used to measure the soil movement with a transparent medium such as perspex or stalinite [59,102, 103, 104, 105]. However, the data obtained are displacement at the model boundary rather than the internal soil deformation. The viewing window of model box imposes a boundary constraint on the soil, which affects the real experimental phenomena. Available techniques able to visualize the inside deformation of soils such as X-Ray radiography tomography, computerized axial tomography (CAT), and magnetic resonance imaging (MRI) have been used [106, 107]. However, routine application of X-Ray Tomography, and MRI in geotechnical research is limited by their high cost and technical difficulties. Recently, several scholars have explored the internal deformation pattern of soil during the process of jacked-pile penetration using the transparent soils technique with particle image velocimetry. Typically, in these methods, spaced markers, painted (opaque) seeded particles or specific fluorescent particles, which can observe the relative displacement inside the transparent soil sample [14, 108, 109, 110, Omidvar et al. 2015]. Previous model tests of transparent soil for jacked-pile penetration have provided useful information relating to the penetration mechanism and displacement pattern [12, 15, 111]. Despite these valuable contributions, there is still a limited amount of data available on induced disturbance effect and displacement pattern during jacked-pile penetration into sandy soil. A new test methodology is adopted for modelling jacked-pile penetration using transparent soil technique with incremental reliability-guided particle image velocimetry (RG-PIV) for it could acquire smaller cumulated displacement measurement error.

The decision to use the incremental RG-PIV method, rather than a traditional PIV method, because it will produce minimized errors for large deformation problems [112].

The visualization research of internal deformation characteristic of jacked-pile penetration still remains insufficient, especially the effect of different shapes of pile tip on soil squeezing effect is not thoroughly considered. Therefore, the objective of the current study is to perform an experimental investigation of measuring the soil movement in sandy soil during jacked-pile penetration using transparent soil technique with the incremental RG-PIV method. The shape effect of pile tip on the governing penetration mechanisms was discussed. The difference between the full-model and half-model test through the comparative experiment was performed. Additionally, the phenomenon of meta-stable sand plug in the process of jacked-pile penetration was studied by means of the method of partial replacement with dyed particles.

11.5.2 Experimental Methodology

11.5.2.1 Transparent Synthetic Soils Preparation

The test soil model is made from a transparent fused quartz and refractive index matching pore fluid [11, 113]. The purity of the fused quartz is 99.99%. In order to simulate sand soil, the grain size of the fused quartz was 0.1–2 mm. The grain size distribution curve for the fused quartz and Fujian (China) standard sand are shown in Fig. 11.52. The properties of this fused quartz are listed in Table 11.6. The curve of the relationship between the refractive index and the volume ratio of White oil 15 to n-Paraffin c12 is shown in Fig. 11.53. The mixed pore fluid comprised of nParaffin C12 and White Oil 15 at a volume ratio of 1:4.75 [23, 58]. The refractive index of this pore fluid is 1.4590. Their properties are listed in Table 11.7 [114]. The reliable mechanical properties of transparent soil are similar to natural sandy soil [22], 32, 109, 115], which can be used as substitute to simulate natural sandy soil in physical model tests.

11.5.2.2 Experimental Apparatus

The schematic diagram of testing apparatus is depicted in Fig. 11.54a. The physical photo of experimental setup is shown in Fig. 11.54b, consisted of a CCD industrial camera, a small-scale loading system, laser device, an optical prism, and an image capturing and processing system [17], 115]. The resolution of the CCD industrial camera is 1028 × 1536 pixels, the focal length of the CCD industrial camera lens is 16 mm, and the CCD industrial camera is set to the fixed focus mode during the test. The classic Proportional-Integral-Derivative (PID) closed-loop control algorithm was implemented in a small-scale loading system. The advantage of the PID

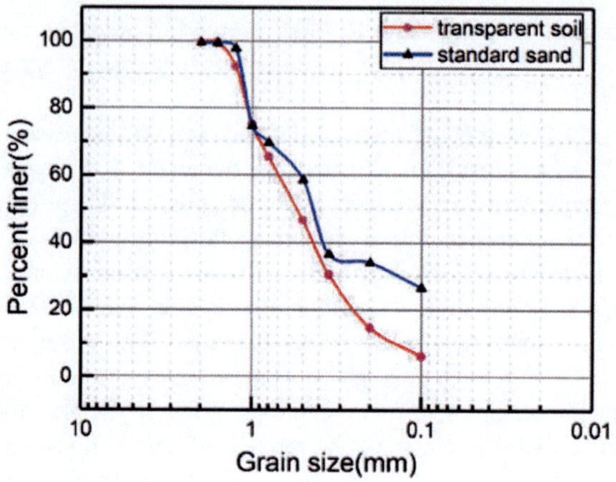

Fig. 11.52 Grain size distribution curve for the fused quartz and Fujian (China) standard sand

Table 11.6 Parameters of the fused quartz

C_u	C_c	ρ_d (g/cm3)	ρ_{dmax} (g/cm^3)	ρ_{dmin} (g/cm^3)	γ (kN/ m^3)
5.45	2.04	1.4	1.56	1.18	2.33

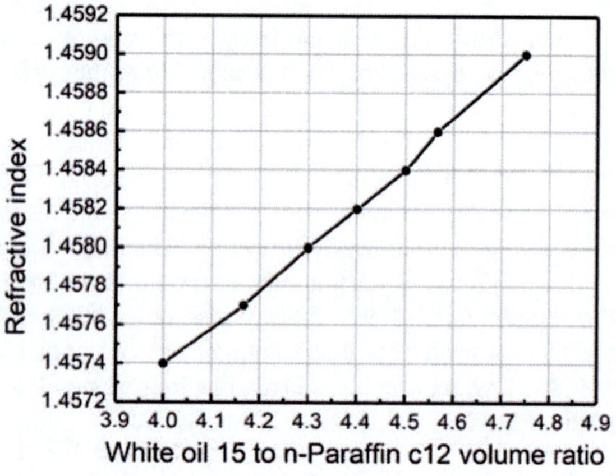

Fig. 11.53 Refraction index versus volume ratio of White oil 15 to n-Paraffin c_{12}

Table 11.7 Properties of n-Paraffin C12 and White Oil 15

	n-Paraffin C_{12}	White oil 15
Appearance	Clear, bright	Clear, bright
Density at 25 °C (Mg/m^3)	0.75	0.849
Viscosity at 40 °C (mm^2/s)	1.45	19.35
Flash point (°C)	71	155
Refractive index at 20 °C[a]	1.4217[b]	1.4660[b]

[a] Obtained by using an Abbe WAY-2S digital refractometer; [b] Average value

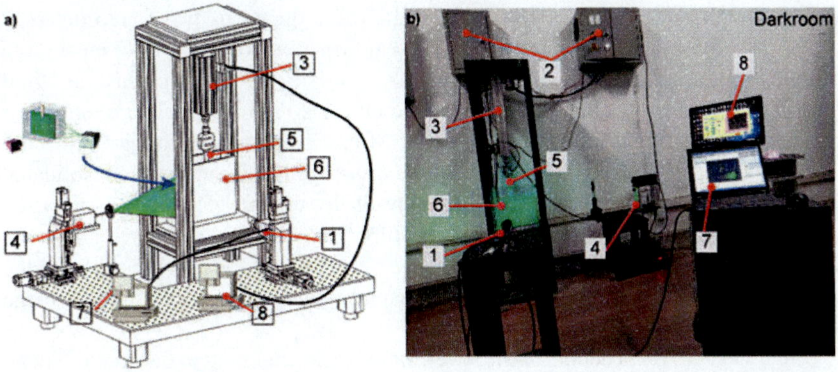

Fig. 11.54 Components of the experimental set up for the model pile penetration test. **a** schematic diagram, **b** physical photo. 1-camera; 2-electric control box; 3-loading device; 4-laser device; 5-model pile; 6-transparent soil; 7-image acquisition system; 8-load control system

controller is that the algorithm is simple, the calculation speed is fast, and the action speed is quick enough, which is suitable for high-precision loading. The small loading system is connected to the pile model by means of the threading nut at the bottom. This testing system was developed in our laboratory.

The tests were performed in a darkroom. By adjusting the laser device and transparent soil model box, the 2 W laser device produced a light beam that spread into a vertical light sheet after passing through a prism and illuminated a plane inside the transparent soil sample. The model pile was penetrated at the centre of the transparent soil sample and the laser sheet should be aligned with the axis of model pile, making sure that the image plane is on the laser sheet [15, 115]. The CCD industrial camera was placed approximately 800 mm away from the model box with its optical axis adjusted perpendicularly to the laser sheet and captured the images during pile penetration. The captured images were stored onto the computer through image processing software. For each model test the same basic test procedure was followed.

11.5.2.3 Model Preparation

In order to avoid the influence of size effect, the model box with a dimension of 300 mm × 170 mm × 260 mm (length × width × height) was finally selected by referring to the influence range of jacked pile penetration on surrounding soil disturbance demonstrated in relevant literature and considering the maximum transparency of the existing preparation method of transparent soil [36, 116, Mei et al. 2008].

For the sample preparation of transparent soils, the fused quartz particles are first cleaned with distilled water and then dried in the oven. After then, fused quartz particles are placed in a model box by 'layered vibrating tamping method'. Then, we insert the hose into the bottom of the model box and fill the model box with pore fluid from the bottom to top slowly. After the fused quartz particles are immersed in the pore fluid, a whisk is used to stir the transparent soil sample to ensure the sample is not stratified. For the transparent soil sample, the height of this pore fluid must be higher than the surface of this fused quartz 20 mm in order to filter the air entering during the pile model penetration. Afterward, a vacuum is applied to de-air the transparent soil sample. The saturated transparent soil sample is consolidated through the loading system by means of the pre-drilled organic glass plate. Finally, the transparent soil sample is prepared successfully (Fig. 11.55a). The pixel equivalent value is converted from the pixel value of the model pile in the speckle image to the actual size of the model pile, and then the displacement of soil in the speckle image is calculated (Fig. 11.55b).

Table 11.8 provides details of the seven model tests presented in this paper. Transparent soil samples were prepared by controlling variables in all tests. Penetration was done at a constant speed of approximately 0.25 mm/s, corresponding to a quasi-static penetration process.

The layout of full-model and half-model tests were shown in Fig. 11.56a. The distance from the full-pile to the model box boundaries was 85 mm. Five types of model piles were designed, as shown in Fig. 11.56b. The length of full-piles is about 125 mm, the length of half-pile is about 170 mm, and their radius, referred to as r, are 11 mm. For the model piles, the ratio of the container size to the pile diameter (L/D = 14) and the ratio of the model pile diameter to the mean grain size of the soil ($D/d50$ = 40) were greater than that suggested by Mo et al. [59] in order to reduce the boundary and grain size effects. The model piles were made of polymethyl methacrylate. In order to minimize reflection of light and consider the effect of friction, the surface of all model piles are roughened.

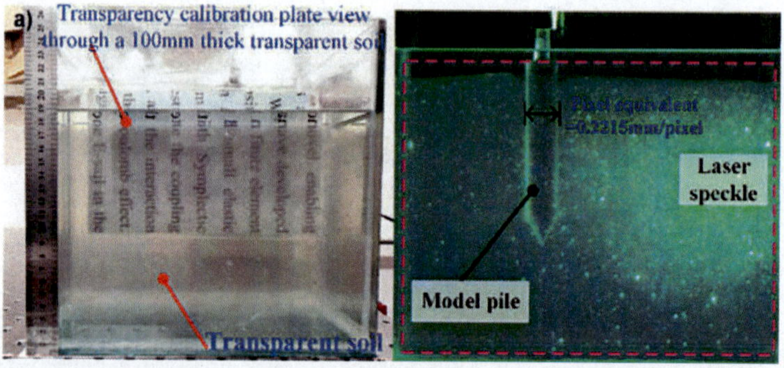

Fig. 11.55 **a** Targets viewed through a 100 mm thick transparent soil model; **b** The speckle image of transparent soil model

Table 11.8 Summary of laboratory tests

Test ID	Shape	Relative density (D_r)	Experiment type	Penetration depth (R)
T1	Flat-ended (half-pile)	D	Full-model	8.5
T2	Flat-ended (half-pile)	D	Half-model	8.5
T3	Flat-ended (full-pile)	D	Full-model	8.5
T4	Pile tip with angle 60	D	Full-model	10
T5	Pile tip with angle 90	D	Full-model	10
T6	Pile tip with angle 120	D	Full-model	10
T7	Flat-ended	D	Full-model	10

'D': dense sand ($D_r = 80\%$)

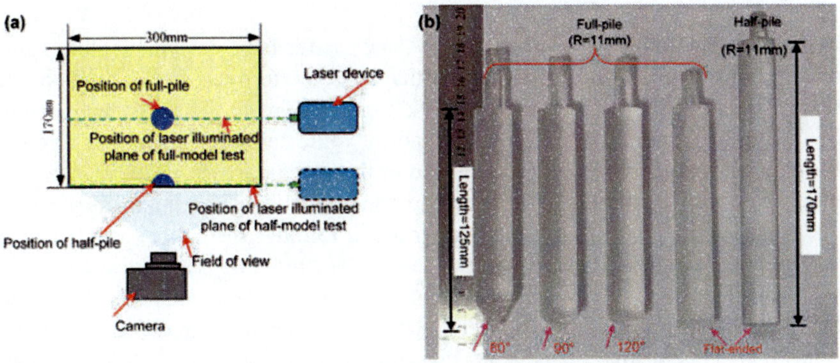

Fig. 11.56 The full-model and half-model tests: **a** top view of test layout, **b** five types of model piles

11.5.2.4 Incremental Reliability-Guided Particle Image Velocimetry

White (2003) reported that particle image velocimetry (PIV) image processing technology was used as a method to measure soil deformation in geotechnical model tests. Since then, PIV technique has become routine experimental practice to obtain deformation patterns in many geotechnical research laboratories [17, 108, 117]. The fundamental principle behind PIV is a classic pattern-recognition technique where two images are compared to obtain the relative displacement between them [Liu et al. 2004, 118]. For tracking the subset between the reference subset and current subset of recorded images, the robust Zero-mean Normalized Sum of Squared Difference (ZNSSD) is commonly used as the correlation criterion. In the conventional PIV algorithm, it basically adopts the 'jumping' method, and the deformed images are correlated with the first initial reference image for image correlation analysis, so as to obtain the deformation information. For the large deformation, using the traditional 'jumping' algorithm to calculate its relative deformation will make the calculation result have a large error. This is the reason for the so-called decorrelation effect. Furthermore, frequently updating reference image will not increase the cumulative error, but also improve the random walk error [112, 119]. Subsequently, an incremental RG-PIV technique, combining the recently developed RG-PIV method and an automatic reference image updating scheme, was proposed to improve the measurement accuracy. The overall flowchart of incremental RG-PIV method is shown in Fig. 11.57. The image post-processing program used in this paper is based on the original MatPIV program, referring to the overarching framework controlling the computation process of the technique of incremental RG-PIV proposed by Pan et al. [112], and adding the function of automatically updating the reference picture. The code was written by ourselves with the help of MATLAB.

11.5.2.5 The Calibration Tests of Accuracy of the PIV Analysis

In order to verify the accuracy of the PIV analysis for large deformation measurement, a translational test of the speckle image was performed. The translation test system is shown in Fig. 11.58a. The speckle image (Fig. 11.58b) was generated in MATLAB by randomly projecting thousands of white dots onto a black background to subpixel positional resolution. The size of the speckle image is 105 mm × 105 mm. The translation test is to fix the standard speckle image on the loading rod of the stepper motor, and control the stepper motor to move the target speckle image vertically downward at a constant rate. Then, the PIV program is used to analyze and calculate the two speckle images before and after the motion, and the accuracy of the PIV program is evaluated by comparing the calculated displacement value with the actual translation displacement value. Figure 11.58c present the vector displacement field obtained by processing the two speckle images before and after translation. Obviously, the displacement vector clearly describes the overall downward translation of the speckle image. It can be clearly found from Fig. 11.58d that the absolute error value of 10 groups of translational calibration tests fluctuates up and down within the range of ± 0.01 mm. The mean absolute errors value of the 10 groups of translational calibration experiment is 0.02102 mm, which is approximately 0.1 pixel equivalent (the pixel equivalent of calibration test is 0.2215 mm/pixel). The results show that the calculated accuracy of the PIV algorithm meets the requirements of image measurement.

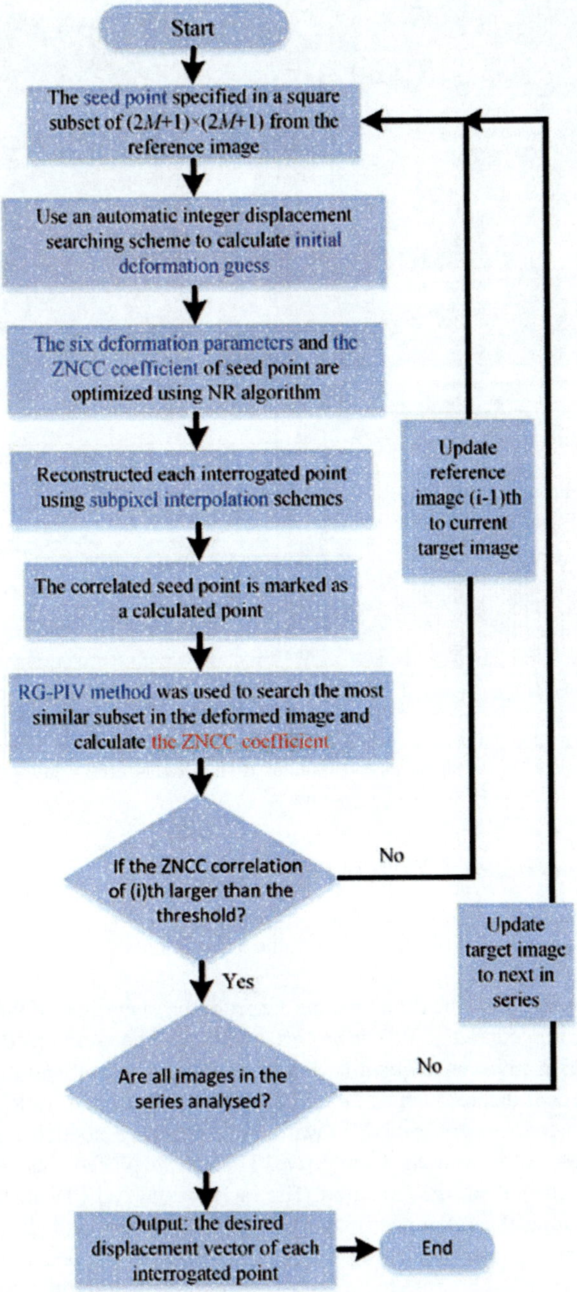

Fig. 11.57 The overall flowchart of incremental RG-PIV method

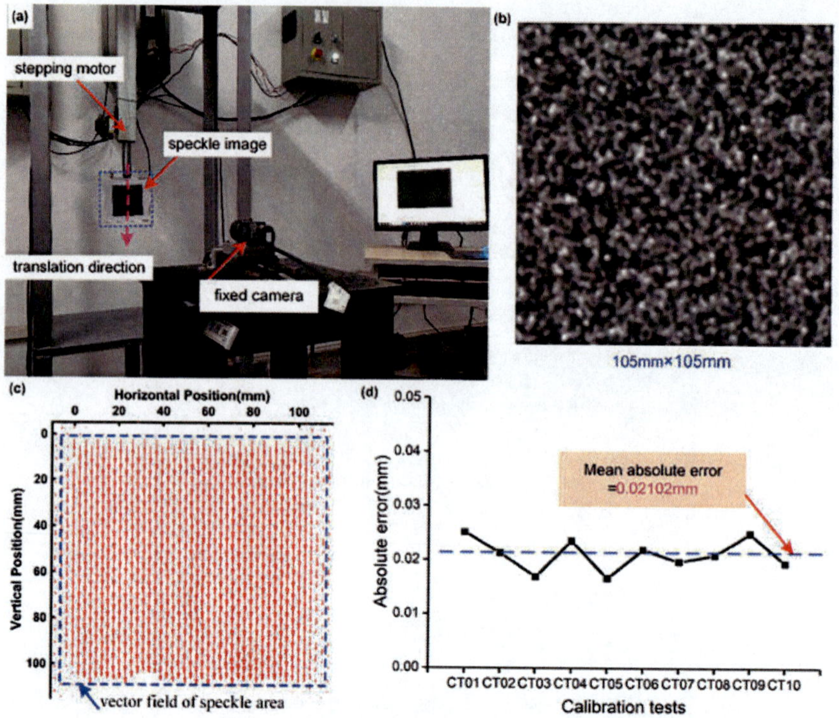

Fig. 11.58 The calibration tests of accuracy of the PIV analysis: **a** The translation test system; **b** The speckle image; **c** The vector displacement vector; **d** The absolute error value of the translational calibration tests

11.5.3 Experimental Results and Analysis

This section presents results obtained from the experiments described in the previous section.

The radial position is the radial distance r from the centerline of the model pile, normalized by the pile radius R. The vertical position is the normalized depth below ground level, z/R. In order to quantify and display the displacement field, the horizontal and vertical displacements, δu and δw, were normalized by R, and contour plot were produced. Outward and downward movements are labelled as positive, and inward and upward movements as negative. The soil displacements were obtained using the incremental RG-PIV method. The incremental RG-PIV method of automatically updating the reference image will reduce the error of the out-of-plane displacement on the post-processing test results. Note that the penetration problem of jacked pile is an axisymmetric problem, and only half of the displacement field were selected for post-processing data analysis.

11.5.3.1 Comparison Between Half-Model and Full-Model Tests

From the measured results of full-model test of half-pile, half-model test of half-pile, and full-model test of full-pile (T1, T2, T3), the relationship curve between penetration resistance and penetration depth are presented in Fig. 11.59. The penetration resistance was obtained from the load cell at the top of model pile. Comparison of the penetration resistance for half-model and full-model tests shows that the maximum penetration resistance at a final penetration depth of 8.5R for three cases is 163 N, 198 N, and 267 N, respectively. It should be noted that White (2004) has confirmed that the interface friction angle between sandy soil and the glass viewing window is significantly lower than the sandy soil friction angle. However, for the penetration resistance of T1 and T2, the penetration resistance of the former is 82% of that of the latter. It shows that the boundary effect increases the penetration resistance of T2. Furthermore, the penetration resistance of T2 is 74% of that of T3. It is because that the contact area between model pile and sandy soil as well as the volume of the soil displaced in half-model test would be different compared to true axisymmetric condition. In addition, for the initial phase of jacked pile penetration, the penetration resistance is always zero. This is attributed to the fact that the sand soil has no cohesive force, and the surface soil cannot withstand pressure. With the increase of penetration depth, the confining pressure around the pile tip increases gradually, so the pile will be affected by penetration resistance when it reaches a certain depth.

The displacement field contour plots of half-model and full-model tests at a penetration depth $L/R = 8.5$, as shown in Fig. 11.60. Figure 11.60a–c show the vertical displacement contour plots of the experimental results with T1, T2, and T3, respectively. It can be seen that the vertical displacement field can be roughly divided into two parts: compression region at the bottom of pile tip and uplifted region around pile body. Nevertheless, there are some noticeable differences. For T1, the influence

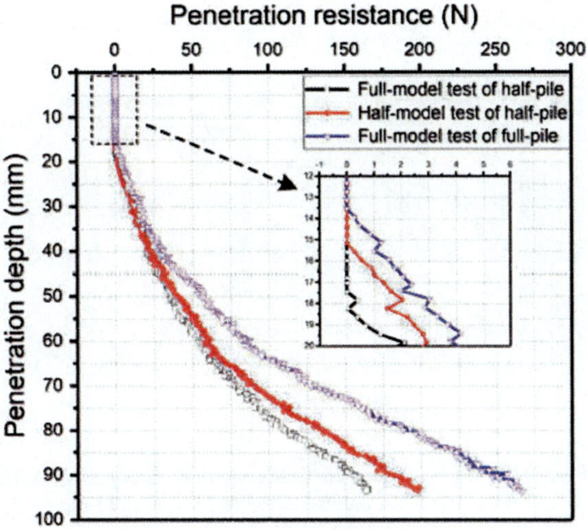

Fig. 11.59 The penetration resistance and penetration depth curves for half-model and full-model tests

range and displacement value of compression region and uplift region are minimal. For T2, the influence range and displacement value of compression region are the largest. The compression displacement of T2 is 160% of that of T3. This is because the glass viewing window limits the movement of the soil particle on the plane and does not produce out-of-plane displacement. Meanwhile, due to the boundary effect, the confining pressure of soil around the pile tip is higher. Therefore, it intensifies the compression of soil around the pile tip. For T3, the influence range and displacement value of value of uplift region are the largest. The uplift displacement of T2 is 65% of that of T3. This disparity is probably due to the volume of soil displaced by T3 is twice that of T1 and T2. Moreover, the problem of jacked pile penetration cannot be simply understood as an ideal axis symmetric problem, and the influence of the squeezing effect of jacked pile penetration is radial, so there is also an interaction effect. Figure 11.60d–f show the horizontal displacement contour plots of the experimental results with T1, T2, and T3, respectively. It is noted that the plane-strain test causes soil horizontal movements to extend further from the pile than would occur in axial symmetry. This further validates the interface friction angle between sandy soil and the glass viewing window is significantly lower than the friction angle of sandy soil. The lateral movements are inversely proportional to distance from the pile for T3, whereas in plane-strain they are constant as observed by White and Bolton [24], and a similar effect was visible in this test.

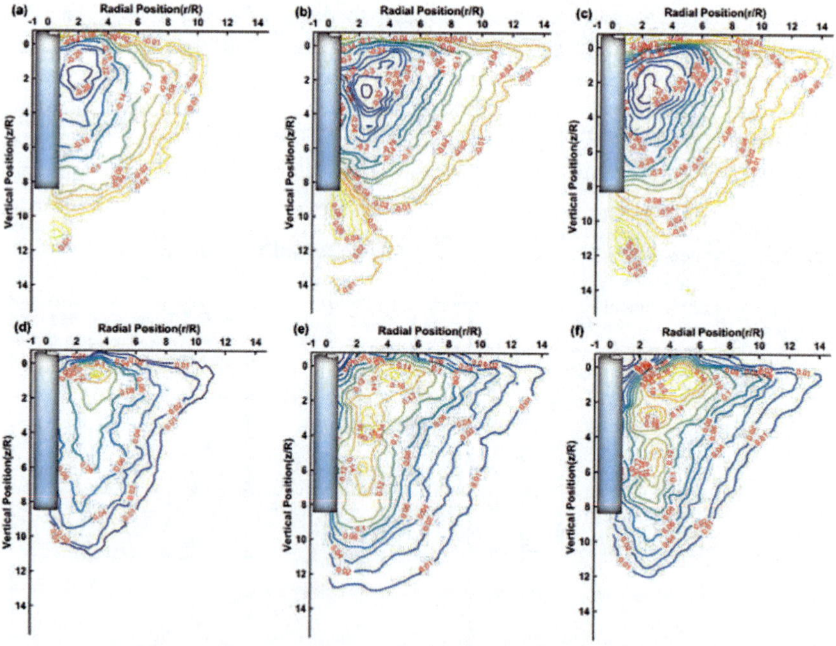

Fig. 11.60 The displacement field contour plots of half-model and full-model tests at a penetration depth L/R = 8.5: **a** T1, δw; **b** T2, δw; **c** T3, δw; **d** T1 δu; **e** T2, δu; **f** T3, δu

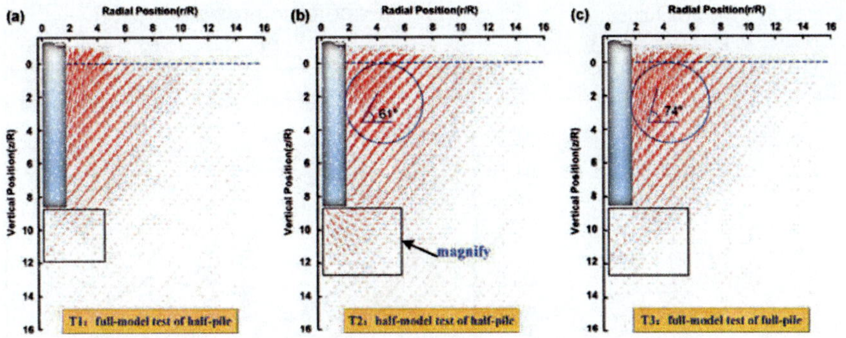

Fig. 11.61 The displacement vectors field plots: **a** full-model test of half-pile; **b** half-model test of half-pile; **c** full-model test of full-pile

The displacement vector field plots of half-model and full-model tests at a penetration depth $L/R = 8.5$, as shown in Fig. 11.61. Figure 11.61a–c reveal the displacement vector fields of the movement of the soil around the piles during the jacked piles penetration test of the three groups of T1, T2, and T3, respectively. It can be seen from the displacement vector field that the disturbance range of T1 is the smallest, and that of T2 and T3 is roughly the same. However, the compaction degree of soil at the bottom of pile is more severe in T2 test, and the angle between the direction of the displacement vector arrow and the horizontal direction is larger in the T3 test. The displacement vector fields can be used to visualize the effect of boundary effect of semi-model test on soil motion amplification.

The practical implications related to observation shown in Figs. 11.59, 11.60 and 11.61 are: (1) the bearing behavior and the penetration mechanism obtained by PIV in surface soil of viewing window would be different compared to true axisymmetric condition. Although the antifriction measures of the viewing window are adopted, the bearing behavior is generally carried out by the full-model test [59, 120]. (2) In order to reduce the boundary and interaction effects, it is of great significance to use the full-model test of transparent soil to further study the penetration of jacked pile.

11.5.3.2 Shape Effect of Pile Tip

To study the shape effect of pile tip, four different full-pile models (60°, 90°, 120°, flat-ended) were considered in this study. For four tests T4, T5, T6, and T7, the relationship curves between penetration resistance and penetration depth are presented in Fig. 11.62 to a final penetration depth of 10R. With each test there is a general trend of no penetration resistance during the first 0–20 mm of penetration, followed by a steady increasing response. In the process of the jacked pile penetration, as the angle of pile tip increases, the volume of soil displaced and the contact area between pile tip and soil also increase correspondingly under the condition of the

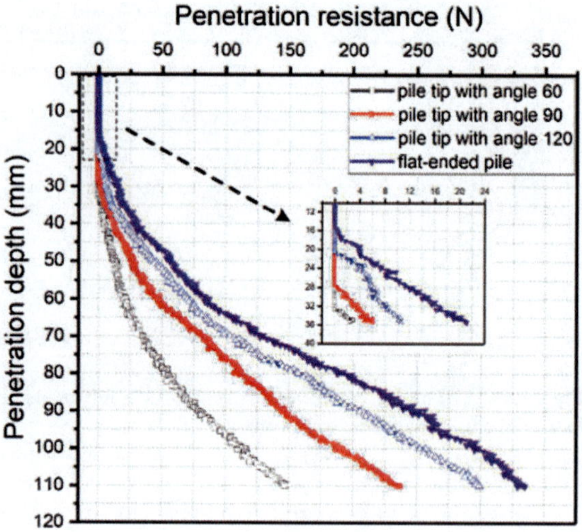

Fig. 11.62 The penetration resistance and penetration depth curves for different shapes of pile tip

same penetration depth. Therefore, the penetration resistance increases gradually, and the compression effect on surrounding soil is more remarkable.

Figure 11.63 exhibited that the contour plot of vertical displacement field for different shapes of pile tip (T4, T5, T6, T7). It can be seen that the vertical displacement patterns of the four model tests are roughly consistent. The vertical displacement contours of the uplift region and the compression region are radially distributed. For the pile tip with angle 60, the maximum vertical displacement in the uplift region is −0.48R, and the maximum vertical displacement in the compression region is 0.04R. As shown in Fig. 11.63, it is clear that the magnitude of the vertical displacement in uplift region and compression region increases with the increase of pile tip angle. Take −0.01R as the minimum quantitative standard, the influence range of the vertical displacement in the radial direction is approximately 14R. Nevertheless, there are some noticeable differences. With the increase of pile tip angle, the influence range of compression region also increases gradually. For the cone-ended pile, the dividing line between uplift region and compression region is at the position of pile tip, while for the flat-ended pile, the dividing line between uplift region and compression region is at the position of 1R at the bottom of pile tip. To some extent this difference is attributable to an entrapped meta-stable sand plug confined below the pile tip. A possible way to account for the effect of the meta-stable sand plug is to assume that it serves as an extension to the pile, and that the flat-ended pile is equivalent to a somewhat longer cone-ended pile [15, 121].

Regarding the horizontal displacement, Fig. 11.64 shows a systematic comparison of horizontal displacement contour plots with different shapes of pile tip at a penetration depth $L/R = 10$. It presents that there is a consistent variation trend of the normalized vertical displacements for different shapes of pile tip. As the upper confining pressure is small and the lower confining pressure is large, the lateral resistance overcome by penetration of jacked pile is different, so the influence range of

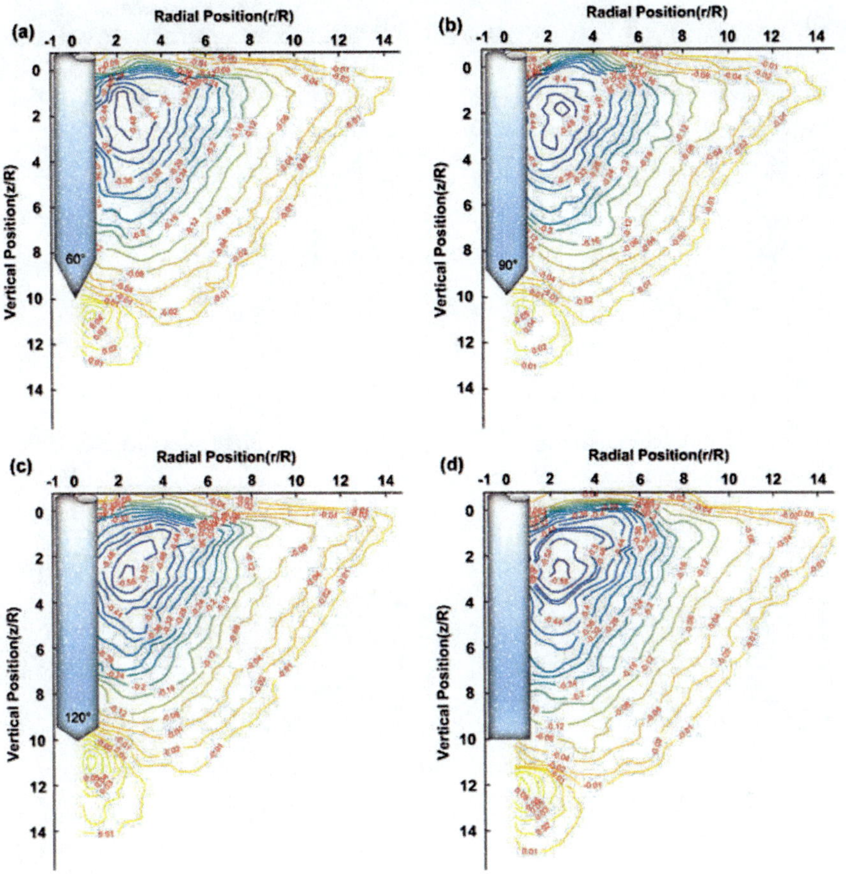

Fig. 11.63 The contour plot of vertical displacement field for different shapes of pile tip at a penetration depth L/R = 10: **a** T4, δw; **b** T5, δw; **c** T6, δw; **d** T7,δw

horizontal displacement gradually decreases in the depth direction. As the angle of pile tip increases, the maximum value of horizontal displacement gradually increases, but the influence range of horizontal displacement is basically the same. Comparing the vertical displacement and the horizontal displacement values in four groups of different pile tip shapes, indicating that the vertical displacement is roughly 2.7 times the horizontal displacement. At the same time, it can be seen from the change trend of displacement contour that the soil moves in inclined upward direction, which further verifies that when the penetration depth of jacked pile is less than 20R, the deformation of soil around the pile is mainly inclined upward uplift [100]. Additionally, a key point to note is that the displacement show maxima developing away from the pile shaft, between $2 < r/R < 4$.

Figure 11.65 presents the displacement vector diagram of soil around piles in the penetration test of four groups of jacked piles with different shapes. The disturbance

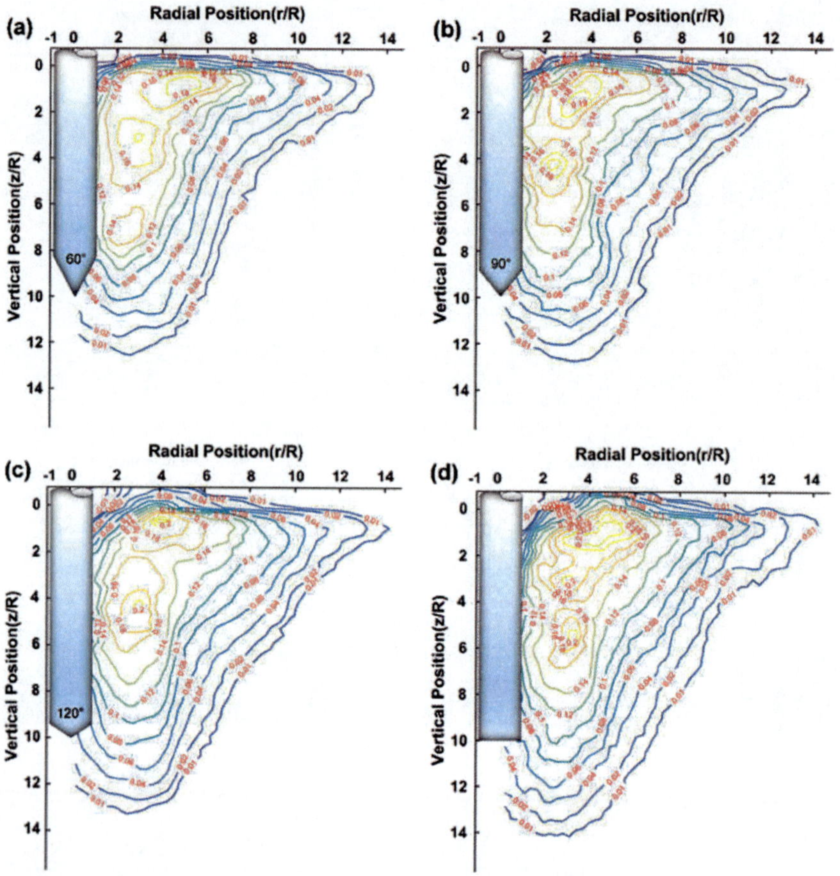

Fig. 11.64 The contour plot of horizontal displacement field for different shape of pile tip at a penetration depth L/R = 10: **a** T4, δu; **b** T5, δu; **c** T6, δu; **d** T7,δu

of jacked pile penetration to the soil around the pile can be simply divided into compression zone, transition zone and uplift zone. The direction of the vector arrow on one side of the pile is inclined upward. With the increase of the angle of pile tip, the influence range of the bottom compression area gradually increases and the length of the vector arrow gradually increases. In addition, with the increase of the angle of pile tip, the surface uplift and the affected range of uplift increase gradually.

In order to quantify and display the disparity of penetration mechanism for different shapes of pile tip, the amount of surface uplift and the schematic diagram of principle analysis were plotted in Fig. 11.66. From Fig. 11.66a, it can be shown that the amount of surface uplift gradually increases with the increase of pile tip angle. In addition, with the increase of the angle of pile tip, the position of maximum surface uplift gradually moves away from the pile shaft. Figure 11.66b schematically shows the stress analysis on soil element "A" around the pile tip during pile penetration.

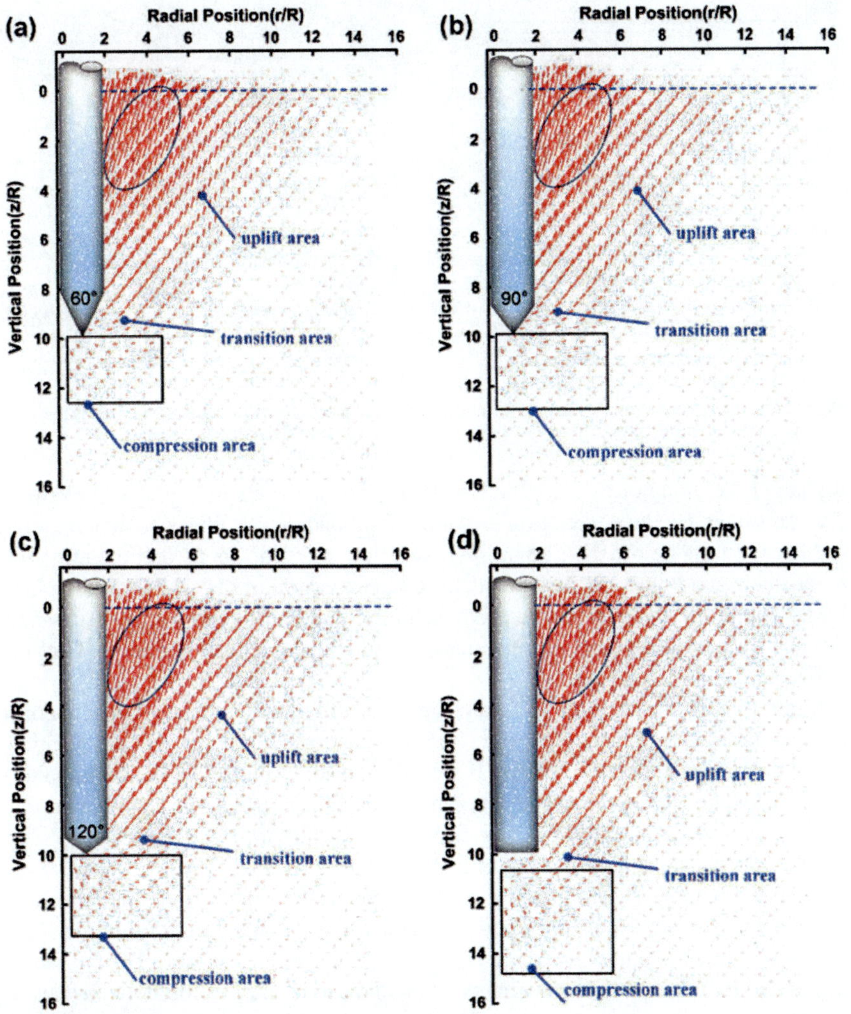

Fig. 11.65 The displacement vector diagram of soil around piles in the jacked piles penetration test with different shapes: **a** T4; **b** T5; **c** T6; **d** T7

For the penetration problem of pile tip angles of 60°, 90° and 120°, the direction of the squeezing force is in the normal direction along the conical plane. The squeezing force can be understood as the resultant force of two components: one is the vertical component, which is used to resist penetration resistance; the other is the horizontal component used to resist passive earth pressure. It has been pointed out in Fig. 11.62 that the larger the angle of pile tip is, the greater the penetration resistance will be. According to the principle of action and reaction, the larger the angle of pile tip is, the greater the squeezing force of pile tip on surrounding soil mass will be. Therefore,

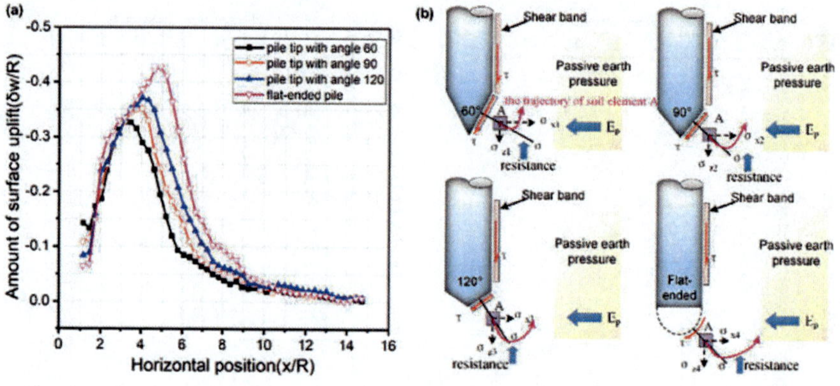

Fig. 11.66 The amount of surface uplift and penetration mechanism of different angles pile tip

the larger the angle of pile tip, the longer the movement trajectory of soil element "A". However, for the penetration problem of flat-ended pile, the existence of meta-stable sand plug below the pile tip increases the squeezing force, and the effect of meta-stable sand plug can be equivalent to the extension of the pile on the original basis. Therefore, compared with the penetration problem of conical pile tip, when the pile tip passes through soil element "A", the quadratic accumulation effect on soil element "A" is also greater.

Figure 11.67 presents the amount of uplift at different positions of the pile body for flat-ended pile. As shown in Fig. 11.67a, it is worthy noticing that the position of the maximum uplift of pile body is roughly z = 2R, which is quite similar to the experiment phenomenon presented in Zeng [122] (Fig. 11.67b). The amount of uplift of each layer is normalized by taking the surface uplift as the standard, it can be seen that the uplift amount of the third layer is the largest, and then gradually decreases with the depth direction. Additionally, the horizontal position of the maximum surface uplift is roughly at x = 5R, which is further away from the center of the pile shaft than that of other depths. The image post-processing analyses in Fig. 11.67b provide physical evidence that the horizontal position of the maximum uplift of the first layer is further away from the pile center. Further detailed illustration of the differential characteristics of uplift presented in Fig. 11.68.

In order to improve the interpretation and understanding of the penetration mechanism of jacked pile, Fig. 11.68 schematically shows the evolution of jacked pile penetration. The penetration process of jacked pile can be divided into two distinct phases: the cavity expansion phase and the primary vertical compression shearing phase. The cavity expansion phase represents the creation of a cylindrical space within the ground as the pile tip pushes the soil away from the path of the pile [123]. The cavity expansion phase is slightly different with the shape of pile tip: for cone-ended pile (Fig. 11.68a, the cavity expansion phase starts from a very small initial cavity radius and ends when the cavity radius becomes equal to the pile radius; for flat-ended pile (Fig. 11.68b), this phase stops when the size of meta-stable sand plug

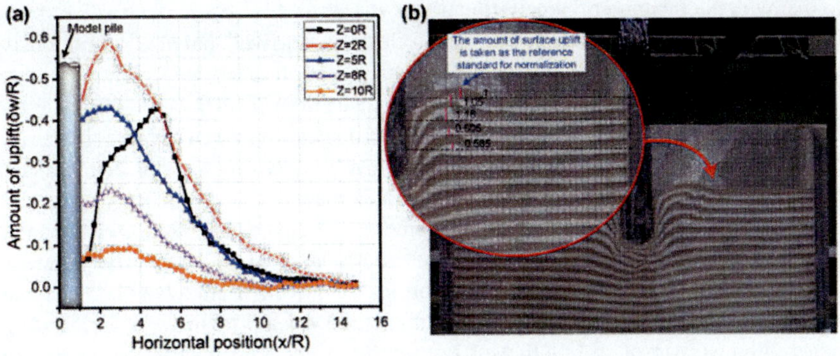

Fig. 11.67 The amount of uplift at different positions of the pile body

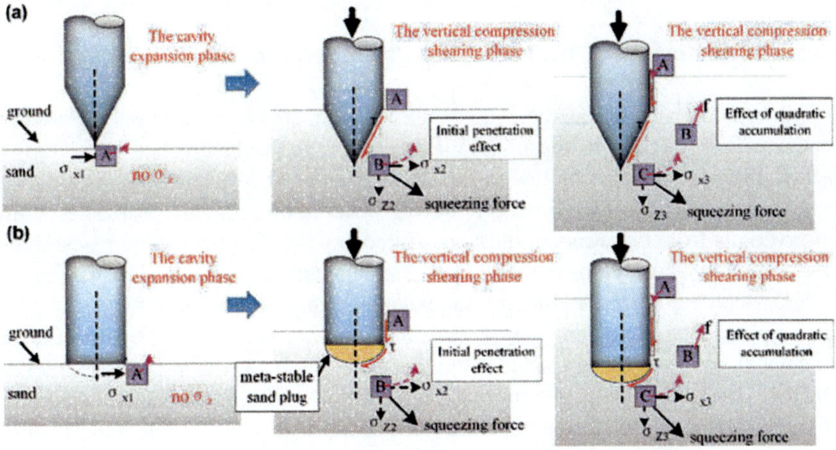

Fig. 11.68 The evolution of jacked pile penetration

reaches its maximum. Since there is no confining pressure on the surface, in the process of cavity expansion, after the pile tip passes through element "A", under the action of squeezing force, element "A" is squeezed to the position aligned with the pile shaft. The cavity expansion is followed by a vertical compression shearing phase. The vertical compression shearing phase refers to the continuous penetration of jacked pile, the upper confining pressure and lateral passive earth pressure limit the movement of soil, the soil at the bottom of pile tip is in the state of high compressive stress, and the soil is continuously sheared along the contact surface between the pile shaft and soil.

'The initial penetration effect' refers to the disturbance effect on the soil around the pile tip when the pile tip first penetrates into soil at a certain depth during the continuous penetration of the jacked pile. The effect of quadratic accumulation refers to that with the increase of penetration depth, the static pile not only has a disturbing effect on the soil at the position where the pile tip just arrived, but also continuously disturbs the soil around the pile shaft that has passed. It is a process in which each subsequent new penetration stage has a cumulative effect on all the previous penetration stages. In the process of progressive vertical shearing, when the pile tip passes through element "B", the initial penetration effect of squeezing force on element "B" is produced. Then the pile tip penetrates into element "C". At this time, the initial penetration effect of squeezing force on element "C" will be generated, and the effect of quadratic accumulation will also be generated on element "B". This is why there is always more uplift at the top than at the bottom. Due to the friction between pile and soil, part of soil around the pile will be dragged downward by the pile shaft during penetration, so part of soil on the surface uplift will roll towards the shear band. In addition, the confining pressure on the upper soil is smaller and the stress can be released more easily. As a result, the amount of surface uplift is not the largest.

11.5.3.3 Disturbance Effect of Jacked Pile Penetration

In order to compare and analyze the disturbance effects of different pile tip shapes on the soil around the pile, the schematic illustration of the disturbance zones during jacked pile penetration is plotted in Fig. 11.69. Figure 11.69ashows the disturbance zones adjacent to the flat-ended pile during penetration into sandy soil, based on observations from these model tests conducted and the results of other researchers [103, 104, 124]. Five regions providing a general, qualitative sense of the disturbance zones, which are of importance for understanding the disturbance effect caused by jacked pile penetration: zone (1)—shear slide zone, which is created by the relative movement of the pile shaft against the adjacent soil. The soil in this area moves downward with the pile penetration. Due to intense shearing, the original structure of sandy soil is completely destructed. The width of this shear slide zone is thin, which was also observed by Yang [159], Massarsch and Wersäll [103] and Arshad et al. [104]. The interface shear band width is controlled initially by grain size alone [125]. Zone (2)—vertical high compression zone, a sector with high compressive which is created during pile progressive penetration. According to the Rankine earth pressure theory, this region is active destruction zone. The zone of intense vertical compression extends 5R from the bottom of flat-ended pile. Considering that the failure mode of the soil at the bottom of the pile is sliding failure, combined with the phenomenon of model test, part A in this zone (2) is defined as meta-stable sand plug, which only produces elastic deformation without shear failure. For sandy soil, the existence of large magnitude vertical stresses at the bottom of pile tip will not only increase the compactness of soil, but also accompany with the phenomenon of particle breakage. Part B in this zone (2) is defined as the compression region. The soil displacement of this zone is purely vertical. Zone (3) —transition zone, the large principal stress in this zone changes gradually from a mainly vertical direction to the inclination upward direction. Zone (4)—stress relaxation zone, this area is in the same stress state as Rankine passive earth pressure, resulting from the expansion of active

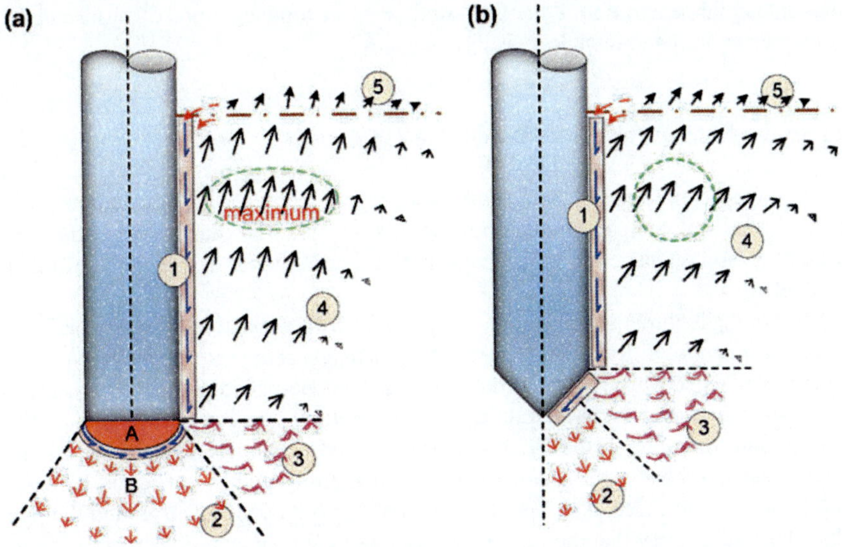

Fig. 11.69 Schematic illustration of the displacement and zones of disturbance during jacked pile penetration: **a** flat-ended pile; **b** cone-ended pile

earth pressure. In zone (4), the soil is displaced primarily in the obliquely upward direction. The lateral movement of soil around the pile shaft decreases sharply with the increase of the distance from the pile axle, the upper influence range is larger than the lower influence range. Zone (5)—surface uplift zone, it is due to the change of stress field in soil. In the process of jacked pile penetration, the horizontal stress is increased due to the extrusion of soil, and the vertical stress is generated. When the stress is greater than the gravity of the overlying soil, the soil will rise.

As for the penetration problem of 60°, 90°, and 120° angle jacked piles, according to the characteristics of its displacement field contour and the similarity of force analysis, it can be uniformly simplified as the penetration problem of cone-ended pile tip. The mechanism of cone-ended pile tip penetration is schematically presented in Fig. 11.69b. In contrast with the penetration mechanism of flat-ended pile, there are some noticeable differences. Five zones have been identified based on the displacement patterns observed. Zone (1)—shear slide zone, the characteristics of the region are consistent with that of the flat-ended pile; Zone (2)—compression expansion zone, because the cone-ended pile penetration problem does not have the phenomenon of meta-stable sand plug, the direction of motion of soil particles is normal along the surface of the conical tip. Zone (3)—transition zone, because the effect of soil compaction of cone-ended pile is weaker than that of flat-ended pile, so the soil displacement field is purely radial, resembling the field that would result from a cavity expansion process. Zone (4)—stress relaxation zone, the angle between the direction of soil particle movement and the horizontal plane is smaller than that of flat-ended pile penetration. Zone (5)—surface uplift zone, the trend of surface uplift is increasing first and then decreasing.

11.5.3.4 Characterisation Around Pile Tip

To describe the existence of meta-stable sand plug more intuitively, fused quartz grain treated with high temperature coloring was used to quantify the size of meta-stable sand plug, and the penetration mechanism of different shapes of pile tip was further studied.

According to the experimental phenomenon of different shapes of pile tip penetration, the images of different stages were selected to reveal the penetrating mechanism of the pile. As shown in Fig. 11.70a and b, as the cone-ended pile penetrates gradually, the friction between the pile and the soil will cause a downward drag on the soil around the piles, so the dyed particles dragged down around the piles can be clearly seen. With the further increase of penetration depth, the friction between piles and soil increases, and there are no longer any dyed particles around the pile tip. The results show that the penetration mode of cone-ended pile is mainly manifested in the way of piercing and squeezing soil on both sides. Figure 11.70c shows the penetration of pile tip with angle 120, it can be seen that a small amount of dyed particles is attached to the surface of the pile tip. However, the meta-stable sand plug beneath the pile tip move progressively downward as the flat-ended pile penetrates into the ground (Fig. 11.70d). The meta-stable sand plug is similar to semi ellipse. This suggests that the friction on the pile–sand interface underneath the pile tip is sufficient such that slippage cannot occur. When the flat-ended pile penetrates from a depth of 6R to 8R, the sand plug's height decreases from 0.71R to 0.62R. As the penetration depth increased, the movement of the dragged dyed particles is restricted by greater confining pressure, so the amount of colored fused quartz on both sides of the pile decreased gradually. After the flat-ended pile penetrates to the depth of 10R, the height of the meta-stable sand plug was 0.6R. By introducing the method of partial replacement by dyed particles, it provided physical evidence that with the change of penetration depth, the confining pressure around the pile tip will change and the shear slip surface beneath the pile tip also changes. The central core of meta-stable sand plug is stationary relative to the pile tip, but the shoulders of the zone are not [24]. In contrast, there is no the phenomenon of meta-stable sand plug beneath the cone-ended pile tip. Therefore, the most important reason for the different mechanism of the flat-ended and cone-ended pile tips is the formation of meta-stable sand plug phenomenon.

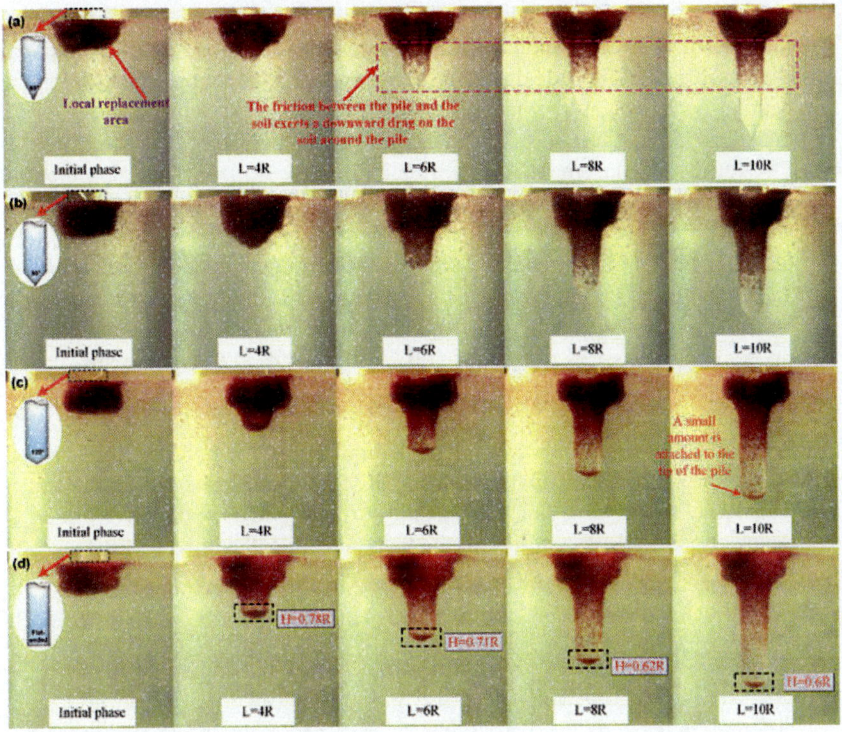

Fig. 11.70 The characterisation around pile tip at different stages: **a** pile tip with angle 60°; **b** pile tip with angle 90°; **c** pile tip with angle 120°; **d** flat-ended pile

11.5.4 Summary and Conclusions

The results of this investigation provide a detailed experimental analysis of the penetration mechanisms of jacked pile and the squeezing effect. A small-scale geotechnical model test, combining transparent soil and digital image measurement system was used to observe soil movement pattern in sandy soil at different stages induced by jacked pile penetration. Interpretation of these experiment data has led to the following conclusions.

(1) The incremental RG-PIV method can improve the serious decorrelation effect in the deformed images due to large deformation in conventional post-processing software by automatically updating the reference image, and can more accurately reflect the real test phenomenon.
(2) Based on three groups of comparison tests (full-model test of half-pile, half-model test of half-pile, and full-model test of full-pile), the differences between the full-model test and the half-model test are quantitatively analyzed. The penetration resistance in the case of a half-model and half-pile is 1.2 times that in the case of a full-model half-pile, and at the same time it is 0.74 times that in

the case of a full-model half-pile. The maximum compressive deformation at the bottom of the pile tip under the half-model and half-pile case is twice that of the full-model full-pile case. The boundary and interaction effect causes the bearing behavior and the penetration mechanism of the half-model test to be different from the true situation of the full-model test.

(3) The vertical displacement is roughly 2.7 times the horizontal displacement in four groups of jacked pile penetration test with different pile tip shapes, which demonstrates that when the penetration depth of jacked pile is less than 20R, the deformation of soil around jacked pile is mainly inclined upward uplift.

(4) The shape effect of pile tip was shown to have a clear impact on the penetration resistance, displacement pattern, and uplift amount. As the angle of pile tip gradually increases from 60° to 90°, 120° and 180°, the maximum vertical disturbance displacement of soil around the pile caused by jacked pile penetration is −0.48R, −0.51R, −0.55R and −0.58R, respectively. Meanwhile, the distance between the position of the maximum surface uplift and the central axis of the pile gradually changed from 2R to 2.7R, 3.2R and 4R. For the cone-ended pile, the dividing line between uplift region and compression region is at the position of pile tips, while for the flat-ended pile, the dividing line between uplift region and compression region is at the position of 1R at the bottom of pile tip.

(5) The evolution of jacked pile penetration can be divided into two distinct phases: the cavity expansion phase and the primary vertical shearing phase. In the cavity expansion phase, the volume of soil mass displaced by different pile tip is different; in the vertical compression shearing phase, the soil around the pile is subjected to different forces for different pile tip. This explains why different pile tip cause different amount of uplift. Additionally, the squeezing effect of jacked pile penetration on surrounding soil should be considered not only the initial penetration effect but also the effect of quadratic accumulation. So the maximum displacement region is between $2 < r/R < 4$.

(6) According to the displacement and deformation modes of sandy soil at different positions during the penetration process of jacked-pile, the soil around the pile affected by the penetration pile can be divided into 5 zones: shear slide zone, compression expansion zone, transition zone, stress relaxation zone, surface uplift zone.

(7) The effect of meta-stable sand plug is the main reason of the different mechanism between flat-ended pile and cone-ended pile. The shape of the meta-stable sand plug is half ellipse. With the continuous penetration, the height of sand plug gradually decreases under the combined action of lateral passive earth pressure and vertical compressive stress, the final height of meta-stable sand plug is 0.6R. The effect of meta-stable sand plug could be interpreted as an extending part of the flat-ended pile, so the flat-ended pile is equivalent to a slightly longer cone-ended pile.

11.6 Visualization of Bulging Development of Geosynthetic-Encased Stone Column

11.6.1 Introduction

Geosynthetic-encased stone column (GESC) is a ground improvement technique for soft soils, which reduces the total settlements and accelerates consolidation. Compared with ordinary stone column (OSC), GESC has greater bearing capacity and undergoes less radial bulging. The behavior of GESC has been investigated through in-situ field tests [126, 127] and model tests [128, 129, 130–137]. Particularly, sand ground was selected as surrounding soil in model tests performed by Lee et al [128]. In these studies, the bulging of OSC and GESC was visualized by carefully removing the stones and casting plaster [136] or pouring concentrated slurry to shape [137]. The operation is complex and only the last bulging state is visible. The study of bulging development is helpful for understanding the bearing mechanism of stone columns. Nevertheless, the internal displacement field, soil movements and radial bulging development during loading are difficult to measure accurately. This can be solved by using innovative techniques such as transparent soil and digital image correlation (DIC).

Transparent soil, made from fused quartz and pore fluid with the same refractive index, has been used to model soil [32, 138]. The visual model technique in combination with the DIC technique and transparent soil have been used in model tests to study internal deformation problems [108, 110, 138]. These studies confirm that the transparent soil technique is an effective method for visualizing the displacement field. Particularly, MacKelvey [72,139] used a transparent soil to allow visual monitoring of OSCs under vertical load. However, the primary purpose of their study was the failure pattern of short and long OSCs. The exact bulging, bulging development process and displacement field of surrounding soil were not measured. The soil was in a passive state when the bulging happens. Model tests using transparent soil like grout permeation [140], plie-deformation induced ground movement [12, 113] proved the feasibility of simulating natural soil by transparent soil in passive state.

In this study, model tests were carried out to investigate the characteristics of GESC using transparent soil and digital image correlation (DIC) technique. The radial bulging development with increasing load was obtained. The displacement field of the surrounding soil was also measured. Furthermore, the deformation results in these tests were compared with those obtained from previous model tests and predicted by existing theory.

11.6.2 Experimental Description

11.6.2.1 Model Test Setup

The visual model test setup, shown in Fig. 11.71a, consisted of a linear laser, a charge-coupled device (CCD) camera, a load system, an optical platform, transparent soil, GESC (or OSC), a plexiglas cylindrical model box and a computer for image processing. The linear laser (EP 532-2 W) can produce a vertical or lateral light sheet. The CCD camera with a resolution of 1280 × 960 was approximately positioned 500 mm away from the laser sheet. A schematic of the load test in a unit cell and loading arrangement is showed in Fig. 11.71b.

The unit-cell idealization is used in this study to simplify the design of the apparatus. Ambily and Gandhi [141, Gniel and Bouazza [133], Murugesan and Rajapopal [136], Nayak et al. [142] have used this concept of a unit cell in their model tests to assess the behavior of a stone column at the centre of a large group. In order to achieve the unit cell model, a plexiglas cylinder, 10 mm thickness, 100 mm inner diameter (d_e) and 230 mm height, was used in the model test. The diameter of the OSC and GESC (d_c) was 25 mm. The dimension of the adopted unit cell is equivalent to stone column installed at a centre-to-centre spacing of 88 mm ($3.5d_c$) in a square pattern or 95 mm ($3.8d_c$) in a triangular pattern. The loading system includes two parts which can apply loads to the column and soil respectively, as shown in Fig. 11.71b. One part is a steel rod of "T" shape with the same diameter as the stone column. It can apply pressure on the GESC or OSC through dead weights placed symmetrically on the hanger. The other part includes a hollow steel pipe with a rectangular steel plate welded on the top and an annular plate welded at the bottom. It can apply pressure to the transparent soil surrounding the column by dead weight on the top plate. Holes were perforated on the annular plate to allow drainage. The casing pipe has an outer diameter of 25 mm with thickness of 0.2 mm. The height of the GESC or OSC is 200 mm.

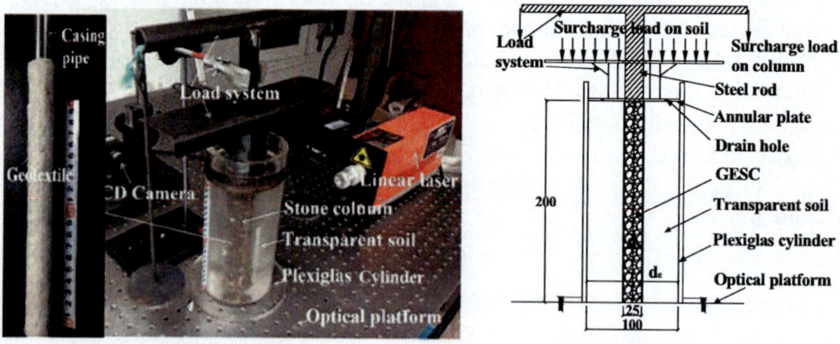

(a) Test setup

(b) Elevation view

Fig. 11.71 Schematic diagram of apparatus

Table 11.9 Chracteristic values used for model to prototype scaling

Dimensionless factor	Characteristic	Scaling factor (prototype/model)	Prototype value	Model value
J/gH^2	J (kN/m)	(=400)	200–4000	0.5–10
T/gH^2	T (kN/m)	(=400)	40–4000	0.1–10
D/H	D (m)	(=20)	0.2–1	0.01–0.05

Table 11.9 shows the scaling factors for the model and prototype with the same stone aggregates density. The model GESC of 0.025 m in diameter was chosen to simulate a prototype column of 0.5 m in diameter, the ratio of prototype column diameter to the model column diameter is 20. Therefore, stresses on a full-scale GESC are 20 times those measured on model GESC.

11.6.2.2 Properties of Materials

Transparent soil, crushed stone aggregate and geotextile were used in this investigation. The transparent soil was manufactured by fused quartz and oil-based solution with the same refractive index of 1.4585. The transparent soil had a uniform gradation and the particle size distribution is shown in Fig. 11.72a, along with those used in previous researches for comparison. As followed the previous studies of stone column or geosynthetic-stone column surrounding by natural soil, the particle size of stone column material was similar to that in these model tests [133, 141, 136, 142] when the transparent soil was similar to natural soil. The properties of the fused quartz are presented in Table 10. The oil-based solution, made of mixed mineral oil and Norpar 12 fluid (4:1 ratio by weight), has a refractive index of 1.4585 at 24°C [23]. More details obtained by consolidated-drained triaxial tests about the transparent soil can be found in Kong et al. [110]. The cohesion was calculated as 0. Hence, the transparent soil was model as a typical sand.

The stone aggregates used to form the stone column were angular granite chips of size 1–10 mm and had uniform gradation. The gradation curve of the stone aggregates is shown in Fig. 11.72a. By controlling the volume of the stone chips and the weight filling in the casing pipe, a column was made with a relative density of approximately 70%. The properties of the stone aggregate were also presented in Table 11.10. Casing geotextile was made of nonwoven geotextile. The tensile force-strain relationship of the tested geotextile samples under standard wide-width tension tests [143] was shown in Fig. 11.72b. Since the encasement was formed by stitching a piece of geotextile into a sleeve, the tensile-strain relationship of the seamed sample of the geotextiles were tested. The tensile strength properties of the geotextiles are listed in Table 11.10.

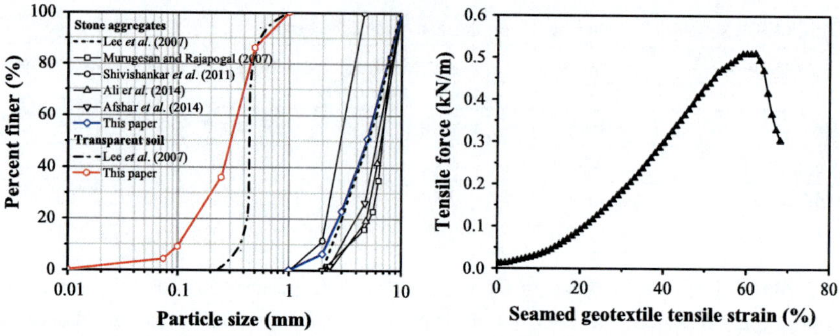

(a) Particle size distributions (b) Tensile-strain curves of geotextile

Fig. 11.72 The properities of the model test materials

Table 11.10 Mechanical properties of the model test materials

Materials	Parameter	Unit	Value
Transparent soil	Coefficient of uniformity		3.38
	Coefficient of curvature		1.19
	Maximum dry unit weight	kN/m^3	13.00
	Minimum dry unit weight	kN/m^3	9.9
	Saturated unit weight $(_s)$	kN/m^3	18.37
	Peak friction angle $(_s)$[a]	°	37
	Cohesion (c)	kPa	0
	One-dimensional constraint modulus $(_s)$	MPa	8.12
	Compression index (C_c)		0.04
Stone column	Coefficient of uniformity		2.62
	Coefficient of curvature		0.96
	Peak friction angle $(_c)$	°	41
	Saturated unit weight $(_c)$	kN/ m^3	19.10
Geotextile	Thickness	mm	1
	Tensile strength with seam (T)	kN/m	0.51
	Strain at peak strength with seam	%	62.09
	Tensile modulus with seam (J)	kN/m	0.82

[a] From consolidation-drained test after Kong et al. [110]

11.6.2.3 Installation of GESC

The installation process was schematically shown in Fig. 11.73. It mainly includes the following steps:

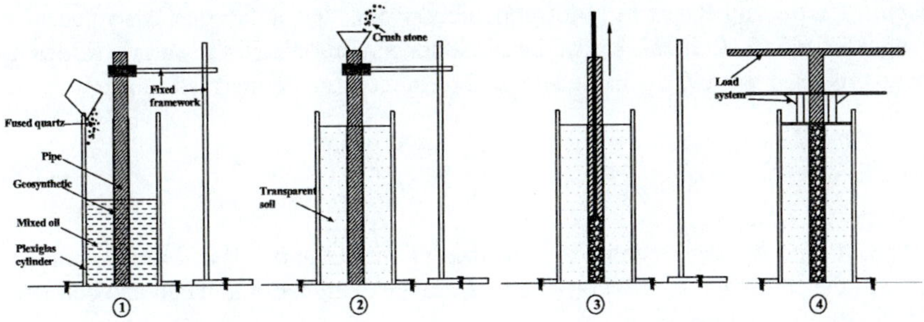

Fig. 11.73 GESC installation process

(1) Preparing casing pipe: The casing pipe was used to install the stone column. It was wrapped with geotextile and fixed at the centre of the plexiglas cylinder, through a fixed framework.
(2) Placing soil and column: First, fused quartz was poured into pore fluid of mixed oil by the sand raining method [144] (400 mm high from the surface). Second, a panel with small holes was placed on the soil sample. The soil sample had a relative density of about 60% after pressure on the panel. Then stone aggregates were poured into the casing pipe in 5 layers.
(3) Constructing of GESC: After placing each layer of the stone aggregate, the stone aggregate was compacted with a tamping rod (10 mm in diameter and 500 mm long). This process was repeated four times until the entire height of the stone column was formed. Then the casing pipe was lifted up gently and the geosynthetic stayed with the stone column. Extra pore fluid was drained out and extracted by a suction bulb.
(4) Setting up the loading and measuring system: Upon completing the stone column installation, the loading system was placed on the top of the soil and column. The measuring system was set around the plexiglas cylinder, with linear laser and CCD camera across.

11.6.2.4 Loading Test on GESC or OSC

As described earlier, the loading system can apply different loads to the column and soil, respectively. Since the GESCs are usually used in column-supported embankment, deformation of column is not same as the one of soil [145]. Hence, the rigid plate loading method seems not proper. The pressure on the GESC was set as the known value based on a given stress concentration ratio in these tests. The pressure applied to the column and soil was determined by

$$q_c = nq_s = \frac{n}{1 + \alpha(n - 1)} q \tag{11.7}$$

where q is the total applied pressure with a range of 0–50 kPa q_c and q_s are the pressure applied to the column and soil, respectively; n $(= q_c/q_s)$ is the stress concentration factor. Castro and Sagaseta [146] concluded from their study that the value of n ranges from 5 to 10. In this study, the stress concentration factor (n) was assumed as 5, as proposed by [147]. α is the area replacement ratio, defined as follow:

$$\alpha = \frac{A_c}{A_e} = \frac{d_c^2}{d_e^2} \tag{11.8}$$

where A_e and A_c are the entire area of the cylindrical unit cell and cross-sectional area of the column, respectively; d_e and $d_c$are the diameter of unit cell and diameter of stone column, respectively.

11.6.2.5 Digital Image Correlation (DIC) Technique

As the interaction between laser light and transparent soil produces a distinctive speckle pattern. Small particle movement can result in the speckle distribution changing in the plane of measurement. Images captured before and after deformation were analyzed by the DIC technique. The images were split into a large number of interrogation windows. For each of these interrogation window, a DIC displacement vector was produced before and after each increment of load pressure with the help of an autocorrecting technique. Performing this for all interrogation regions produces a vector map of the average particle displacements. The discrete form of the standard cross-correlation function is as follows:

$$C(\Delta x, \Delta y) = \frac{1}{MN} \sum_{m=0}^{M-1} \sum_{n=0}^{N-1} f(m, n) g(m + \Delta x, n + \Delta y) \tag{11.9}$$

where M and N are the dimensions of interrogated images, and f and g are the grayscale intensities of the two images being interrogated. The spatial displacement field can be obtained from the displacement in each interrogated window.

11.6.2.6 Warping Effect and Modification

According Snell's law from curved surfaces, warping effect will make difference between position in DIC images and in actual. Plexiglas cylinder is like a magnifying glass when the objects in the cylinder are photographed. In order to eliminate this effect, it is necessary to modify the displacement results obtained from DIC technique. The displacements were calculated by cross-correlating images from same cross section on the physical coordinate plane. The distance between two outside marines of cylinder is 120 mm (diameter and thickness of cylinder). The surface is linear along vertical direction because of the cylinder surface, rather than spherical surface. On account of cylinder is only radial-curved, the warping effect along vertical orientation is ignored. A simple test was using to measure the effect in this study.

As shown in Fig. 11.74a, metal-grid with width of 100 mm was inserted into transparent soil in cylinder. The width of every grid is 9.5 mm. The center of the metal-grid is set as the starting point of the physical coordinate at horizontal orientation. The position of each grid joint was measured by photo and compared with metal-grid in actual. The details of linear difference are shown in Fig. 11.75b.

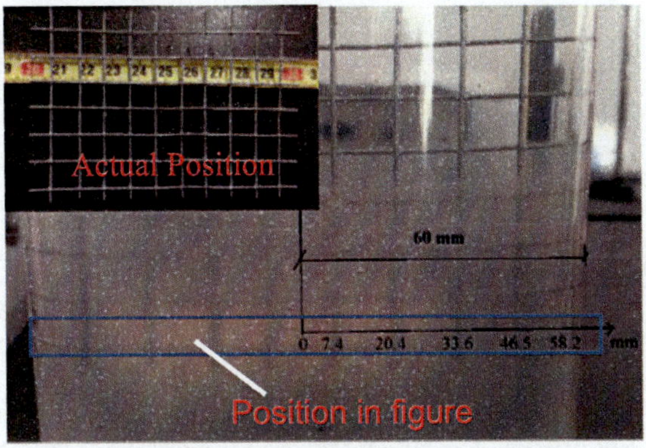

(a) Diagram of modification method

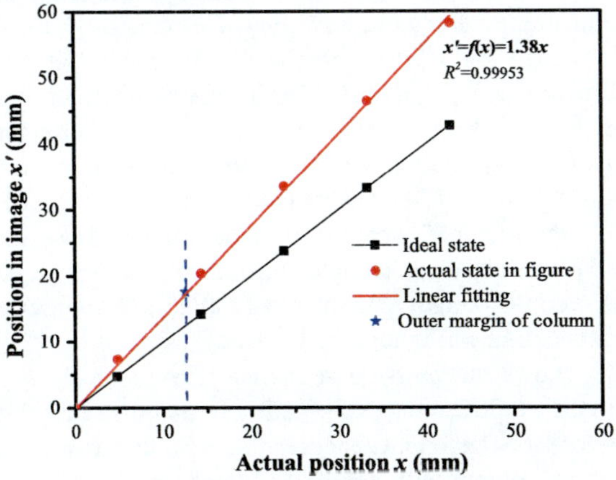

(b) Results of modification method

Fig. 11.74 Modification for horizontal position between actual and image

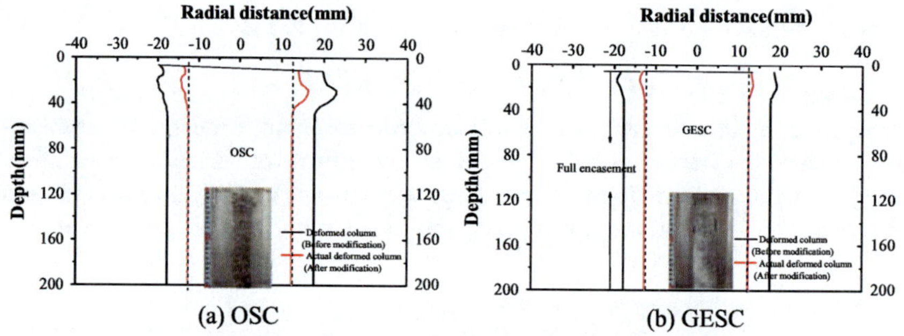

Fig. 11.75 Deformed column under pressure of 50 kPa

11.6.3 Results and Discussion

11.6.3.1 Deformation of GESC/OSC

The detailed test parameters were list in Table 11.11. The pressure-deformation responses observed for OSC and GESC are shown in Figs. 11.75 and 11.76, respectively. Figure 11.75 shows the physical photos and the bulging distribution at a total applied pressure of 50 kPa. The bulging distribution of GESC or OSC obtained by DIC technique and then modified were plotted in this figure. Figure 11.76 plotted the deformation of GESC or OSC (settlement and bulging). The radial bulges of GESC or OSC before and after modification are S'_c and d_c respectively. Figure 11.76a plots the applied pressure (q) against the normalized settlement (S_c/d_c). The test results obtained by Murugesan and Rajagopal [136] using clay soil and by Ali et al. [131] using Kaolinite clay are also presented for comparison. In the figure, the general variation pattern of q-S_c/d_c relationship was similar. Furthermore, at $q = 50$ kPa, the settlement of GESC ($S_c/d_c = 0.244$) is about 50% of that of the OSC ($S_c/d_c = 0.479$). This demonstrates the use of transparent soil is feasible in GESC model tests. The bulking of columns was usually taken into consideration for the super-long column. In the study, the ratio that length to diameters (200/25 = 8) is not so large. Hence, the possibility of buckling was ignored in the study.

Figure 11.75b shows the pressure settlement curves of the surrounding soil. Compared with OSC, GESC reduces the soil deformation nearly by 30% (S/d_c = 0.198 and 0.279 for GESC and OSC encased under 50 kPa, respectively). It clearly shows that the stone column with geosynthetic encasement has undergone minor vertical settlement (S_c) and radial bulging deformation (δ_c) at the top of the column.

Figure 11.76b presents the normalized radial bulging (δ_c/d_c) development process observed along the normalized length of the column (z_c/d_c) with and without geosynthetic encasement under 10, 20, 30, 40 and 50 kPa. It can be seen that the deformation shape of the GESC is similar to that of the OSC, but the maximum bulging, the depth of maximum bulging and bulging depth varies. Bulging depth (z_b), the

Table 11.11 Detailed test parameters

Test description	Diameters of columns (mm)	Lengths of columns (mm)	Applied load (kPa)	Encasement length (mm)
OSC	25	200	0	0
OSC	25	200	10	0
OSC	25	200	20	0
OSC	25	200	30	0
OSC	25	200	40	0
OSC	25	200	50	0
GESC	25	200	0	200
GESC	25	200	10	200
GESC	25	200	20	200
GESC	25	200	30	200
GESC	25	200	40	200
GESC	25	200	50	200

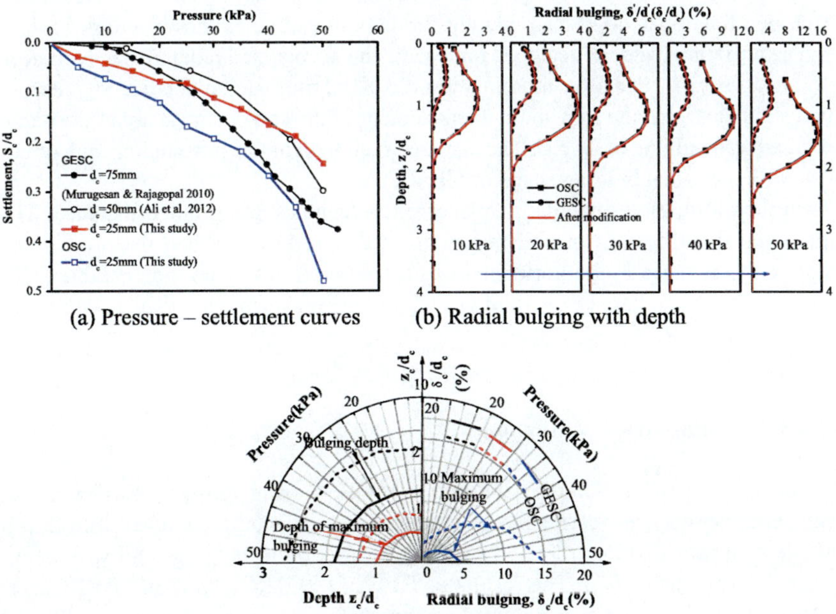

(a) Pressure – settlement curves (b) Radial bulging with depth

(c) Bulging development

Fig. 11.76 Deformation development of column

maximum radial bulging and the depth of the maximum bulging were three important measurements characterizing radial bulging. Figure 11.76c plots development of these characteristic measurements, specifically. The maximum radial bulging of GESC of 0.83%, 1.41%, 2.33%, 3.49% and 4.86% occurs at depth z_c/d_c of 0.45–0.75. By contrast, the maximum bulging (δ_c/d_c) of OSC of 2.63%, 3.88%, 6.83%, 11.19%, and 15.54% occurs at depth z_c/d_c of 0.7–1.2. The maximum radial bulging of GESC reduces by 70% compared with that of OSC. The bulging depth of GESC (z_c/d_c) ranges from 1.05–1.4 to 1.72–2.16 for GESC and OSC, respectively. This is consistent with the findings of Lee et al. [128], who performed model tests and observed that bulging occurs at a depth 1–2.5 times the diameter of OSC.

11.6.3.2 Displacement of Surrounding Soil

The displacement of soil was measured before and after each increment of pressure by DIC technique and then modified by modification coefficient, as describe before. Under various pressures, the pattern of vector and contours (contains radial direction and vertical direction) are similar to each other. The vector and contours before and after modification at 40 kPa are plotted for demonstration, as shown in Fig. 11.77.

The horizontal and vertical axis represents the normalized radial distance (before modification r/d'_c and after modification r/d_c) and normalized depth z_c/d_c, respectively. It's clear that the maximum displacement of the soil occurs at about the depth of maximum bulging of GESC. The displacement field of the surrounding soil agrees well with the radial bulging shape of GESC.

Furthermore, the horizontal displacement attenuates along radial distance. The horizontal displacement of soil ranges from 0 to 1.57 d_c. When the normalized depth z_c/d_c reaches 3, the horizontal displacement is nearly not appreciable. This demonstrates the influence depth of GESC bulging on surrounding soil is 0–3 times diameter of GESC.

11.6.3.3 Comparison Validation

In order to compare the bulging deformation of GESC using transparent soil with that using ordinary clay, the test results are further analyzed. A bulging ratio is introduced and defined as δ_c/S_c. Figure 11.78 shows the variation of bulging ratio with depth (after deformation z'_c/d_c). The results of ET1X (J = 0.34 kN/m) and ET2X (J = 0.48 kN/m) in [134] are also analyzed and presented. The applied pressure is 30–40 kPa when the deformation of the stone column was measured in Hong et al. [134]. Hence bulging ratios at pressures of 30 and 40 kPa were chosen for comparison. The bulging shapes are closely matched in this figure. The maximum bulging ratios in this study are about 24% and are close to the 23% (ET2X) and the 17.5% (ET1X). The depth of maximum bulging is 0.4 and 0.35 times the diameter of the GESC in Hong et al. [134], consistent with the results in the current study. The comparison demonstrates the validity of bulging deformation shape for GESC in transparent soil.

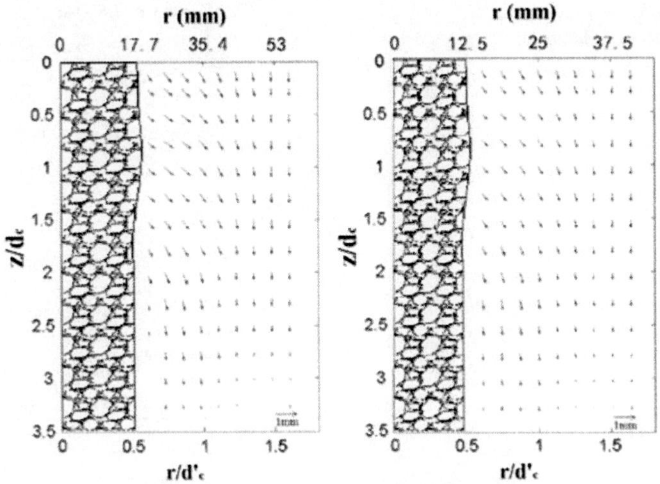

(a) Vectors before and after modification

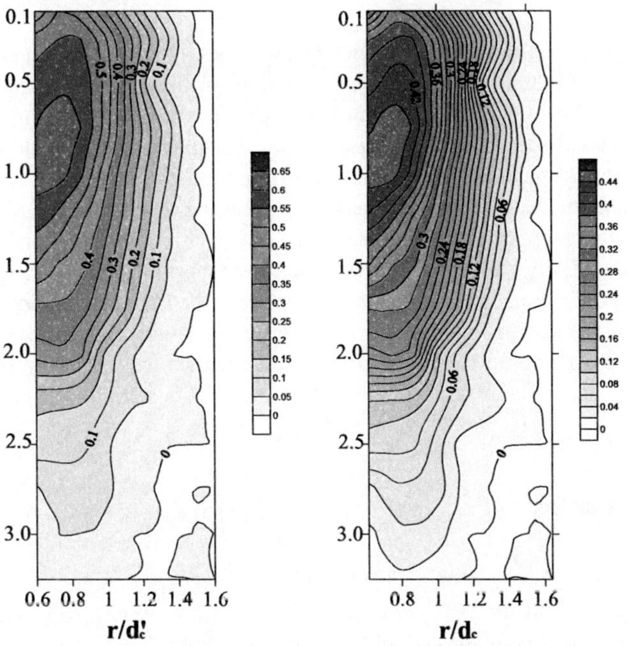

(b) Radial contours before and after modification

Fig. 11.77 Normalized displacement field of surrounding soil under pressure of 40 kPa

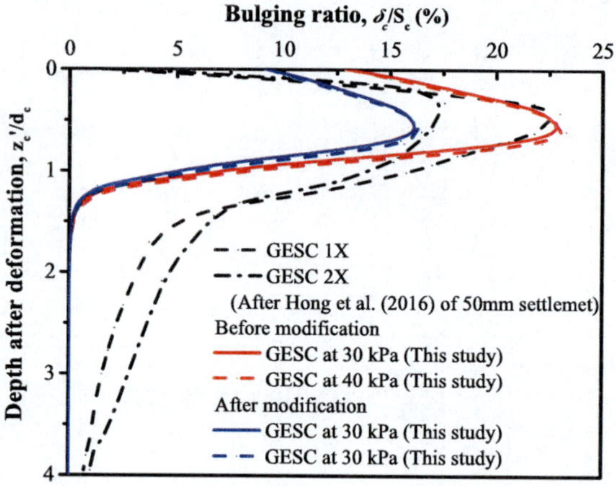

Fig. 11.78 Variation of bulging ratio with depth

It is worth noting that the bulging depth is quite different from that of Hong et al. [134], which may be due to the higher friction angle of transparent soil compared to clay.

The measured deformation behavior of the single GESC at any depth was also predicted using a simple analytical model based on the unit-cell theory proposed by Kong et al. [148]. In their theoretical model, it was assumed that the geotextile material behaves as a nonlinear elastic material. The approximate calculation model were shown in Appendix. The detailed description could be referred to Kong et al. [148]. Settlement (S_c/d_c) and maximum radial bulging (δ'_c/d_c) are meaningful focus in engineering cases. In order to quantify the deformation, the comparison analysis were conduct. Figure 11.79 plots the deformation against the total applied pressure. A reasonably good agreement is obtained between the theoretic calculation and model test results for settlement under relative low pressure. The results of maximum radial bulging (δ'_c/d_c) are slightly smaller than that obtained by theoretical predicted. That might be caused by the assumption that stone material is a nonlinear material.

11.6.4 Conclusions

The bulging development of a GESC was measured by conducting model tests using transparent soil and DIC technique. With an innovative experimental apparatus developed, major conclusions could be drawn as follows:

- The continuous bulging development of GESC (including bulging depth, the maximum bulging and the depth of maximum bulging) and the displacement field of surrounding soil can be obtained by using transparent soil and DIC techniques.

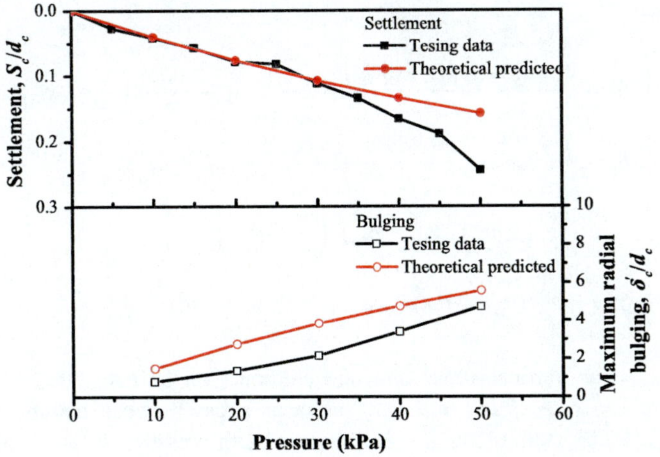

Fig. 11.79 Comparison of deformation with pressure

- Compared with OSC, the settlement and bulging of GESC reduce by 50% to 70%. The bulging depth and the depth of maximum bulging of GESC ranges from 1.05–1.40 to 0.45–0.75 times the diameter of GESC.
- The displacement field of the surrounding soil agrees well with the radial bulging shape of GESC. The radial displacement of soil ranges from 0 to 1.57 times the diameter of GESC. The influence depth of GESC bulging on surrounding soil is 0–3 the times diameter of GESC.

Appendix: Theoretical Predicted

In the theoretical model [148], it was assumed that the geotextile material behaves as an elastic material with a constant stiffness modulus and the stone column rests on a hard stratum. The confining stress acting on the column is derived via two approaches: the lateral confining stress provided by the surrounding soil and the additional confining stress provided by geotextile. The deformation (δ_c and S_c) of the stone column can be derived as:

$$\delta_c = \frac{-B \pm \sqrt{B^2 - 4AC}}{2A} \tag{A.10}$$

$$S_c = \int \left[\frac{2\delta_c}{r_c + 2\delta_c} \right] dz_c \tag{A.11}$$

In which,

$$\begin{cases} A = \dfrac{2\sin\varphi_c + R_f(1 - \sin\varphi_c)}{r_c}\left(\dfrac{J}{r_c^2} + k_h\right) + \dfrac{\sin\varphi_c}{E_c}\left(\dfrac{J}{r_c^2} + k_h\right)^2 \\ B = \dfrac{2}{r_c}(P_{s0}\sin\varphi_c) - \dfrac{R_f(1 - \sin\varphi_c)}{r_c}(\sigma_{zc} - P_{s0}) \\ \quad + \dfrac{1}{E_c}(+\sin\varphi_c(2P_{s0} - \sigma_{zc}))\left(\dfrac{J}{r_c^2} + k_h\right) \\ C = -\dfrac{1}{E_c}(\sigma_{zc} - P_{s0})P_{s0}\sin\varphi_c \end{cases}$$

where σ_{zc} is the vertical stress of stone column, can be calculated by loading transfer $\sigma_{zc} = q_c - \int 2\tau_{cs}/r_c dz$; τ_{cs} is shear stress between column and soil, $\tau_{cs} = (p_{s0} + k_h\delta_c)\tan\varphi_{cs}$; $(p_{s0} + k_h\delta_c)$ is the earth pressure; k_h is the horizontal modulus of soil reaction, determined by $(p_{su}-p_{s0})/\Delta r_{su}$; p_{su} and p_{s0} are passive earth pressure and earth pressure at rest respectively; Δr_{su} is the relative displacement when passive earth pressure achieves the maximum and can be taken as 0.05 m; Poisson's ratio v_c can be taken as 0.25. R_f is the parameter of stone materials, taken as 0.7. The radial bulging (δ_c) and settlement (S_c) can be obtained from Eqs. (A.10) and (A.11), respectively. Hence, the radial bulging along the depth can be calculated through this method for comparison with the data obtained in this study.

References

1. Kong G, Yang Q, Liu H, Robert L (2013) Numerical study of a new belled wedge pile type under different loading modes. European J Environ Civil Eng 17(Sup 1):65–82
2. Spronken JT (1998) Bearing capacity of tapered piles. PhD thesis. University of Calgary, Alberta, Canada
3. EI-Naggar MHE (2004) The 2002 Canadian geotechnical colloquium: the role of soil-pile interaction in foundation engineering. Canad Geotech J 41(3):485–509
4. Manandhar S, Yasufuku N, Omine K, Kobayashi T (2010) Response of tapered piles in cohesionless soil based on model tests. J Nepal Geological Soc 40(10):85–92
5. Ismael NF (2010) Behavior of step tapered bored piles in sand under static lateral loading. J Geotech Geoenviron Eng 136(5):669–676
6. Khan MK, Naggar MHE, Elkasabgy M (2008) Compression testing and analysis of drilled concrete tapered piles in cohesive frictional soil. Canad Geotech J 45(3):377–392
7. Metellia G, Beschia C, Rivab P (2011) Cyclic behaviour of a column to foundation joint for concrete precast structures. Eur J Environ Civ Eng 15(9):1297–1318
8. Lee J, Paik K, Kim D, Hwang S (2009) Estimation of axial load capacity for bored tapered piles using CPT results in sand. J Geotech Geoenviron Eng 135(9):1285–1294
9. Ilamparuthi K, Dickin EA (2001) Predictions of the uplift response of model belled piles in geogrid-cell-reinforced sand. Geotextiles Geomembranes 19(2):89–109
10. Ilamparuthi K, Dickin EA (2001) The influence of soil reinforcement on the uplift behaviour of belled piles embedded in sand. Geotextiles Geomembranes 19(1):1–22
11. Iskander MG, Liu JY, Sadek S (2002) Transparent amorphous silica to model clay. J Geotech Geoenviron Eng 128(3):262–273

12. Hird CC, Ni Q, Guymer I (2011) Physical modelling of deformations around piling augers in clay. Geotechnique 61(11):993–999. https://doi.org/10.1680/geot.9.T.028
13. Hover EQ, Ni Q, Guymer I (2013) Investigation of centreline strain path during tube penetration using transparent soil and particle image velocimetry. Geotechnique Lett 3(2):37–41
14. Lehane BM, Gill DR (2004) Displacement fields induced by penetrometer installation in an artificial soil. Int J Phy Model Geo 4(1):25–36. https://doi.org/10.1680/ijpmg.2004.040103
15. Ni QCCI, Hird CC, Guymer I (2010) Physical modelling of pile penetration in clay using transparent soil and particle image velocimetry. Géotechnique 60(2):121. https://doi.org/10.1680/geot.8.P.052
16. Iskander MG, Sadek S, Liu J (2003) Optical measurement of deformation using transparent silica to model sand. Int J Phys Model Geotech 2(4):13–26
17. Iskander M, Liu J (2010) Spatial deformation measurement using transparent soil. Geotechnical Testing J 33(4):1–8, 314–321. https://doi.org/10.1520/GTJ102745
18. Liu J, Iskander MG (2010) Modelling capacity of transparent soil. Can Geotech J 47(4):451–460. https://doi.org/10.1139/T09-116
19. Ahmed M, Iskander MG (2011) Analysis of tunneling-induced ground movements using transparent soil models. J Geotech Geoenviron Eng 137(5):525–535
20. Lo HC, Tabe K, Iskander MG, Yoon SH (2010) A transparent water-based polymer for simulating multiphase flow. Geotech Test J 33(1):1–13
21. Serrano FR, Iskander MG, Tabe K (2011) 3D contaminant flow imaging in transparent granular porous media. Geotechnique Letters 1:71–78
22. Ezzein FM, Bathurst RJ (2011) A transparent sand for geotechnical laboratory modeling. Geotech Test J 34(6):1–12, 590–601. https://doi.org/10.1520/GTJ103808
23. Zhao H, Ge L, Luna R (2010) Low viscosity pore fluid to manufacture transparent soil. Geotech Test J 33(6):463–468. https://doi.org/10.1520/GTJ102607
24. White DJ, Bolton MD (2004) Displacement and strain paths during plane-strain model pile installation in sand. Géotechnique 54(6):375–397. https://doi.org/10.1680/geot.2004.54.6.375
25. Au SK, Yeung AT, Soga K (2006) Pressure-controlled cavity expansion in clay. Canad Geotech J 43(7):714–725
26. Chen JY, Gilbert RB, Puskar FJ, Verret S (2013) Case study of offshore pile system failure in hurricane ike. J Geotech Geoenviron Eng, ASCE 139(10):1699–1708
27. Choo Y, Kim D, Park J, Kwak K, Kim J, Kim D (2014) Lateral response of large-diameter monopiles for offshore wind turbines from centrifuge model tests. Geotech Testing J, ASTM 37(1):107–120
28. Fan HJ, Liang RY (2013) Performance-based reliability analysis of laterally loaded drilled shafts. J Geotech Geoenviron Eng, ASCE 139(12):2020–2027
29. Gurpersaud N, Vanapalli S, Sivathayalan S (2013) Semiempirical method for estimation of pulling capacity of grouted soil nails in saturated and unsaturated soil environments. J Geotech Geoenviron Eng, ASCE 139(11):1934–1943
30. Pisanò F, Di Prisco CG, Lancellotta R (2014) Soil–foundation modelling in laterally loaded historical towers. Geotechnique 64(1):1-15
31. Zhou H, White DJ, Randolph MF (2008) Physical and numerical simulation of shallow penetration of a cylindrical object into soft clay. Paper presented at GeoCongress 2008: Characterization, Monitoring, and Modeling of GeoSystems, 108–117. https://doi.org/10.1061/40972(311)14
32. Guzman I, Iskander M, Suescun E, Omidvar M (2014) A transparent aqueous -saturated sand-surrogate for use in physical modeling. Acta Geotech 9(2):187–206
33. Peters SB, Siemens G, Take WA (2011) Characterization of transparent soil for unsaturated applications. Geotech Testing J, ASTM 34(5):1–12
34. Siemens GA, Peters SB, Take WA (2013) Comparison of confined and unconfined in transparent porous media. Water Resour Res 49(2):851–863

35. Stanier SA, Black JA, Hird CC (2012) Enhancing accuracy and precision of transparent synthetic soil modelling. Int J Phys Modell Geotech 12(4):162–175
36. Cao ZH, Kong GQ, Liu HL (2014) Model test on deformation characteristic of pile driving in sand using PIV technique. Eng Mech 31(8):168–174. https://doi.org/10.6052/j.issn.1000-4750.2013.03.0217
37. Patra NR, Pise PJ (2006) Model pile groups under oblique pullout loads—an investigation. Geotech Geol Eng 24(2):265–282
38. Krishna B, Patra NR (2006) Effect of compressive load on oblique pull-out capacity of model piles in sand. Geotech Geol Eng 24(3):593–614
39. Randolph MF, Leong EC, Houlsby GT (1991) One-dimensional analysis of soil plugs in pipe pile. Géotechnique 41(4):587–598. https://doi.org/10.1680/geot.1991.41.4.587
40. Paik K-H, Lee S-R (1993) Behavior of soil plugs in open-ended model piles driven into sands. Mar Georesour Geotechnol 11(4):353–373. https://doi.org/10.1080/10641199309379929
41. Jeong S, Ko J, Won J et al (2015) Bearing capacity analysis of open-ended piles considering the degree of soil plugging. Soils Found 55(5):1001–1014. https://doi.org/10.1016/j.sandf.2015.06.007
42. Liu J, Duan N, Cui L et al (2019) DEM investigation of installation responses of jacked open-ended piles. Acta Geotech 14(6):1805–1819. https://doi.org/10.1007/s11440-019-00817-7. doi:10.1007/s11440-019-00817-7
43. Liu J, Guo N, Han (2019) Load transfer of offshore open-ended pipe piles considering the effect of soil plugging. J Marine Science Eng 7(9):313. https://doi.org/10.3390/jmse7090313
44. Kikuchi Y, Sato T, Morikawa Y, et al (2007) Visualization of plugging phenomena in vertically loaded open-ended piles. Geo-Congress, 118–124. https://doi.org/10.1061/40972(311)15
45. Liu JW, Zhang ZM, Yu F, et al (2012) Case history of installing instrumented jacked open-ended piles. J Geotechnical Geoenvironmental Eng, 138(7):810–820. https://doi.org/10.1061/(ASCE)GT.1943-5606.0000638
46. Ko J, Jeong S (2015) Plugging effect of open-ended piles in sandy soil. Can Geotech J 52(5):664–667. https://doi.org/10.1139/cgj-2014-0041
47. Murthy DS, Robinson RG, Rajagopal K (2018) Formation of soil plug in open-ended pipe piles in sandy soils. Int J Geotech Eng, 1–11. https://doi.org/10.1080/19386362.2018.1465742
48. Kou HL, Li W, Chu J, et al (2019) Model tests on open-ended concrete pipe piles jacked in sand. Marine Georesources and Geotechnology, 1–8. https://doi.org/10.1080/1064119X.2019.1642972
49. Kumara JJ, Kikuchi Y, Kurashina T (2016) The effectiveness of thickened wall at the pile base of open-ended piles in increasing soil plugging. Japanese Geotech Soc Special Publ 4(6):138–143. https://doi.org/10.3208/jgssp.v04.j29
50. Kumara JJ, Kikuchi Y, Kurashina T et al (2017) Influence of relative density on the degree of soil plugging of pipe piles driven in sandy soils. Japanese Geotech Soc Special Publ 5(2):28–33. https://doi.org/10.3208/jgssp.v05.007
51. Kumara, JJ, Kikuchi Y, Kurashina T, et al (2016) Effects of inner sleeves on the inner frictional resistance of open-ended piles driven into sand. Frontiers Structural Civil Eng 10(4). https://doi.org/10.1007/s11709-016-0335-0
52. Kumara JJ, Kurashina T, Yajima T et al (2016) Understanding inner friction mechanism of open-ended piles - an experimental study. Japanese Geotech Soc Special Publ 2(37):1333–1338. https://doi.org/10.3208/jgssp.LKA-01
53. Han F, Ganju E, Salgado R, et al (2019) Comparison of the load response of closed-ended and open-ended pipe piles driven in gravelly sand. Acta Geotechnica 14(6):1785–1803. https://doi.org/10.1007/s11440-019-00863-1
54. Yu F, Yang J (2012) Base capacity of open-ended steel pipe piles in sand. J Geotech & Geoenviron Eng 138(9):1116–1128. https://doi.org/10.1061/(ASCE)GT.1943-5606.0000667
55. Seo H, Kim M (2018) Soil plug behaviour of open-ended pipe piles during installation. J Deep Found Institute, 1–9. https://doi.org/10.1080/19375247.2018.1448552

56. Liu, C, Tang, X, Wei, H, et al (2020) Model tests of jacked-pile penetration into sand using transparent soil and incremental particle image velocimetry. KSCE J Civ Eng 24:1128–1145. https://doi.org/10.1007/s12205-020-1643-4
57. Zhou J, Chen XL, Wang GY, Zhou KM (2012) Experimental and numerical analysis of open-ended pipe piles during jacking into sand. J Tongji Univ (Nat Sci) 40(2):173–178. https://doi.org/10.3969/j.issn.0253-374x.2012.02.003
58. Zhao H, Ge L (2014) Investigation on the shear moduli and damping ratios of silica gel. Granular Matter 16(4):449–456. https://doi.org/10.1007/s10035-014-0495-3
59. Mo PQ, Marshall AM, Yu HS (2015) Centrifuge modelling of cone penetration tests in layered soils. Géotechnique 65(6):468–481. https://doi.org/10.1680/geot.14.P.176
60. Wang T, Zhang Y, Bao X et al (2020) Mechanisms of soil plug formation of open-ended jacked pipe pile in clay. Comput Geotech 118:103334. https://doi.org/10.1016/j.compgeo.2019.103334
61. Doherty P, Gavin K (2011) Shaft capacity of open-ended piles in clay. J Geotech Geoenviron Eng 137(11):1090–1102. https://doi.org/10.1061/(ASCE)GT.1943-5606.0000528
62. Paikowsky SG (1990) The mechanism of pile plugging in sand. In Proc 22nd Offshore Technology Conf 1, 593–603. https://doi.org/10.4043/6490-MS
63. Ko J, Jeong S, Lee JK (2016) Large deformation FE analysis of driven steel pipe piles with soil plugging. Computers & Geotech 71(1):82–97. https://doi.org/10.1016/j.compgeo.2015.08.005
64. Chen F, Lin Y, Dong Y et al (2020) Numerical investigations of soil plugging effect inside large-diameter, open-ended wind turbine monopiles driven by vibratory hammers. Mar Geotechnol 38(1):83–96. https://doi.org/10.1080/1064119X.2018.1553081
65. Paik K, Salgado R (2003) Determination of bearing capacity of open-ended piles in sand. J Geotech Geoenviron Eng 129(1):46–57. https://doi.org/10.1061/(ASCE)1090-0241
66. Terzaghi K (1943) Theoretical soil mechanics. John Wiley & Sons Inc., New York
67. Meyerhof GG (1963) Some recent research on the bearing capacity of foundations. Can Geotech J 1(a):16–26.
68. Vesic AS (1972) Expansion of cavities in infinite soil mass. J Soil Mech Founds Div 98(SM3):265–290
69. Poulos HG, Davis EH (1968) The settlement behaviour of single axially loaded incompressible piles and piers. Geotechnique 18:351–371
70. Jardine RJ, Potts DM, Fourie AB, Burland JB (1986) Studies of the influence of non-linear stress-strain characteristics in soil-structure interaction. Geotechnique 36(3):377–396
71. Poulos HG (1989) Pile behaviour-theory and application. Geotechnique 39(3):365–415
72. MacKelvey D (2002) Performance of virbo stone column reinforced foundations in deep soft clay. Ph.D. thesis, Queen's Univ. Belfast, UK
73. Han F, Salgado R, Prezzi M, Lim J (2017) Shaft and base resistance of non-displacement piles in sand. Comput Geotech 83:184–197
74. Poulos HG, Aust MIE (1968) The influence of a rigid pile cap on the settlement behaviour of an axially-loaded pile. Transactions of the Institution of Engineers, Australia. Civil Eng 10(2):206–208
75. Chen Q, Zhang Y, Hong M (1992) Experimental research on the interaction between cap and short pile. Building Science (in Chinese) 2:30–36
76. Lou X-M, Fang W-X, Fei P-Y, Ye W-Y (2005) Comparative testing on characteristic of load transfer between single pile and single pile with cap. Rock Soil Mech 26(9):1399–1402. https://doi.org/10.3969/j.issn.1000-7598.2005.09.010
77. Lou X-M, Yao H-Y, Ye W-Y (2007) Load transfer tests of single capped friction piles in situ. J Tongji University (Natural Science) 35(1):15–20
78. Wang H, Zhou J, Deng Z (2006) Model tests on pile-soil-cap interaction. Chinese J Geotech Eng 28(10):1253–1258
79. Li X, Jiang G, Zai J, Wang X (2007) Model tests of pile-soil-cap nonlinear working mechanism. Industrial Const 37(11):57–60

80. Iskander M, Lai J, Oswald C, Mannheimer R (1994) Development of a transparent material to model the geotechnical properties of soil. Geotech Test J 17(4):425–433
81. Mannheimer RJ, Oswald CJ (1993) Development of transparent porous media with permeabilities of soils and reservoir materials. Ground Water 31(5):781–788
82. Iskander M (2010) Modeling with transparent soils, visualizing soil structure interaction and multi phase flow, non-intrusively. Springer, New York
83. Cao Z, Liu J, Liu H (2011) Transparent fused silica to model natural sand. Proceedings of Pan American CGS Conference (CD-ROM). Toronto, Canada, October 2–6, 2011, Canadian Geotechnical Society, Richmond, British Columbia, Canada
84. Liu JY, Iskander M (2004) Adaptive cross correlation for imaging displacements in soils. J Comput Civ Eng 18(1):46–57
85. Iai S (1989) Similitude for shaking table tests on soil-structure-fluid model in 1g gravitational field. Soils Found 29(1):105–118
86. Sveen JK (2004) An Introduction to MatPIV, v.1.6.1. Eprint No.2, ISSN 0809-4403, 2004, Dept. of Math., University of Oslo.
87. Bhasi A, Rajagopal K (2015) Geosynthetic-reinforced piled embankments: comparison of numerical and analytical methods. Int J Geomech 15(5):04014074-1-12
88. Baligh MM (1985) Strain path method. J Geotechnical Eng 111(9):1108–1136. https://doi.org/10.1061/(ASCE)0733-9410(1985)111:9(1108)
89. Sagaseta C, Whittle AJ, Santagata M (1997) Deformation analysis of shallow penetration in clay. Int J Numer Anal Meth Geomech 21(10):687–719. https://doi.org/10.1680/geot.1979.29.4.361
90. Chow YK, The CI (1990) A theoretical study of pile heave. Geotechnique 40(1):1–14. https://doi.org/10.1680/geot.1990.40.1.1
91. Zhang QQ, Liu SW, Zhang SM, Zhang J, Wang K (2016) Simplified non-linear approaches for response of a single pile and pile groups considering progressive deformation of pile–soil system. Soils Found 56(3):473–484. https://doi.org/10.1016/j.sandf.2016.04.013
92. Bozozuk M, Fellenius BH, Samson L (1978) Soil disturbance from pile driving in sensitive clay. Can Geotech J 15(3):346–361. https://doi.org/10.1139/t78-032
93. Massarsch KR (1976) Soil Movements Caused by Pile Driving in Clay. Dept. *Soil and Rock Mechanics*, Royal Institute of Technology (KTH), Stockholm, Sweden, Thesis in partial fulfillment of the requirements for the Degree Doctor of Engineering, Job-Rapport No 6, 261
94. Poulos HG (1994) Effect of pile driving on adjacent piles in clay. Can Geotech J 31(6):856–867. https://doi.org/10.1139/t94-102
95. Hu P, Stanier SA, Cassidy MJ, Wang D (2013) Predicting peak resistance of spudcan penetrating sand overlying clay. J Geotech Geoenviron Eng 140(2):04013009. https://doi.org/10.1061/(ASCE)GT.1943-5606.0001016
96. Jiang M, Dai Y, Cui L, Shen Z, Wang X (2014) Investigating mechanism of inclined CPT in granular ground using DEM. Granular Matter 16(5):785–796. https://doi.org/10.1007/s10035-014-0508-2
97. Gill DR, Lehane BM (2001) An optical technique for investigating soil displacement patterns. Geotech Test J 24(3):324–329. https://doi.org/10.1520/GTJ11351J
98. Lundberg AB, Dijkstra J, Tol AFV (2013) Displacement pile installation effects in sand. Paper presented at the International Conference on Installation Effects in Geotechnical Engineering, 79–85. https://doi.org/10.4233/uuid:01d8943f-e3eb-4051-8b44-32097e18c4da
99. Randolph MF, Wroth CP (1979) An analytical solution for the consolidation around a driven pile. Int J Numer Anal Meth Geomech 3(3):217–229. https://doi.org/10.1002/nag.1610030302
100. Cooke RW, Price G, Tarr K (1979) Jacked piles in London clay: a study of load transfer and settlement under working conditions. Geotechnique 29(2):113–147. https://doi.org/10.1680/geot.1979.29.2.113
101. Hwang JH, Liang N, Chen CH (2001) Ground response during pile driving. J Geotech Geoenviron Eng 127(11):939–949. https://doi.org/10.1061/(ASCE)1090-0241(2001)127:11(939)

102. Randolph MF, Carter JP, Wroth CP (1979) Driven piles in clay—the effects of installation and subsequent consolidation. Geotechnique 29(4):361–393. https://doi.org/10.1680/geot.1979.29.4.361
103. Massarsch KR, Wersäll C (2013) Cumulative lateral soil displacement due to pile driving in soft clay. Paper presented at Sound Geotechnical Research to Practice: Honoring Robert D. Holtz II:462–479. https://doi.org/10.1061/9780784412770.031
104. Arshad MI, Tehrani FS, Prezzi M (2014) Experimental study of cone penetration in silica sand using digital image correlation. Géotechnique 64(7):551–569. https://doi.org/10.1680/geot.13.P.179
105. Mo PQ, Marshall AM, Yu HS (2017) Layered effects on soil displacement around a penetrometer. Soils Found 57(4):669–678. https://doi.org/10.1016/j.sandf.2017.04.007
106. Paniagua P, Andó E, Silva M, Emdal A, Nordal S, Viggiani G (2013) Soil deformation around a penetrating cone in silt. Géotechnique Letters 3(4):185–191. https://doi.org/10.1680/geolett.13.00067
107. Huang B, Zhang Y, Fu X, Zhang B (2018) Study on Visualization and Failure Mode of Model Test of Rock-Socketed Pile in Soft Rock. Geotechnical Testing J 42(6). https://doi.org/10.1520/GTJ20170348. ISSN 0149-6115
108. Chen Z, Omidvar M, Iskander M (2014) Modelling of projectile penetration using transparent soils. Int J Phy Model Geotech 14(3):68–79. https://doi.org/10.1680/ijpmg.14.00003
109. Cao ZH, Liu HL, Kong GQ, Zhou H (2015) Physical modelling of pipe piles under oblique pullout loads using transparent soil and particle image velocimetry. J Central South University 22(11):4329–4336. https://doi.org/10.1007/s11771-015-2981-0
110. Kong GQ, Cao ZH, Zhou H, Sun XJ (2015) Analysis of piles under oblique pullout load using transparent-soil models. Geotech Test J 38(5):725–738. https://doi.org/10.1520/GTJ20140109
111. Qi CG, Zheng JH, Zuo DJ, Chen G (2017) Measurement on soil deformation caused by expanded-base pile in transparent soil using particle image velocimetry (PIV). J Mt Sci 14(8):1655–1665. https://doi.org/10.1007/s11629-016-4025-0
112. Pan B, Dafang W, Yong X (2012) Incremental calculation for large deformation measurement using reliability-guided digital image correlation. Opt Lasers Eng 50(4):586–592. https://doi.org/10.1016/j.optlaseng.2011.05.005
113. Liu JY, Liu ML, Gao HM (2010) Influence of pile geometry on internal sand displacement around a laterally loaded pile using transparent soil. Geoshanghai, 104–110
114. Xiao Y, Yin F, Liu HL, Chu J, Zhang W (2016) Model tests on soil movement during the installation of piles in transparent granular soil. Int J Geomech 17(4):06016027. https://doi.org/10.1061/(ASCE)GM.1943-5622.0000788
115. Sang Y, Wang Z, Yu S, Zhao H (2019) The loading test on the singe pile with pile cap in transparent soil model. Geotech Test J 42(2):385–406. https://doi.org/10.1520/GTJ20170153
116. Xu JP, Zhou J, Xu CY, Zou CY (2000) Model test research on pile driving effect of squeezing against soil. Rock and Soil mechanics 21(3):235–238. https://doi.org/10.16285/j.rsm.2000.03.011
117. Toiya M, Hettinga J, Losert W (2007) 3D imaging of particle motion during penetrometer testing. Granular Matter 9(5):323–329. https://doi.org/10.1007/s10035-007-0044-4
118. Gao Y, Cheng T, Su Y, Xu X, Zhang Y, Zhang Q (2015) High-efficiency and high-accuracy digital image correlation for three-dimensional measurement. Opt Lasers Eng 65:73–80. https://doi.org/10.1016/j.optlaseng.2014.05.013
119. Stanier SA, Blaber J, Take WA, White DJ (2015) Improved image-based deformation measurement for geotechnical applications. Can Geotech J 53(5):727–739. https://doi.org/10.1139/cgj-2015-0253
120. Faizi K, Armaghani DJ, Sohaei H, Rashid ASA, Nazir R (2015) Deformation model of sand around short piles under pullout test. Measurement 63:110–119. https://doi.org/10.1016/j.measurement.2014.11.028
121. Hu P, Stanier SA, Wang D, Cassidy MJ (2015) Effect of footing shape on penetration in sand overlying clay. Int J Phy Model Geotech 16(3):119–133. https://doi.org/10.1680/jphmg.15.00013

122. Zeng Z (2013) Deformation and bearing behavior of piles with different tip angle in sandy soils. MSc Thesis, Shanghai Jiao Tong University, Shanghai, China (in Chinese)
123. Basu P, Loukidis D, Prezzi M (2014) The mechanics of friction fatigue in jacked piles installed in sand. From Soil Behavior Fundamentals to Innovations in Geotechnical Engineering: Honoring Roy E Olson, 546–557. https://doi.org/10.1061/9780784413265.044
124. Yang ZX, Jardine RJ, Zhu BT, Rimoy S (2013) Stresses developed around displacement piles penetration in sand. J Geotech Geoenviron Eng 140(3):04013027. https://doi.org/10.1061/(ASCE)GT.1943-5606.0001022
125. Jardine RJ, Zhu BT, Foray P, Yang ZX (2013) Interpretation of stress measurements made around closed-ended displacement piles in sand. Géotechnique 63(8):613–627. https://doi.org/10.1680/geot.9.P.138
126. Yoo C (2010) Performance of geosynthetic-encased stone columns in embankment construction: numerical investigation. J Geotech Geoenviron Eng 136(8):1148–1160
127. Hosseinpour I, Almeida MSS, Riccio M (2015) Full-scale load test and finite-element analysis of soft ground improved by geosynthetic-encased granular columns. Geosynth Int 22(6):428–438
128. Lee D, Yoo C, Park S (2007) Model tests for analysis of load carrying capacity of geogrid encased stone column. 17th Int. International Society of Offshore and Polar Engineers (ISOPE), Cupertino, CA, Offshore and Polar Engineering Conf., pp 1632–1635
129. Black JA, Sivakumar V, Madhav MR, Hamill GA (2007) Reinforced stone columns in weak deposits: laboratory model study. J Geotech Geoenviron Eng 133(9):1154–1161
130. Afshar JN, Ghazavi M (2014) Experimental studies on bearing capacity of geosynthetic reinforced stone columns. Arab J Sci Eng 39(3):1559–1571
131. Ali K, Shahu JT, Sharma KG (2012) Model tests on geosynthetic-reinforced stone columns: a comparative study. Geosynth Int 19(4):292–305
132. Demir A, Sarici T (2017) Bearing capacity of footing supported by geogrid encased stone columns on soft soil. Geomech. Eng. 12(3):417–439
133. Gniel J, Bouazza A (2009) Improvement of soft soils using geogrid encased stone columns. Geotext Geomembr 27(3):167–175
134. Hong YS, Wu CS, Yu YS (2016) Model tests on geotextile-encased granular columns under 1-g and undrained conditions. Geotext Geomembr 44(1):13–17
135. Malarvizhi SN, Ilamparuthi K (2007) Comparative study on the behavior of encased stone column and conventional stone column. Soils Found 47(5):873–885
136. Murugesan S, Rajagopal K (2010) Studies on the behavior of single and group of geosynthetic-encased stone columns. J Geotech Geoenviron Eng 136(1):129–139
137. Shivashankar R, Badu MRD, Nayak S, Manjunath R (2010) Stone columns with vertical circumferential nails-laboratory model study. Geotech Geol Eng 28(4):695–706
138. Ezzein FM, Bathurst RJ (2014) A new approach to evaluate soil–geosynthetic interaction using a novel pullout test apparatus and transparent granular soil. Geotext Geomembr 42(3):246–255
139. MacKelvey D, Sivakumar V, Bell A, Graham J (2004) Modelling vibrated stone columns in soft clay. Proc ICE Geotech Eng 157(3):137–149
140. Liu J, Gao Y, Sui W (2013) Visualization of grout permeation inside transparent soil. In IACGE 2013: Challenges and Recent Advances in Geotechnical and Seismic Research and Practices, pp. 188–194
141. Ambily AP, Gandhi SR (2007) Behavior of stone columns based on experimental and FEM analysis. J Geotech Geoenviron Eng 133(4):405–415
142. Nayak S, Shivashankar R, Ramanaidu M (2011) Performance of stone columns with circumferential nails. Proc ICE Ground Improvement 164(12):97–106
143. ASTM D4595 (1986) Standard test method for tensile properties of geotextiles by wide-width strip method. Annual Book of ASTM Standards, ASTM International, West Conshohocken, PA
144. Madabhushi SPG, Houghton NE, Haigh SK (2006) A new automatic sand pourer for model preparation at University of Cambridge. In: Proceedings of the 6th International Conference on Physical Modelling in Geotechnics, pp: 217–222

145. Alamgir M, Miura N, Poorooshasb HB, Madhav MR (1996) Deformation analysis of soft ground reinforced by columnar inclusions. Comput Geotech 18(4):267–290
146. Castro J, Sagaseta C (2011) Deformation and consolidation around encased stone columns. Geotext Geomembr 29(3):268–276
147. Zhang L, Zhao MH (2015) Deformation analysis of geotextile-encased stone columns. Int J Geomech 15(3):04014053
148. Kong GQ, Zhou Y, Liu HL (2018) 2018, Nonlinear model analysis of radial bulging deformation of geosynthetic-encased stone columns". Int J Geomech 18(10):06018022
149. Seo H, Kim M (2018) Soil plug behaviour of open-ended pipe piles during installation. DFI J-J Deep Foundations Institute 11(2–3):128–136
150. O'Neill MW, Raines RD (1991) Load transfer for pipe piles in highly pressured dense sand. J Geotech Eng 117(8):1208–1226
151. Liu J, Duan N, Cui L, Zhu, N (2019) DEM investigation of installation responses of jacked open-ended piles. Acta Geotechnica 14(6):1805–1819
152. Jardine R, Chow F, Overy R, Standing J (2005) ICP design methods for driven piles in sands and clays, vol 112. Thomas Telford, London
153. Lehane BM, Schneider JA, Xu X (2005) The UWA-05 method for prediction of axial capacity of driven piles in sand. Front Offshore Geotech ISFOG, 683–689
154. Kikuchi Y, Sato T, Morikawa Y, Kigata K (2008) Visualization of plugging phenomena in vertically loaded open-ended piles. In: GeoCongress 2008: Characterization, Monitoring, and Modeling of GeoSystems, pp 118–124
155. Kikuchi Y, Sato T, Mizutani T, Morikawa Y (2013) Plugging mechanism of open-ended piles. Advances in computed tomography for geomaterials. John Wiley & Sons
156. Zhang QQ (2018) Pile foundations engineering. China Architecture & Building Press
157. Liu JW, Zhang ZM, Yu F, Xie ZZ (2012) Case history of installing instrumented jacked open-ended piles. J Geotech Geoenviron Eng 138(7):810–820
158. Kou HL, Li W, Chu J, Yang DL (2020) Model tests on open-ended concrete pipe piles jacked in sand. Marine Georesource Geotech 38(8):939–946
159. Yang ZX, Gao YY, Jardine RJ, Guo WB, Wang D (2020) Large deformation finite-element simulation of displacement-pile installation experiments in sand. J Geotech and Geoenviron Eng 146(6):04020044
160. Sagaseta C (1978) Analysis of undrained soil deformation due to ground loss[J]. Geotecech 37(3):301–302

Chapter 12
Application of Transparent Soil Modeling Technique to Grouting

Abstract Rock grouting has been widely used in geotechnical and mining engineering. The grouts may be adhesives, chemical solutions, or have a variety of resinous origins. Penetrating, fracturing and compaction grouting have been studied by using theoretical analysis, experimental investigation, numerical simulation and in situ measurement. This chapter first presented a study of modeling grout propagation in transparent replica of rock fractures. Another modelling study on the chemical grout column permeated by water in transparent soil was also presented, which acts grouted barrier walls to reduce the permeability and/or strengthen the formation to prevent water inrush in the coal mining. Detailed experimental studies and finite element simulation are reported.

12.1 Modeling of Grout Propagation in Transparent Replica of Rock Fractures

12.1.1 Introduction

Rock grouting has been widely used in geotechnical and mining engineering. The grouts may be adhesives, chemical solutions, or have a variety of resinous origins [1]. Penetrating, fracturing and compaction grouting have been studied by using theoretical analysis, experimental investigation, numerical simulation and in situ measurement [2–6]. For example, [7] experimentally investigated the propagation of a Bingham fluid in rock discontinuities and proposed a method to calculate its penetration length. Dai and Bird [8] theoretically deduced formulations for slit flow, volume flow rate and location of plug-flow regions of a Bingham fluid between parallel plates. Grout propagation in rock mass has also been investigated by imaging fracture and pores filled with grout in laboratory and in situ studies. The fluorescent penetrant method was first used to study the development of fractures in rock specimens deformed in the laboratory [9]. This approach was then modified and used to identify grout injected into rock fractures and pores by using a borehole television system or a microscope [10]. This kind of method can only present the final state or phased process of grout propagation. Moreover, excavation, cutting or drilling

H. Zhao et al., *Transparent Soil Modelling Technique and Its Application*, Springer Tracts in Civil Engineering, https://doi.org/10.1007/978-981-19-6825-9_12

after grouting is required to allow observation or imaging of the grouted fractures and pores. Therefore, there is the need for a real-time non-interfering visualization approach to investigate grout propagation during rock grouting.

In terms of a real-time non-interfering visualization approach, transparent soil modeling [11, 12, 29] has been applied in many geotechnical and geological studies to examine internal displacement, strain, stress, and failure and seepage, such as in pile displacement [13, 14], shield tunneling [15, 16], grout propagation [17] and contaminant transportation [18, 19].

In this study, preliminarily attempts are made to further extend transparent soil experiment techniques to the area of rock fracture grouting experiments. Impermeable fused silica blocks in a cubic form are chosen to simulate rock mass and fused silica grains are used to fill the spaces between these blocks to simulate fractures with a certain aperture. As well, a mixed mineral oil that has the same refractive index as that of the fused silica blocks and grains is used to simulate fluid and groundwater. Grout propagation images are subsequently captured and analyzed, and the feasibility of using a transparent replica to simulate grouting in rock is verified by this study.

12.1.2 Materials

Fused silica blocks with a dimension of $2 \times 2 \times 2$ cm^3 were used to construct the replica of fractured rock in this experimental investigation. The blocks are made of silicon dioxide which has stable physical and chemical properties and cannot be dissolved by oil, water or grout. The silica blocks were placed uniformly and the apertures of the blocks were considered to be the fractures in this experiment. The blocks themselves are almost impermeable; therefore, the pore fluid and grout will not be able to penetrate them. Instead, the fractures of the blocks that were filled with fused silica grains constituted the pathways for the penetration of the pore fluid and grout. The grains have a particle size between 0.1 and 1 mm. The refractive index (RI) of the fused silica is 1.4585. The physical properties of the fused silica are shown in Table 12.1 and compared with those of Chinese standard sand. Mineral Oils A and B were mixed and applied as the pore fluid. Some of the physical properties of the two oils are listed in Table 12.2. The viscosity of the oils varies with temperature. The RI of Mineral Oils A and B was 1.4664 and 1.4396 respectively for testing at a temperature of 30 °C. The oils were mixed at a volume ratio of 2:1. The RI of the mixed oil was equal to the RI of the fused silica at 1.4585 when the temperature reached 30 °C.

A modified urea–formaldehyde (UF) resin and an oxalic acid solution with a mass concentration of 5% were mixed together and used as the injection solution to simulate Bingham grout. The density of this grout is 1.17 g/cm^3 and its other properties are shown in Fig. 12.1. The two solutions gelled when they were mixed together and the grout gel time varied, depending on the liquid concentration and the volume ratio of the two liquids (Table 12.3). The gel time was tested in air, because

Table 12.1 Physical properties of fused silica and Chinese standard sand [17]

	Gs	e_{max}	e_{min}	D_r (%)	e	K (10^{-3} cm/s)
Chinese standard sand	2.65	0.76	0.51	30	0.68	7.00
				50	0.63	6.73
				70	0.58	6.52
Fused silica	2.21	0.8	0.51	30	0.71	8.32
				50	0.65	7.41
				70	0.59	6.39

Table 12.2 Physical properties of mineral oils A and B, and mixed oil

	Specific gravity (10^3 N/m^3)	Viscosity index	Kinematic viscosity (10^{-6} m^2/s)	Flash point (°C)	Pour point (°C)	Refractive index (30 °C)
Mineral oil A	0.812	100	5–6	150 (close)	< − 10	1.4663
Mineral oil B	0.844	> 98	10	180 (open)	− 15	1.4397
Mixed oil	0.823	–	7	–	–	1.4585

it was taken into consideration the difficulty of determining whether the gelling had taken place in the mixed oil. There could be a difference between the gel time in air and the mixed oil. Therefore, the gel time is used here for reference purposes with respect to the properties of the modified UF resin and oxalic acid solution. The mixture with a gel time of 45.2 s was chosen in this experiment, with a volume ratio of 1:3 for the acid solution and modified UF resin.

12.1.3 Experimental Set Up and Procedure

12.1.3.1 Experimental Set Up

Figure 12.2 shows the experiment setup and a detailed schematic of the experiment. The experimental devices are placed on a precision optical table. The experimental setup consists of a transparent model of a fractured rock, a stress-load frame, air pressure detectors, a grouting system, an optical detecting system and data collectors.

A transparent Plexiglas mold with inside dimensions of 10 cm × 10 cm × 12 cm was used to accommodate the transparent rock model. An air compressor with a pneumatic reducing valve which was connected to a gas pressure gauge with a measurement range of 0–400 kPa was utilized to provide the grouting pressure, which was detected by pressure sensors that could detect a pressure range of 0–100 kPa and connected to a digital acquisition instrument.

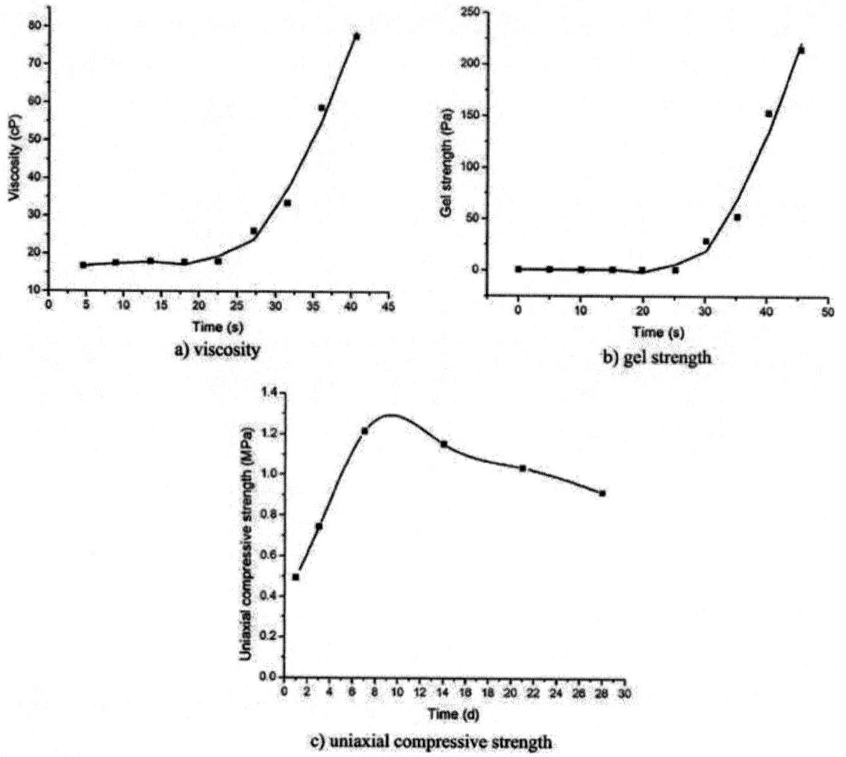

Fig. 12.1 Variation of **a** grout viscosity, **b** gel strength and **c** uniaxial compressive strength with time

Table 12.3 Gel time for grout with different volume ratios of modified urea-formaldehyde resin and solid oxalic acid

Volume ratio oxalic acid: modified UF resin	Average gel time (s)
1:1	20.5
1:1.5	26.8
1:2	32.1
1:3	45.2

The Plexiglas mold was partially filled with the fused silica mixed with the pore fluid. Sixteen silica blocks were evenly and symmetrically packed in blocks of 4 × 4 to form a layer of the rock replica. Then, the procedure was repeated until 4 layers of silica blocks were built inside the Plexiglas mold, and the fractures were filled with fused silica and pore fluid. A total of 125 g of fused silica was used to fill the mold, and the relative density of the fused sand in the mold was determined to be approximately 70%. The model was de-aerated by using a vacuum pump until

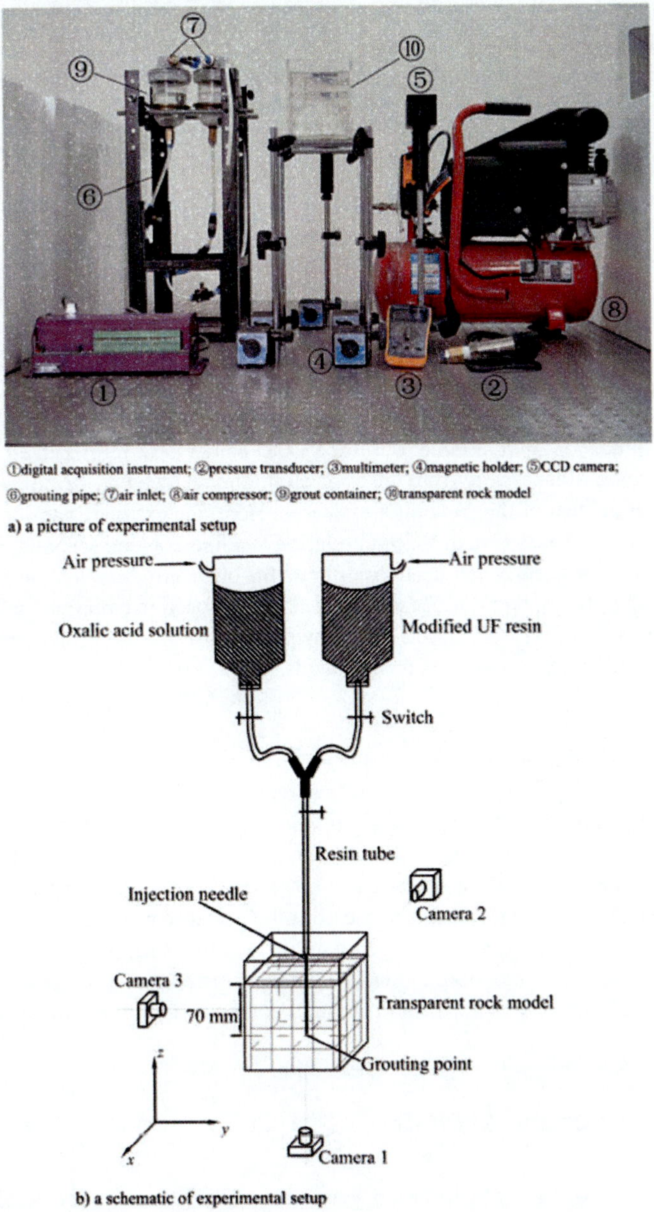

Fig. 12.2 **a** Photo and **b** schematic of experimental set up

the best transparency was obtained. The aperture of the fractures between the silica blocks was around 2.5 mm.

The point source of the grouting was simulated by using an injection needle with an inner diameter of 0.9 mm which was inserted into the model at a depth of 70 mm below the sample surface. Figure 12.3 is a photo and schematic of the placement of the fused silica blocks used in this study. Figure 12.4 is a photo of the transparent rock model which shows the transparency of the fused silica blocks, grains and pore fluid after saturation.

The modified UF resin and oxalic acid solution were kept in separate containers. The mixture was injected into the transparent rock model through air-pressure.

Three CCD cameras and two lights with a power of 1 kW for illuminating purposes were used in the experiment. Each CCD camera has a resolution of 1024 × 768 pixels and a maximum frame rate of 30 frames per second (fps). The CCD camera was controlled by a computer.

On the precision optical table, the three CCD cameras were placed so that they were the same distance away from the transparent model (see Fig. 12.2b), so that the extrinsic parameters of the cameras were same. Here, the extrinsic parameters refer to the location of the camera in space in relation to a fixed object. Meanwhile, a right angle laser line projection was used to calibrate the distortion between the object and camera. A calibration board [20] (see Fig. 12.5a) was used to calibrate the accuracy of the cameras (Fig. 12.5). Figure 12.5c shows that most of pixel errors are less than 2 pixels and the maximal error is near 3 pixels.

12.1.3.2 Test Procedure

After the experimental set up was prepared, the chemical grout was injected into the transparent fractured rock replica through the point source which was simulated by an injection needle. The grouting pressure was 60 kPa in the first stage, and then changed to 80 kPa at about 80 s. The change in the grouting pressure was manually controlled by taking into consideration that the grout might gel after 80 s. A background image was taken before the injection. A series of images were then captured during the injection at a frame rate of 15 fps and simultaneously recorded into the computer.

12.1.4 Results and Analysis

Figure 12.6 shows a sample of the images captured by Camera 1 at different times during the experimental investigation, in which the grouting process lasted for 140 s and approximately 2100 images were captured. The grouting point is marked with a black dot and the projection of the grout propagation path is depicted by an almost round white area.

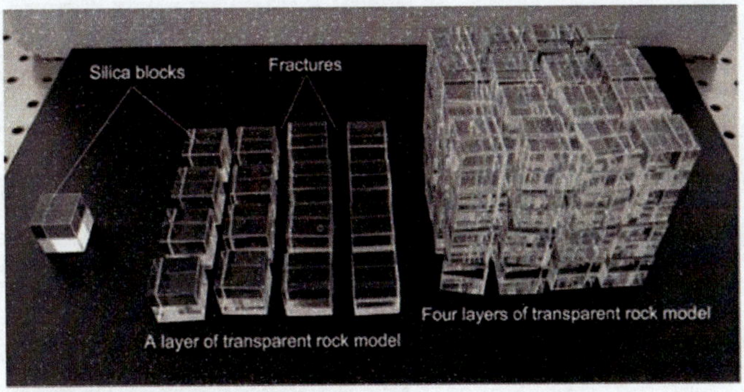

a) a picture of transparent silica block

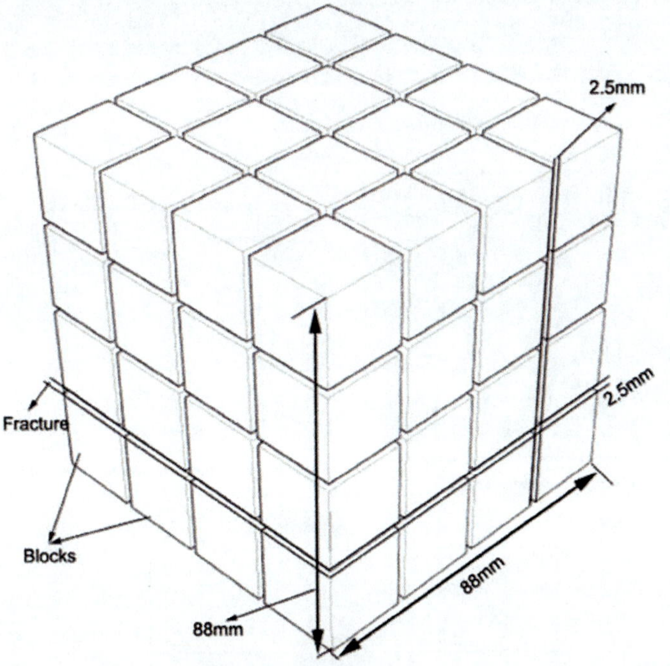

b) a schematic of the placement of blocks

Fig. 12.3 **a** Picture and **b** schematic of the placement of transparent silica blocks

12.1.4.1 Propagation of Grout Front

In the majority of the images shown in Fig. 12.6, the grout propagates almost symmetrically. Since the silica blocks are impermeable, the grout first penetrates through different fractures and then spreads in a circular shape inside the fractures in the same direction and plane (Fig. 12.6a–f). However, the grout did not propagate in

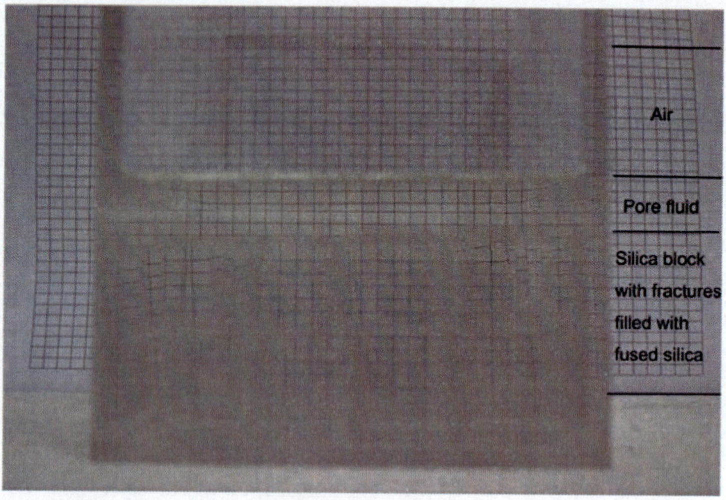

Fig. 12.4 Transparent effect of transparent rock model

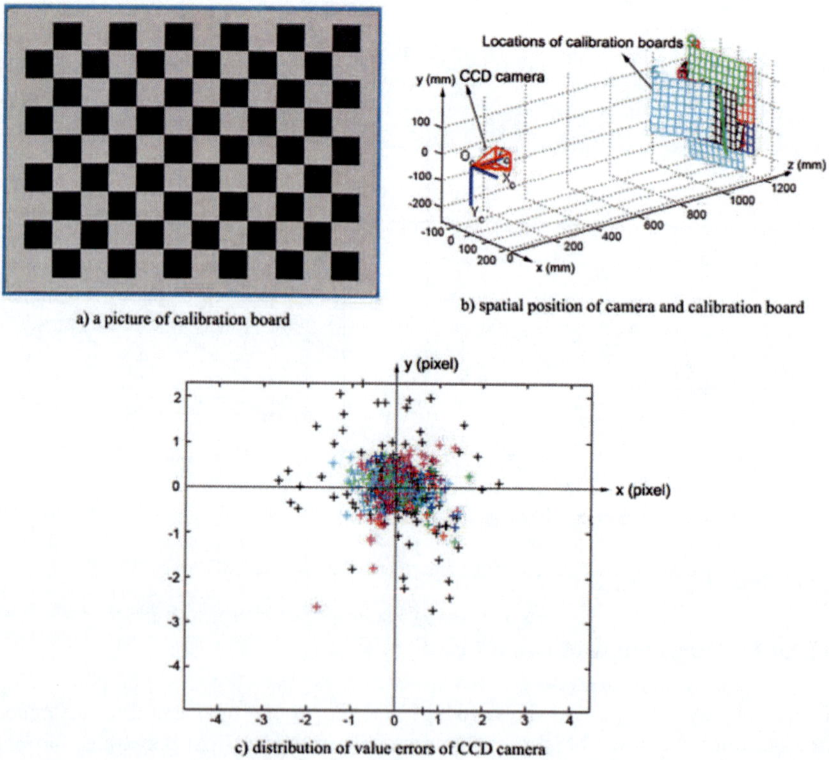

Fig. 12.5 Calibration of CCD camera

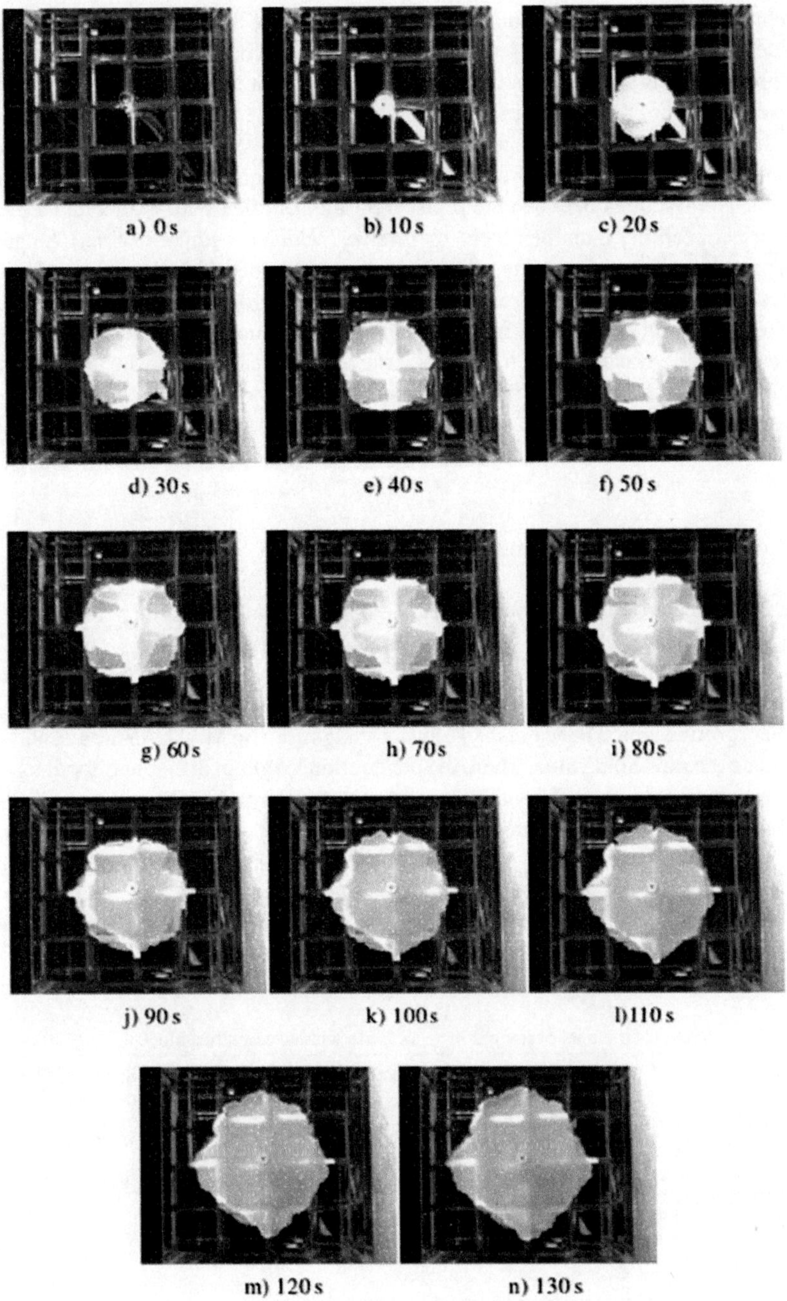

Fig. 12.6 Grout propagation images captured by camera 1 at different times during grouting

an entirely circular shape. Figure 12.6l–n show some of the grout front with fingers inside the fractures where the penetration was faster. With a longer propagation time, the grout boundary became more irregular. Therefore, the propagation of the front is uneven in most of the images in Fig. 12.6 rather than having a smooth circular shape.

The structure of the transparent replica was spatially symmetrical to the grouting point at the beginning of the grouting. The grout propagation was almost symmetrical at first. The fractures between the transparent blocks filled with grains had a much higher permeability than the blocks themselves, which are impermeable. The grout first propagated into the fractures, then along the surface of the blocks and was slightly adhesive until there was another fracture. During this process, there was interaction between the grout and the rock fracture. This interaction might have changed the inner structure of the fractures, such as the aperture, angle, and shape when the internal forces between the grout and fracture reached a certain value. When the grout began to gel, the internal force increased. The grout then propagated in an irregular circular shape and thus resulted in the uneven edges of the front propagation.

12.1.4.2 Grout Penetration Length and Area

The image scale coefficients were calculated in accordance with the pixel value and real size of the silica blocks. The results are shown in Table 12.4.

The border of the propagation for every image was automatically and approximately traced according to the different grayscale value between the grout bulb and the background area. The watershed transform algorithm in MATLAB was applied to calculate the threshold value. Then, the penetration length in different directions and the propagation area at different times were calculated by using the scale coefficients in Table 12.4 and the measurement of the pixel value corresponding to the grout penetration and propagation. Figure 12.7a shows the maximum penetration length of the grout in the *X* and *Y* axis directions in the images captured by the Camera 1. Figure 12.7b shows the variation of the grout propagation areas in three projection planes with grouting time, captured by the three cameras.

Table 12.4 Scale coefficients of images in accordance with size of silica block

Location	Item	Pixel value	Real size (cm^2)	Scale coefficient ($pixels/cm^2$)
Camera 1, underneath the model	Area of silica block	22,500	2 × 2	75 × 75
Camera 2, behind the model		22,500		75 × 75
Camera 3, on the left side of the model		24,300		78 × 78

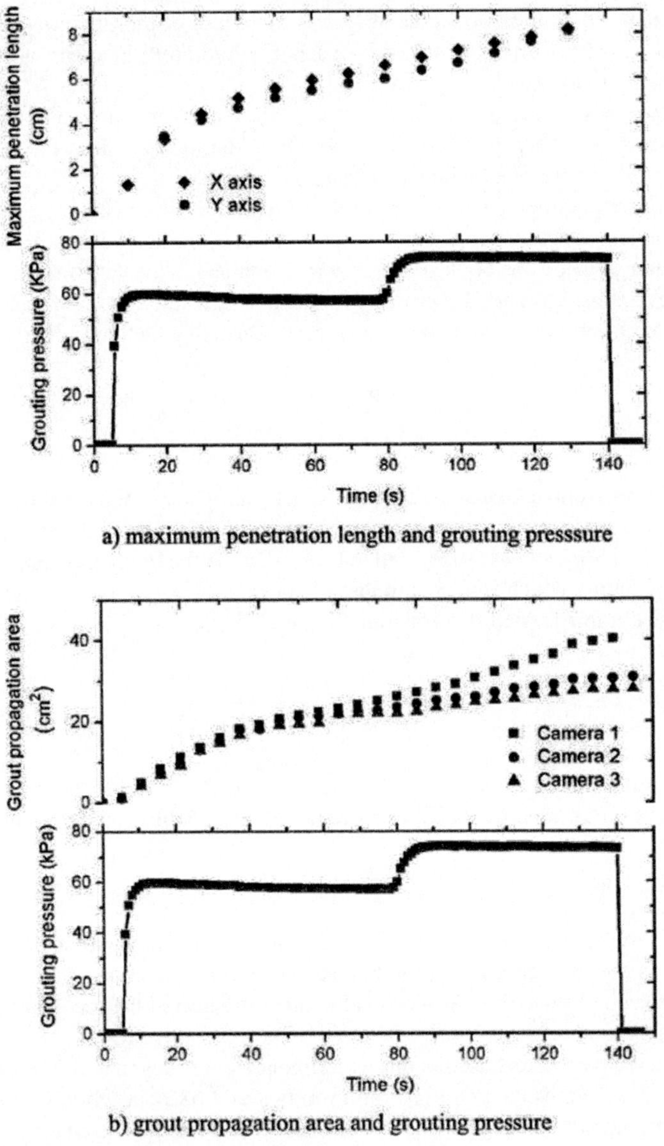

Fig. 12.7 Variation of **a** maximum penetration length of grout and **b** propagation area with time

The cubic law is considered to be valid for laminar flow through parallel wall fractures with smooth surfaces:

$$q = \frac{gb^3}{12\upsilon} J \tag{12.1}$$

where q is the flow in the fracture (m^3), b is the width of the fracture (m), υ is the kinematic viscosity of the water (m^2/s), J is the hydraulic gradient, and g is the acceleration of gravity.

The flow in the silica grain filled fractures different from that in porous media or unfilled fractures. Therefore, Eq. (12.1) needs to be modified since the fractures in the present study are filled with silica grains.

The images acquired from any one of the cameras can be considered as the projection of the largest grout propagation area on a fracture plane. Therefore, the grout propagation taken by one of the cameras was referenced as the flow in a fracture filled with silica grains. Su et al. [21] proposed a seepage flow rate in a filled fracture based on a modification of the cubic law and Kozeny-Carman equation [22] as follows:

$$v = \frac{gn^3b^2}{20.4\upsilon\alpha^2\left[1+3(1-n)\frac{b}{d}\right]^2}J = K_f J \quad (12.2)$$

where v is the average flow velocity (m/s), n is the porosity of the filled fracture, α is the particle shape factor, and d is the grain size of the fillings. If the fillings are not homogeneous, $d_{10} \sim d_{20}$ is suggested to be used. K_f is the hydraulic conductivity of the filled fracture, which is related to the fracture aperture, particle size and porosity of the silica grains in the filled fracture.

$$K_f = \frac{gn^3b^2}{20.4\nu\alpha^2\left[1+3(1-n)\frac{b}{d}\right]^2} \quad (12.3)$$

The grout propagation needs to be divided into o periods in accordance with the changes in grout pressure. Then, the grout penetration radius can be calculated by:

$$r_c = v_1t_1 + v_2t_2 \quad (12.4)$$

where r_c is the calculated penetration radius (cm), v_1 is the flow rate in the first stage (cm/s), t_1 is the time duration of the first stage grout propagation (s), v_2 is the flow rate in the second stage (cm/s), and t_2 is the time duration of the second stage of the grout propagation (s).

The related parameters used in this experiment and the calculated result are given in Table 12.5. The calculated penetration radius is 4.586 cm while the measured maximum penetration length is 4.07 cm. The reason for the difference is that the penetration in the rock fractures in this study is actually not a flow between a parallel filled fracture, but a flow inside a fractured network. Moreover, the influence of the grout boundary could be a significant factor during the last stage of the grouting. The experiment was terminated once the grout propagated to the boundary of the model mold.

Table 12.5 Value of parameters and calculation result

Test stage (s)	Gravitational acceleration g (cm/s^2)	Porosity n	Width of fracture b (cm)	Kinematic viscosity υ (cm^2/s)	Particle shape factor α
10–80	981	0.371	0.25	0.17	1.5
80–140	981	0.371	0.25	0.68	1.5
0–140	981	0.371	0.25	–	1.5
d_{20} (cm)	b/d	Hydraulic conductivity K_f (10^{-3} cm/s)	Hydraulic gradient J	Flow velocity υ (cm/s)	Penetration radius r_c (cm)
0.025	10	1.016	51.28	0.05212	3.648
0.025	10	0.2541	61.54	0.01564	0.938
0.025	10	–	–	–	4.586

12.1.5 Conclusions

This study has presented an experimental investigation in which grout propagation is observed in real-time in a transparent replica of fractured rock filled with sands through visualization with CCD cameras. A series of grout propagation images have been captured and analyzed to identify the propagation boundary, and penetration length and area.

The results show that the grout propagates in a roughly circular shape, and the diffusion of the grouting front is uneven and irregular. The results from the image processing also indicate that the grout propagation follows the modified cubic law which takes into consideration fractures that are fillings with silica grains.

This study verifies the feasibility of using a transparent replica of rock and soil for the simulation of grouting. However, there are still many factors that need to be considered in further studies, such as the boundary effect, fracture surface friction, dip angle, shape and complex networks. In addition, pore water pressure, interaction between fractures and grout, etc. also warrant further investigation.

12.2 Modeling of Chemical Grout Column Permeated by Water in Transparent Soil

12.2.1 Introduction

Water inrush is a typical hydrogeological hazard which threatens the safety of workers in the coal mining process. To address this issue, grouted barrier walls are now primarily used to reduce the permeability and/or strengthen the formation to prevent water inrush. Vertical walls are typically formed by overlapping grout columns which form a single row. Additional rows can be added to ensure the effectiveness of the wall. Grout columns are typically constructed through two methods: jet and permeation grouting. Most of the coal mines in east China use permeation grouting to prevent water inrush from the shaft walls [23–26]. Chemical grouts that are low in viscosity can be injected into soil at low pressure to form the grout columns. Yet despite the improvement in methods for controlling grouting operations, there is still the need for a better understanding of the conditions that govern the distribution of chemical grouts that are injected into the soil. During the past two decades, there have been a number of studies that focus on the effect of groundwater flow on the distribution of chemical grouts [27, 28]. In their development of better technologies to address this issue, they have carried out laboratory testing with physical modeling.

In the modeling, transparent soil which is made of aquabeads or silica powder has been used to understand the flow process in soil. However, there are limitations to using aquabeads or silica powder in large-scale models due to their poor transparency [11, 29]. Thus, fused quartz has been proposed for modeling sand [30]. This material is adequately transparent to visualize the process of chemical grout permeation in a 2D model [31]. However, this study proceeds further and adds to previous experiments by proposing the3D modeling of the permeation of a chemical grout column by water in transparent soil made of fused quartz. An equation based on the fluid dynamics in porous media and Darcy's law is proposed to predict the displacement of the groundwater-grout interface in flowing water. A numerical model based on finite element modeling is also developed to simulate this process.

12.2.2 Materials

12.2.2.1 Chemical Grout

Urea–formaldehyde resin (UFR) is a white chemical grout and widely used in coal mines in the east of China [32]. The UFR was mixed with a catalyst, or oxalic acid, in the experiment, and then foamed in place for block filling. The foam can be forced through small openings and delivered to the entire area of any cavity prior to curing. The viscosity of the chemical grout was measured by using a Brookfield viscometer in accordance with ASTM standard D4016-08. The initial viscosity of the chemical grout was around 5 mPa s. Figure 12.8 shows the typical gel time curves of the UFR with various concentrations of oxalic acid. The gel times decrease as the concentration of oxalic acid increases from 3 to 12%. The UFR has a low initial viscosity and a wide range of controllable gel times.

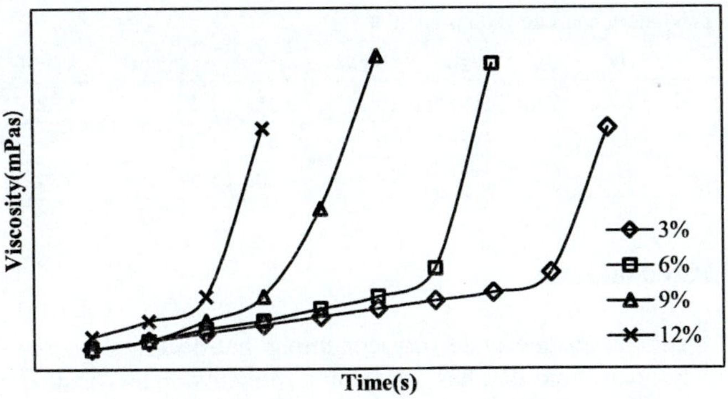

Fig. 12.8 Typical gel time curves of UFR with various concentrations of oxalic acid

12.2.2.2 Fused Quartz for Modeling Sand

Fused quartz was used to model natural sand in the experiment. The fused quartz was delivered in particle form from Jiangsu Kaida Silica Co., Ltd in China. The particle-size distribution graph of the fused quartz is shown in Fig. 12.9. The geotechnical properties of the fused quartz are summarized in Table 12.6. Compared to the silica gel used in a previous study [33], the fused quartz has solid particles without internal pores, and thus demonstrates better capacity to model natural sand. The hydraulic conductivity of the fused quartz was investigated by using the constant head method in ASTM-D2434.

Fig. 12.9 Grain-size distribution of fused quartz

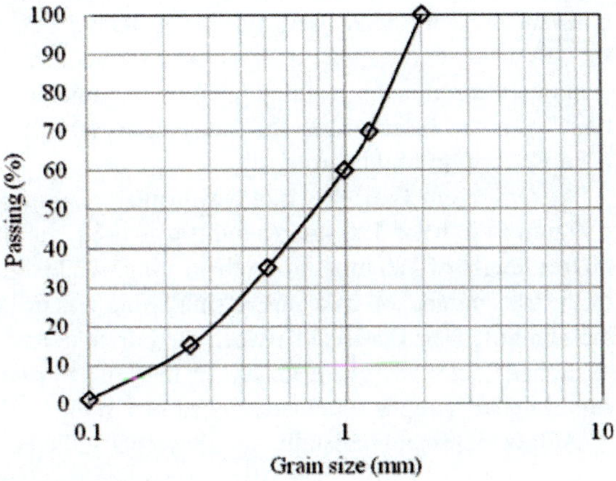

Table 12.6 Laboratory measured soil properties

	Gs	e_{max}	e_{min}	Dr (%)	e	ρ_d (g/cm^3)	$k \times 10^{-3}$ (cm/s)
Transparent soil	2.25	0.79	0.55	30	0.71	1.29	8.32
				50	0.65	1.34	7.41
				70	0.60	1.39	6.39

12.2.2.3 Pore Fluid

The fused quartz is opaque under dry conditions but becomes transparent when saturated with a pore fluid that has a matching refractive index. That is, when the solid (fused quartz) and liquid (pore fluid) refractive indexes are matched, the former becomes transparent. A blend of two different mineral oils was used as the pore fluid for modeling groundwater. The mixed mineral oil is a 1:1.2 blend by weight of the #2 and #10 Mineral Oils provided by the China National Petroleum Corporation. The refractive index, viscosity, and density of the blend of the two different mineral oils at room temperature(25 °C) were 1.458, 3.5 mPa s, and 840 kg/m^3, respectively.

12.2.3 Physical Modeling Experiments

12.2.3.1 Grouting Equipment Setup

A grout injection apparatus that uses air pressure to discharge grout was set up with a pressure capacity from 10 to 500 kPa. Compared to the grouting device with a constant pressure head used by [34] in their investigation on chemical grouting, this grout injection apparatus that uses air pressure is easier to operate and more accurately controls the grouting pressure. An injection needle with an inner diameter of 1.8 mm was inserted into the transparent soil at a depth of 45 mm. Figure 12.10a is an illustration of this grout injection apparatus.

A transparent Plexiglas mold with internal dimensions of 300 mm (length) × 150 mm (width) × 120 mm (height) was used, and filled with the transparent soil up to a height of 120 mm during the testing. A Plexiglas plate was placed on the top of the transparent soil as a cap. A small hole was drilled in the middle of the cap as the borehole. The distance between the borehole and the inflow hole was 100 mm. A rubber circle was glued on top of the hole to prevent any leakages. The entire experimental setup is schematically shown in Fig. 12.10b.

A flow meter and peristaltic pump were installed on the inflow and outflow pipes as shown in Fig. 12.10b. The peristaltic pump was used to control the flow rate and the flow meter was used to monitor and record the volumetric flow rate.

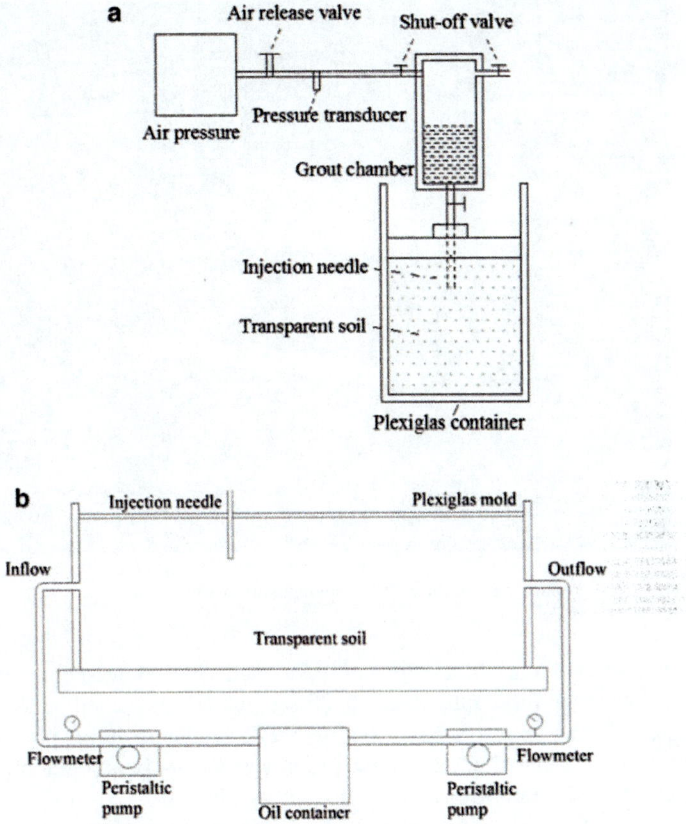

Fig. 12.10 Schematic diagram of experimental setup (not to scale). **a** Grout injection device [31]. **b** Experimental setupto permeate grout column with water

12.2.3.2 Sample Preparation

The preparation of the samples was carried out as follows. First, two filters that consisted of five layers of steel screens with an opening size that ranged from 0.5 to 2 mm were placed at both ends of the Plexiglas model tank in order to distribute the flow vertically along the depth of the model.

Then, the model tank was partially filled with the blend of the two different mineral oils. The fused quartz was slowly poured into the tank with the help of a spoon. At the same time, the fused quartz was stirred to release air bubbles entrapped during pouring. When a certain amount of fused quartz was immersed into the blend of the two different mineral oils, vacuum was applied to de-air the mixture until it became transparent. The time required for vacuuming depended on the depth of the fused quartz and the power of the vacuum pump. Approximately 2 h was required to fully saturate 30 mm of fused quartz in height. Following that, the fused quartz

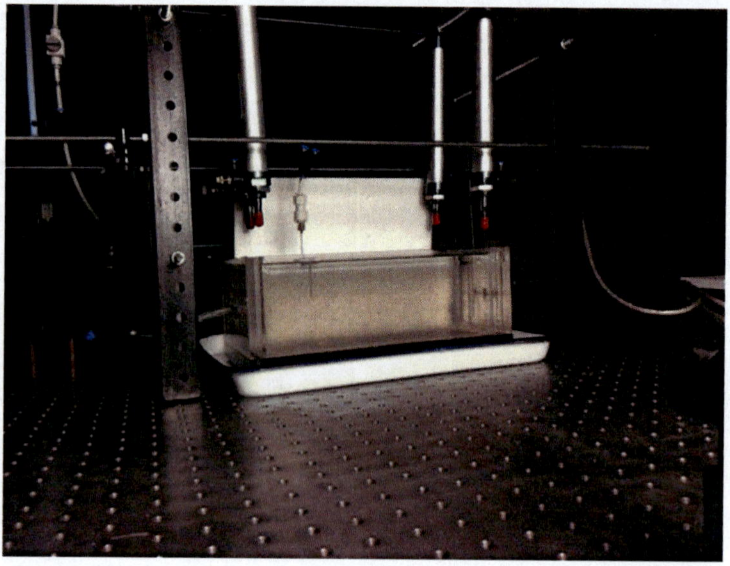

Fig. 12.11 White board viewed through transparent soil model

was compacted to form the transparent soil model. The tank was vibrated during the compaction of the fused quartz, and the fused quartz was placed in 30 mm lifts. A piezoelectric vibrator was used to compact the fused quartz during placement in the model tank. The magnitude of the vibration was 0.1 mm and duration was 2 min for each lift. Once the fused quartz filled the model tank, a cover plate was tightly screwed onto the tank with a gasket placed between the plate and the tank. The transparency of the model is shown in Fig. 12.11, in which a white board with the width of 150 mm can be viewed through the transparent soil.

12.2.3.3 Experimental Procedure

Two black and white charge-coupled device (CCD) cameras were mounted in an orthogonal position as shown in Fig. 12.12. They have a resolution of 1024 × 768 pixels and a maximum frame rate of 30 frames per second (fps). The CCD cameras were controlled by using a PC through an in-house developed driver that uses MATLAB® to acquire images at the same time [35]. They were placed onto an optical table and calibrated by using a standard calibration board to correct lens distortion.

Prior to the grout injection, the model was flushed with the blend of two different mineral oils for 1 h to saturate the transparent soil as much as possible. The blend of the two different mineral oils was slowly introduced into the model from the inflow pipe by using the peristaltic pump. Once a steady outflow of the blend of two different

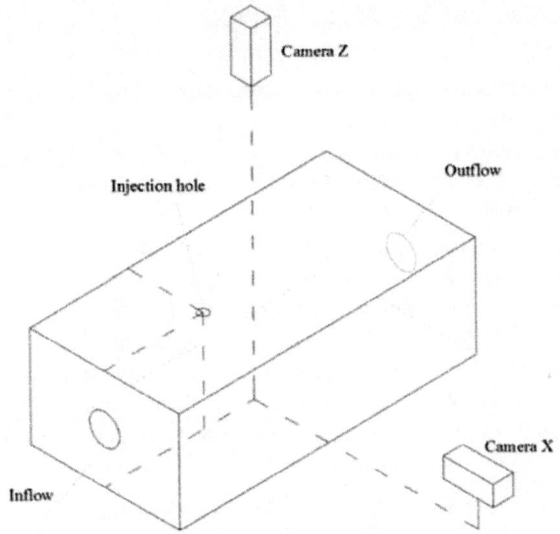

Fig. 12.12 An optical system for groundwater-grout interface measurement (not to scale)

mineral oils was realized, the flow rate was recorded by using the flow meter on the outflow pipe. As a result, the velocity of flow can be determined. The flow rate was adjusted to meet the requirements of the designed value. Details of these individual tests are summarized in Table 12.7.

Several background images were taken before the grout injection started. Once the switch on the injection pipe was turned on, a series of images were taken during the injection with the time data acquired simultaneously into the computer. The injection pressure (air pressure applied on the grout chamber) was a constant value (40 kPa) in the experiment.

After the predetermined time passed, the switch on the injection pipe was turned off. Then the chemical grout was allowed to gel. After gelation, the cameras were turned off.

Table 12.7 Summary of individual tests

Property	Relative density Dr	Hydraulic conductivity $k \times 10^{3}$ (cm/s)	Vertical pressure σ (kPa)	Velocity of flow v (cm/s)	Grouting pressure P (kPa)	Gel time t (s)
Transparent soil model	0.50	7.41	150	0.06	40	120
Numerical model	0.50	7.41	150	0.06	40	120

12.2.4 Transparent Soil Model Results

A total of 1200 grayscale digital images were acquired by the two cameras during the injection process. Several typical images are selected for presentation in this manuscript as examples to illustrate how the groundwater-grout interface is obtained. The grayscale digital images acquired by Cameras X and Z are provided in Figs. 12.13 and 12.14, respectively.

The observed penetration length (L_f) along the flow direction obtained by Camera X during the grouting process was normalized by using the diameter (Di, 2 mm) of the injection needle. The lateral penetration length (L_l) was normalized in the same manner. Figure 12.15 shows that L_f and L_l change with time. L_f and L_l increase with time at the same rate in the first stage. However, L_f sharply increases after 80 s when the flowing water controls the distribution of the chemical grout. The reduction in seepage velocity from the progression of the UFR through the transparent soil may have contributed to this phenomenon.

12.2.5 3D FEM Model

12.2.5.1 Model Development

In order to better understand the mechanism of the permeation of the grout column by the water, the process of the grout injection was simulated by using a commercial 3D finite element code, COMSOL Multiphysics®.

A 3D model with 10,978 elements was developed by using COMSOL Multiphysics®.

The properties of this numerical model are listed in Table 12.7. The model boundaries and mesh grid are schematized in Fig. 12.16.

UFR is basically a Newtonian fluid with a constant viscosity, where the addition of oxalic acid rapidly increases its viscosity after a certain time has passed. Therefore, Darcy's law can be applied for the flow of the chemical grout in pore spaces. The groundwater-grout interface was considered to be a vertical plane as shown in Fig. 12.17. The actual seepage velocity v for the groundwater-grout interface can be estimated as follows.

$$v = \frac{q}{n} \tag{12.5}$$

where

q flow rate per unit width, m^3/s, and
n porosity, %.

Based on several assumptions, including that the chemical grout and groundwater are immiscible and interface is impervious, and under steady groundwater flow in

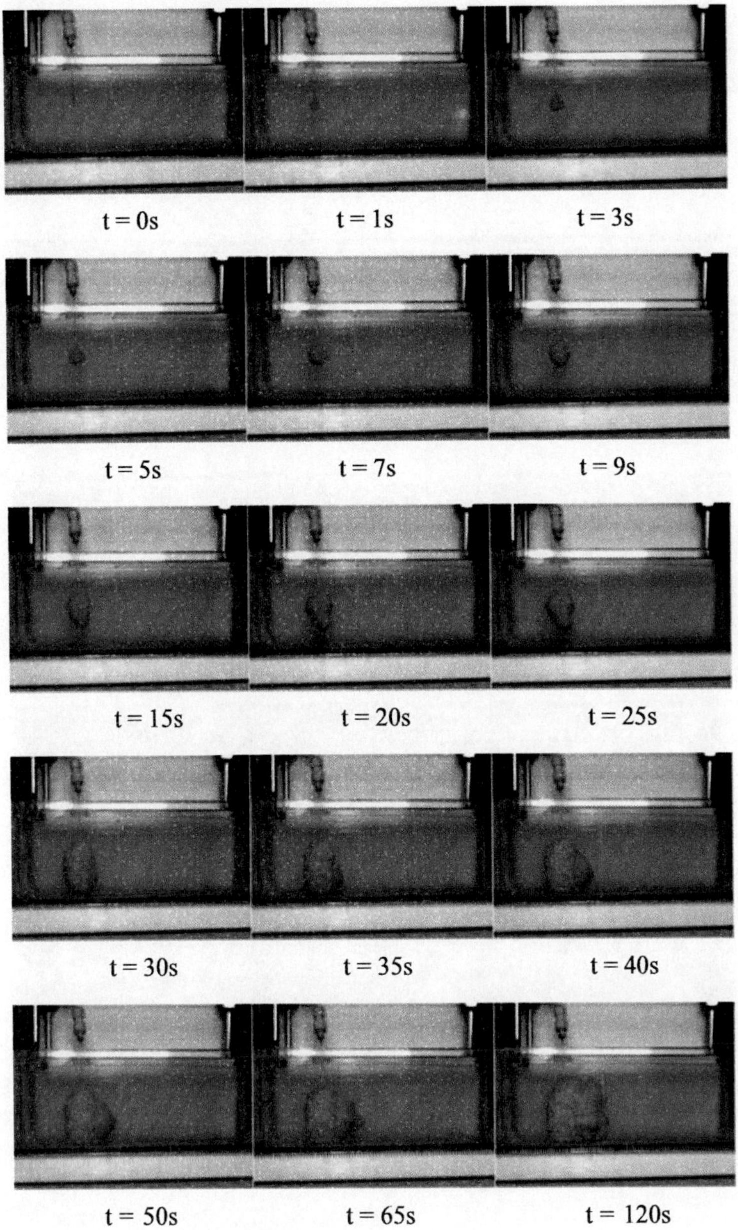

Fig. 12.13 Images acquired by Camera X

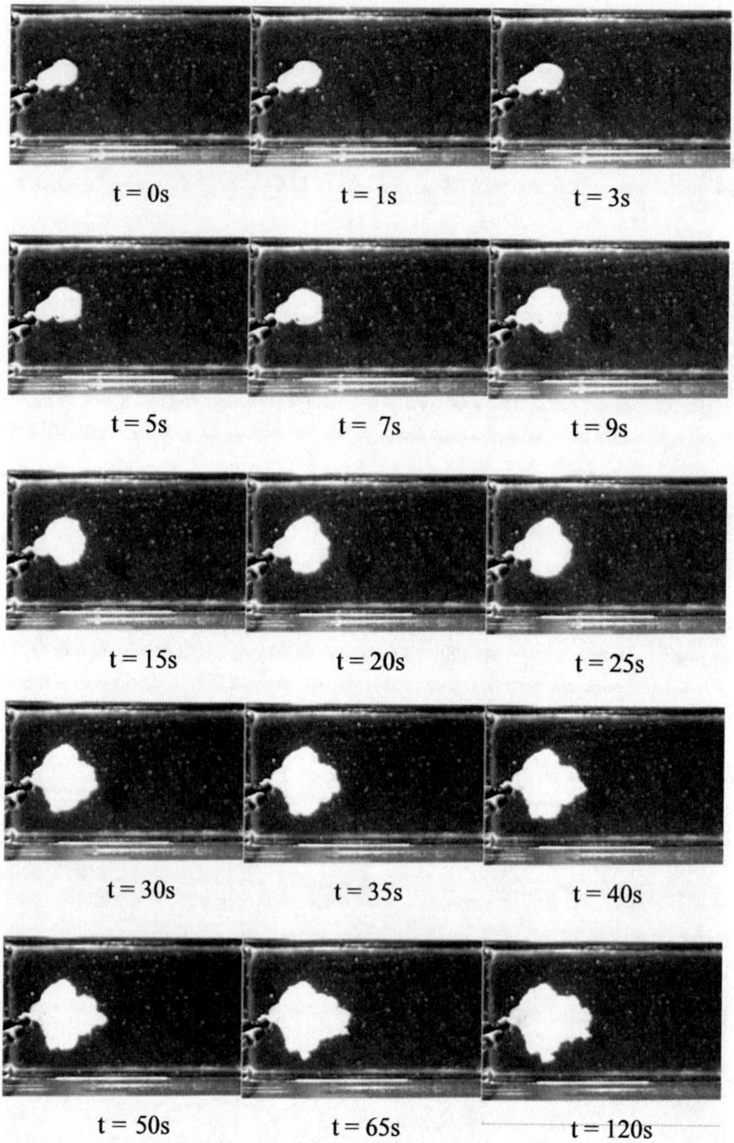

Fig. 12.14 Images acquired by Camera Z

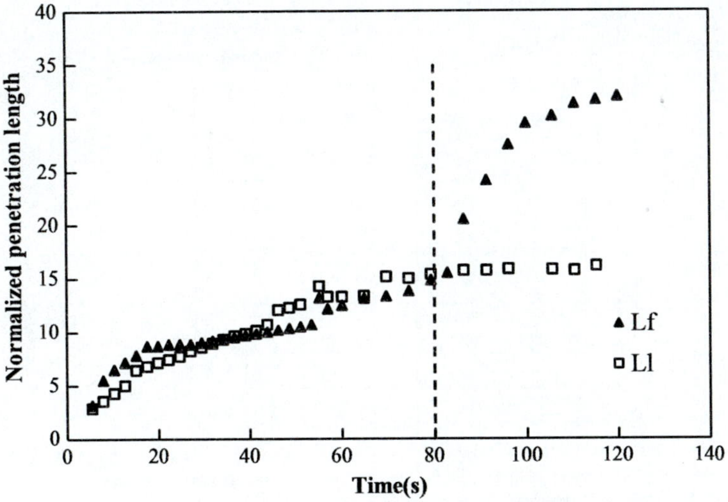

Fig. 12.15 Normalized penetration length (L_f, L_l) observed in transparent soil model

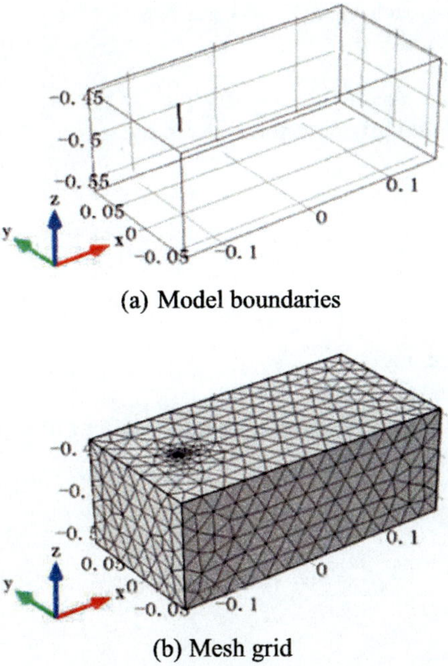

Fig. 12.16 Numerical model boundaries and mesh grid: **a** model boundaries, **b** mesh grid

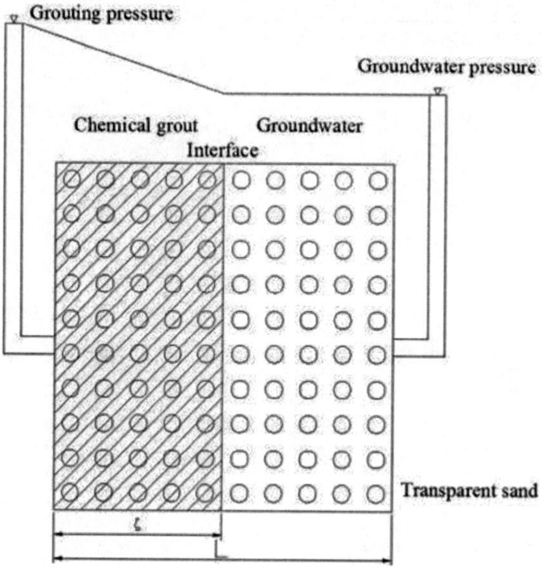

Fig. 12.17 Groundwater-grout interface in two phase flow

the horizontal direction, the continuity equation for the two immiscible fluids, that is, the chemical grout and groundwater, is given by:

$$\left.\begin{array}{l} \frac{\partial^2 P_G}{\partial x^2} = 0\ 0 < x \le \zeta(t) \\ \frac{\partial^2 P_w}{\partial x^2} = 0\ \ \zeta(t) < x \le L \end{array}\right\} \quad (12.6)$$

where

P_G grouting pressure, m,
P_W groundwater pressure, m,
$\zeta(t)$ displacement of groundwater-grout interface after t time, m, and
L length of porous media, m.

The outer boundary conditions can be written as:

$$\left.\begin{array}{l} x = 0,\ P_G = p' \\ x = L,\ P_w = p'' \end{array}\right\} \quad (12.7)$$

The boundary conditions on the interface are given by:

$$\left.\begin{array}{ll} P_G = P_w, & \zeta = \zeta(t) \\ \frac{k_G}{\mu_G} \cdot \frac{\partial P_G}{\partial x} = \frac{k_w}{\mu_w} \cdot \frac{\partial P_w}{\partial x} = -q_x, & \zeta = \zeta(t) \end{array}\right\} \quad (12.8)$$

where

k_G intrinsic permeability of the chemical grout, m^2,
μ_G dynamic viscosity, mPa s,
k_W intrinsic permeability of groundwater, m^2,
μ_W dynamic viscosity, mPa s, and
q_x flow rate per unit width, m^3/s.

Continuity Eq. 12.6 can be calculated by using Eqs. 12.7 and 12.4. The integral results are given by:

$$t = \frac{\mu_G n L^2}{k_G(p' - p'')}\left[M\left(\frac{\zeta}{L}\right) + \frac{1}{2}(1 - M)\left(\frac{\zeta}{L}\right)^2\right] \tag{12.9}$$

where

$$M = \left(\frac{k_G}{\mu_G}\right) / \left(\frac{k_w}{\mu_w}\right)$$

12.2.5.2 Analysis of the Experimental Results and Comparison with FEM Simulations

The 3D FEM simulated the process of the permeation of the chemical grout column by water based on the mechanisms of fluids dynamics in porous media as discussed in the previous section. The numerical experimental results are provided in Fig. 12.18. The process of the permeation of chemical grout in flowing water, which was obtained by numerical simulation, is similar to that in the laboratory experiments. The general trend is a reduction in the pressure as shown in Fig. 12.17. This trend is attributed to the reduction in seepage velocity from the progression of the UFR through the transparent soil as discussed in the previous section. As a result, the pressure and hydrodynamic advection both decrease with increasing distance from the injection point. The groundwater-grout interface is forced to move along the flow direction after a certain time has passed. This trend was observed in both the physical modeling (around 80 s) and the numerical modeling (around 40 s).

More importantly, the analysis with numerical modeling was performed under idealized conditions which assumed a homogeneous porous medium. This may have contributed to the discrepancy in the time.

12.2.6 Limitations and Discussion

The test results provided important information for understanding the process of the permeation of chemical grout in flowing water. This understanding is shown to have

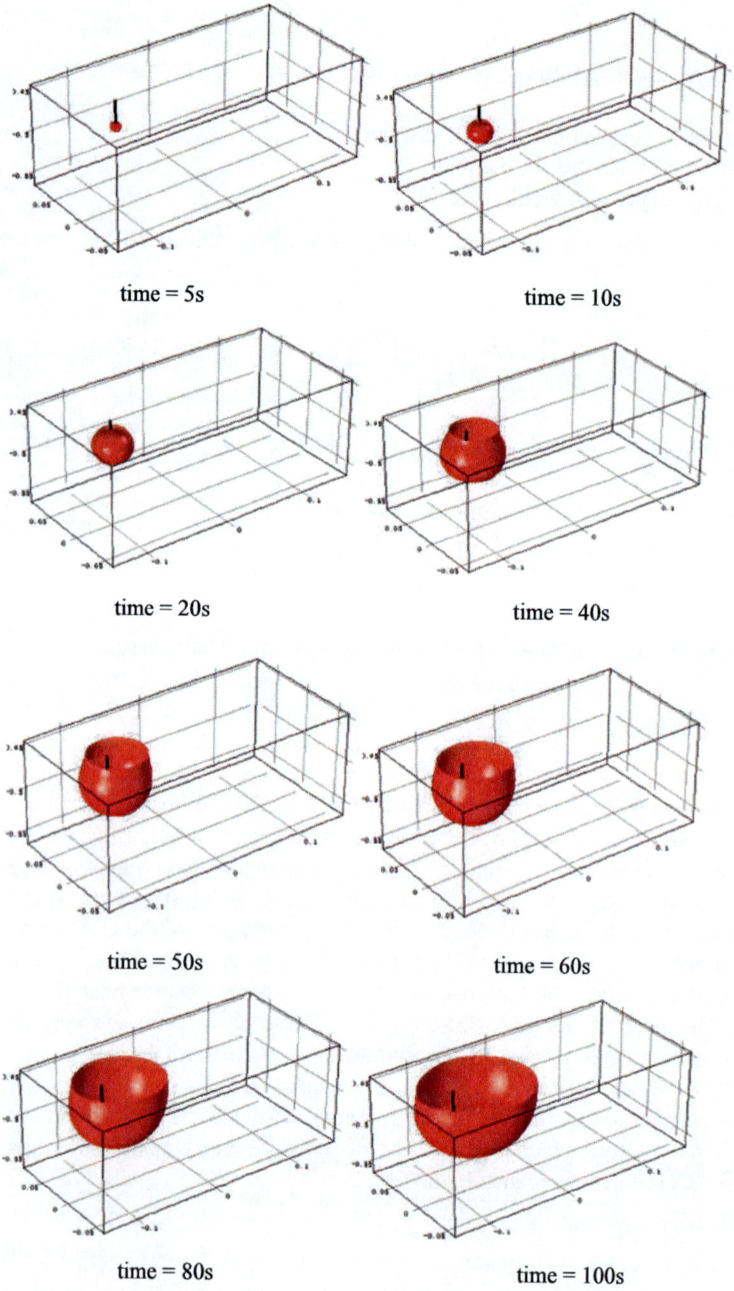

Fig. 12.18 Results of numerical experiment

practical value for reducing environmental impacts. For example, grouted barrier walls are considered to be an effective way to control the discharge of polluted water into coal mine goafs. The boreholes for chemical grout columns can be designed to provide greater distance to the water flow as the grout-water interface will move in the direction of the flow. It is possible to optimize injection spacing so that fewer boreholes are needed in coal mine goafs.

Aside from reducing environmental impacts, this research also contributes to mine safety. First, since the process of chemical grout permeation in flowing water should be ideally carried out in two stages, a very high injection pressure should be applied in the first stage to overcome the seepage pressure in the flowing water. The effectiveness of the grouted barrier wall can therefore be ensured.

A numerical simulation is therefore conducted to demonstrate that in ideal conditions, that is, assuming that there is a homogeneous porous medium, these two stages are feasible.

Finally, a limitation of this study is that the intergranular and the discontinuity components of flow are not considered in the equation, which deserves further rigorous research.

12.2.7 Conclusions

This investigation presents the results of transparent soil modeling which is used to visualize the process of the permeation of chemical grout columns by water. The laboratory work consists of both physical and numerical modeling. The following conclusions are drawn after carrying out the work.

There are two stages in the permeation of UFR which is used as the chemical grout in porous media inflowing water. First, the chemical grout is injected into the porous media by using air pressure. The groundwater-grout interface is sphere shaped. After a certain time, the grouting pressure is reduced with increasing distance from the injection point. The groundwater-grout interface is then forced to move in the flow direction. The groundwater-grout interface is primarily influenced by the advection of the flowing water with decreases in the grouting pressure.

An equation based on fluid dynamics in porous media and Darcy's law is then proposed to predict the displacement of the groundwater-grout interface in flowing water. The numerical modeling is performed under an idealized condition which assumes that the porous medium is homogeneous. If a heterogeneous medium is used, it would be quite difficult to obtain the same correlations found with a homogeneous medium as presented above. A heterogeneous porous medium will make the application of the technology for the chemical grouting of columns more difficult and unpredictable. This problem is not considered here because it is beyond the scope of this study.

References

1. Bruce DA (2005) Glossary of grouting terminology. J Geotech Geoenviron Eng 131(12):1534–1542
2. Carter TG, Dershowitz W, Shuttle D, Jefferies M (2012) Improved methods of design for grouting fractured rock. In: Grouting and deep mixing. Geotechnical Special Publications (GSP), United States, pp 1472–1483
3. El Tani M (2012) Grouting rock fractures with cement grout. Rock Mech Rock Eng 45(4):547–561
4. Funehag J, Gustafson G (2008) Design of grouting with silica sol in hard rock—new methods for calculation of penetration length. Part I. Tunnell Undergr Space Technol Incorp Trenchless Technol Res 23(1):1–8
5. Hässler L, Håkansson U, Håkan S (1992) Computer-simulated flow of grouts in jointed rock. Tunnell Undergr Space Technol Incorp Trenchless Technol Res 7(4):441–446
6. Yang M, Yue Z, Lee P, Su B, Tham L (2002) Prediction of grout penetration in fractured rocks by numerical simulation. Can Geotech J 39:1384–1394
7. Wallner M (1976) Propagation of sedimentation stable cement pastes in jointed rock. Rock Mech Waterways Constr 2:49–165
8. Dai G, Bird RB (1981) Radial flow of a Bingham fluid between two fixed circular disks. J Nonnewton Fluid Mech 8(3):349–355
9. Gardner RD, Pincus HJ (1968) Fluorescent dye penetrants applied to rock fractures. Int J Rock Mech Min Sci 5:155–158
10. Chen Y, Nishiyama T, Terada M, Iwamoto Y (2000) A fluorescent approach to the identification of grout injected into fissures and pore spaces. Eng Geol 56:395–401
11. Iskander M (2010) Modelling with transparent soils. Polytechnic Institute of NYU
12. Mannheimer RJ, Oswald C (1993) Development of transparent porous media with permeabilities and porosities comparable to soils, aquifers, and petroleum reservoirs. Ground Water 31(5):781–788
13. Liu J, Liu M, Zhu Z (2012) Sand deformation around an uplift plate anchor. J Geotech Geoenviron Eng 138(6):728–737
14. Yuan B, Liu J, Chen W, Xia K (2012) Development of a robust stereo-PIV system for 3-d soil deformation measurement. J Test Eval 40(2):58–66
15. Ahmed M, Iskander M (2011a) Transparent soil model tests and FE analyses on tunneling induced ground settlement. In: Geo-Frontiers. United States, pp 3381–3390
16. Ahmed M, Iskander M (2011) Analysis of tunneling-induced ground movements using transparent soil models. J Geotech Geoenviron Eng 137(5):525–535
17. Liu J, Gao Y, Sui W (2013) Visualization of grout permeation inside transparent soil. In: IACGE 2013. China, pp 188–194
18. Fernández Serrano R, Iskander M, Tabe K (2011) 3D contaminant flow imaging in transparent granular porous media. Geotech Lett 1(3):71–78
19. Lo H-C, Tabe K, Iskander M, Yoon SH (2008) Modeling of multi-phase flow and surfactant flushing using transparent aquabeads. In: Geo congress. United States, pp 846–853
20. Zhang Z (1999) Flexible camera calibration by viewing a plane from unknown orientations. In: Computer vision. The proceedings of the seventh IEEE international conference on computer vision, vol 1. Corfu, Greece, pp 666–673
21. Su B, Zhan M, Zhang Z (1994) Experimental research of seepage characteristic for filled fractured. Rock Soil Mech 15(4):46–52
22. Miyazaki T (2006) Water flow in soils. CRC Press Taylor & Francis Group, pp 377–378
23. Brill GT, Burke GK, Ringen AR (2003) A ten year perspective of jet grouting: advancements in applications and technology. In: Proceedings of grouting and ground treatment, New Orleans, LA, USA
24. Burke G (2007) Vertical and horizontal groundwater barriers using jet grout panels and columns. ASCE Geotechnical Special Publication Geo-Denver. Proceedings. Denver, Colorado, USA

25. Pearlman L (1999) Subsurface containment and monitoring systems: barriers and beyond. National Network of Environmental Management Studies Fellow for US Environmental Protection Agency
26. Sui W, Liu J, Yang S, Chen Z, Hu Y (2011) Hydrogeological analysis and salvage of a deep coalmine after a groundwater inrush. Environ Earth Sci 62(4):735–749
27. Krizek RJ, Perez T (1985) Chemical grouting in soils permeated by water. J Geotech Eng 111(7):898–915
28. Moseley MP, Kirsch K (2004) Ground improvement. CRC Press
29. Iskander M (2010) Modelling with transparent soils: visualizing soil structure interaction and multi-phase flow, non-intrusively. Springer Science and Business Media
30. Ezzein FM, Bathurst RJ (2011) A transparent sand for geotechnical laboratory modeling. Geotech Test J 34(6):1–6
31. Gao Y, Sui WH, Liu JY (2015) Visualization of chemical grout permeation in transparent soil. Geotech Test J 38(5):1–13
32. Cao CM, Feng ZQ (2009) Development and application of low viscosity urea-formaldehyde resin grouting material. J China Coal Soc 34(4):482–486
33. Iskander M, Sadek S, Liu J (2002) Optical measurement of deformation using transparent silica gel to model sand. Int J Phys Model Geotech 2(4):13–26
34. Hayashi S, Chai XJ, Matsunaga K, Toki A (2006) Drip injection of chemical grouts: a new apparatus. Geotech Test J 29(2):108–116
35. Gonzalez RC, Woods RE, Eddins SL (2009) Digital image processing using MATLAB, 2nd edn. Gatesmark Publishing, Tennessee

Chapter 13
Application of Transparent Soil Modeling Technique to Rapid Penetration of Objects

Abstract Penetration into granular media is a complicated process. This chapter presented a study on the penetration of different cylindrical rods into dry fused quartz and transparent soil at an impact velocity of 5 m/s and the measurement of internal flow fields. The influences of rod nose shape and rod size on the penetration were discussed. The penetration depth and the collision time were investigated for dry fused quartz sand and transparent soil. The flow field of the soil during cylindrical penetration process were obtained.

13.1 Introduction

Penetration of projectiles into soils entails a complicated process involving a variety of soil physical–mechanical properties. The projectile encounters soil resistance and friction during penetration. Dynamic compaction, penetration tests, offshore foundation installation, and projectile penetration into soil are commonly encountered in engineering practice [1–3]. During impact and penetration, a cavity can be formed in the soil and surrounding soil particles can be disturbed, which can alter the mechanical behavior of surrounding soils. The forces encountered during penetration generally include two components, i.e., collisional and frictional forces, which are depth-dependent forces [4]. Different forms of equations have been reported to describe the penetration process [5]. Clark and Behringer [6] formulated a kinetic energy-based equation to describe penetration phenomena [6].

The velocity of a given object gradually declines during penetration due to the resistance of soil particles. Different methods are available to measure the object velocity variation during penetration, such as the interferometric method [7] and inverse experiments [8]. These methods are effective but not very convenient. The digital image correlation (DIC) [9] and digital speckle-radiography methods [10] have been adopted to obtain the displacement of sand particles. The transparent soil modeling technique has been applied to investigate the soil response during rapid penetration of solid objects [11]. Collins et al. [12] employed X-ray technology and lead particles with the DIC method to measure the internal flow fields during high-speed penetration of long-rod projectiles.

H. Zhao et al., *Transparent Soil Modelling Technique and Its Application*, Springer Tracts in Civil Engineering, https://doi.org/10.1007/978-981-19-6825-9_13

The final penetration depth is an important topic investigated in considerable works [13–15]. The final penetration depth has been found to be related to the initial impact velocity of the solid object [14], the shape of the solid object, and the sand density [15]. Hinch [16] found that the penetration depth increases with the logarithm of the impact velocity, while the stopping time decreases with the inverse of the square of the logarithm of the impact velocity. Guo [17] presented an exact solution for the depth of the impact crater created by a sphere in a granular bed [17].

The influences of the projectile shape on penetration into granular media have been investigated by a few researchers. Bless et al. [18] applied cylindrical rods with three head shapes in penetration tests, but no clear conclusions were given. Chian et al. [19] employed projectiles with five head shapes to investigate nose shape effects and found that a blunt flat head requires the most energy for penetration into sand blocks, while the projectile with an ogival head exhibited the lowest ballistic limit. Glößner et al. [20] also employed cylindrical rods with three head shapes to perform a series of tests on sand at a speed of approximately 380 m/s, and a similar finding was obtained for the cylindrical rods with the different head shapes. Omidvar et al. [21] allowed cylindrical objects to penetrate into a transparent soil model and found that an opaque region was formed at the early stage of penetration, which was wider for projectiles with a blunt nose than that for projectiles with a cone nose. Similar observations were also reported by Guzman et al. [11], in which the velocity magnitude immediately ahead of the blunt projectile was larger. Newhall and Durian [22] reported experimental results for the shape effect on low-speed granular impact. They found that during shallow penetration, the actual depth depends on the tip shape, while during deep penetration, the actual depth does not depend on the tip shape. Clark and Behringer [6] determined that the collision force of a spherical head at impact is high, while for an elongated head, this force is reduced. The available data pertaining to head shape effects on penetration remain limited. More research is still needed to investigate the penetration dependence on the projectile head shape.

Soil properties such as saturation, relative density, and friction coefficient are the primary factors involved in the penetration process. Takita and Sumita [23] investigated the shape of the impact crater in moist sand with different saturations. Chian et al. [19] studied the penetration of cylindrical objects into well-graded sand for $D_r = 65\%$, 75% and 90%. Seguin et al. [24] simulated the penetration process into frictionless granular media with molecular dynamics. Bragov et al. [25] assessed the impact and penetration of cylindrical bodies into dry and water-saturated sand models.

The purpose of this study was to investigate the penetration of cylindrical rods into sand, which is a simplification of problems such as anchor foundation installation and penetration tests. Since foundations and exploration tools can exhibit different head shapes, cylindrical rods with different head shapes were selected to conduct penetration tests. Penetration tests with cylindrical rods of different diameters were further conducted to investigate the size effect. In addition, transparent soil was employed as a substitute for saturated sand to visualize the penetration phenomena produced by cylindrical rods. Penetration results into dry fused quartz and transparent soil models are presented. Saturation influences on penetration were explored by comparing the obtained test results between dry fused quartz and transparent soil. A kinetic energy-based equation was extended to describe cylindrical rod movement after impact. Moreover, the peak acceleration and collision time were investigated.

13.2 Experimental Program

The experimental setup consists of an electromagnetic projectile accelerator, soil model box, laser sheet setup, and image-acquisition system, as described below. To compare the differences in penetration process between dry and saturated granular materials, dry and saturated fused quartz (transparent soil) were adopted in this study. The density of dry fused quartz is 2.1 g/cm^3, and the bulk densities of dry fused quartz and transparent soil are 1.44 g/cm^3 and 2.38 g/cm^3, respectively. The fused quartz particles are angular, as shown in Fig. 13.1. The pore fluid used to make the transparent soil is a mixture of white oil and dodecane.

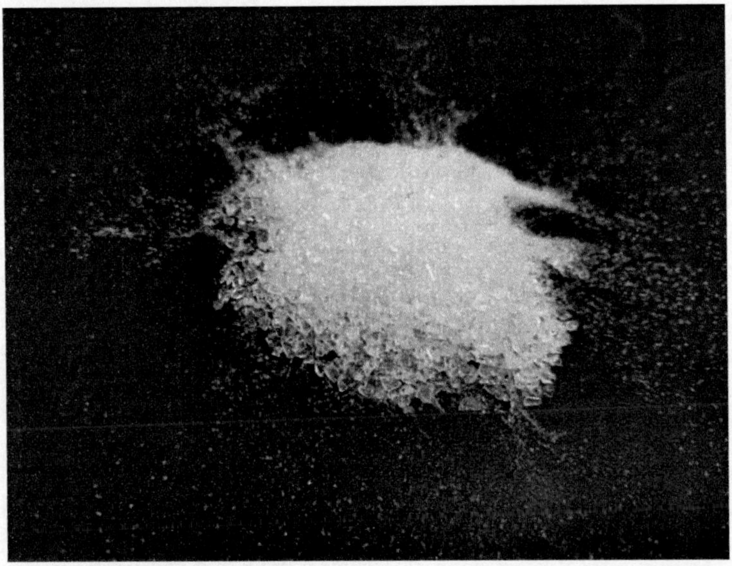

Fig. 13.1 Fused quartz applied in the test

13.2.1 Projectile Accelerator and Projectile

To perform the test, a projectile accelerator is designed based on solenoid coils similar to an electromagnetic cannon. Specially, the projectiles consist of iron. If a solenoid coil is powered, an electromagnetic field can be generated. When an iron steel bar is placed within the generated electric magnetic field, it is accelerated and moves quickly. The accelerator setup is shown in Fig. 13.2. Solenoid coils surround a guiding pipe, which is firmly maintained vertical with two pincers and a steel column. A cylindrical steel rod is held immediately above the guiding pipe head by a magnetic plate. Upon release, the cylindrical steel rod passes through the solenoid coil part of the guiding pipe, the solenoid coils are then powered, and the cylindrical steel rod is accelerated. Finally, the cylindrical steel rod penetrates the transparent soil model located below the guiding pipe. This setup is easy to accomplish in the laboratory. The obtained velocity can exceed 10 m/s. By adjusting the magnitude of the electric current, the intensity of the magnetic field can be varied, and the cylindrical steel rod can obtain different velocities upon acceleration.

In regard to the cylindrical rods, considering the head shape, length and diameter as variables, twelve rods were prepared in this study. Three different shapes, including flat, cone, and hemisphere heads, were considered. The chosen diameters of the cylindrical rods were 10 mm and 15 mm. The lengths of the rods were selected as 60 mm and 100 mm. Geometry and mass information on all the cylindrical rods is summarized in Table 13.1. The cylindrical rod consists of cast iron, and the density is 7.78 g/cm^3. The rods are named based on the length, diameter and head shape. For example, S10-F indicates a short (60-mm length) flat-head cylindrical rod with a diameter of 10 mm (D = 10 mm). The entry velocities of the cylindrical rods were measured through a photoelectric timer.

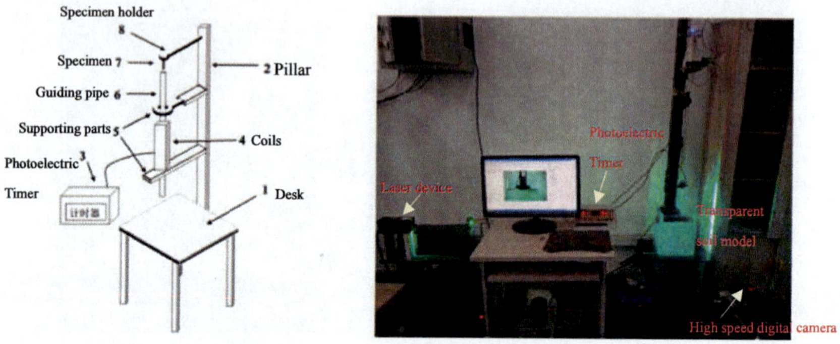

(a) Schematic sketch (b) Physical picture

Fig. 13.2 Test setup: **a** Schematic sketch; **b** physical picture

Table 13.1 Test cylindrical specimens

	Head shape	Length/mm	Diameter/mm	Mass/g	Cone angle	Head radius/mm
S10-F	Flat	60	10	35.251	–	–
S15-F		60	15	82.495	–	–
L15-F		100	15	123.84	–	–
L10-F		100	10	58.812	–	–
S10-C	Cone	60	10	31.767	53.13°	–
S15-C		60	15	68.953		–
L15-C		100	15	109.619		–
L10-C		100	10	55.234		–
S10-Sp	Hemisphere	60	10	35.599	–	5
S15-Sp		60	15	80.81	–	7.5
L15-Sp		100	15	121.15	–	7.5
L10-Sp		100	10	60.596	–	5

13.2.2 Transparent Soil Model

The transparent soil model box consists of glass and exhibits dimensions of 200mm × 150mm × 250mm. The transparent soil comprises fused quartz, and the refractive index matches that of the pore fluid consisting of dodecane and white oil (volume ratio = 1:4). Dry clean fused quartz is emplaced in layers in the glass model box. Each layer is compacted before the next layer. A relative density higher than 70% is achieved. The box is gradually filled with the pore fluid through a hose placed at the bottom of the model box. This approach displaces air and obtains a good transparency. After the pore fluid rises above the fused quartz particles, the transparent soil model box is vacuumed for 30 min in a large vacuum chamber. This further expels any air trapped in the model and improves the transparency. After vacuuming, a stress of 10 kPa is applied to the transparent soil model through a plastic plate with holes, and the transparent soil model is further consolidated. In this way, the transparent soil model becomes relatively dense.

The grain size distribution of the fused quartz applied in the model is shown in Fig. 13.3. The grain size ranges from 0.1 mm to 3 mm. The dry fused quartz model is prepared similarly to the transparent soil model, as described above, without pore fluid. The porosity n of the dry fused quartz model is approximately 0.75.

Before the penetration test, the bottom guiding pipe was located 10 cm above the center of the transparent soil model, which ensured that the cylindrical rod penetrated the soil vertically. Adjustment of the laser sheet allowed coincidence with the center plane of the transparent soil model. Laser speckle images of the center plane of the transparent soil model during penetration could be obtained with a high-speed digital camera set in front of the model. Tests were conducted on both saturated fused quartz (transparent soil) and dry fused quartz with 12 cylindrical rods, as listed in Table 13.1.

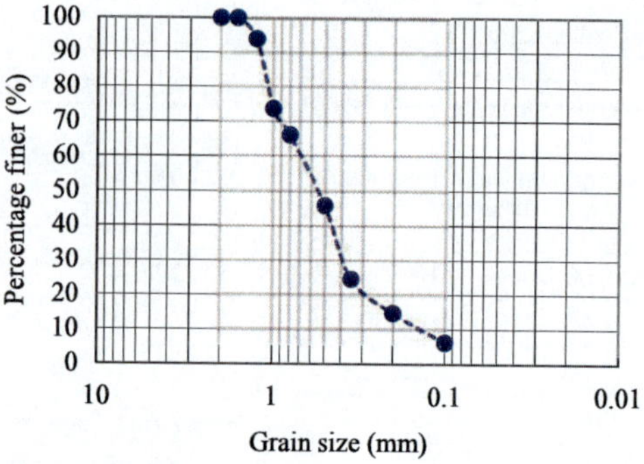

Fig. 13.3 Grain size distribution of the fused quartz

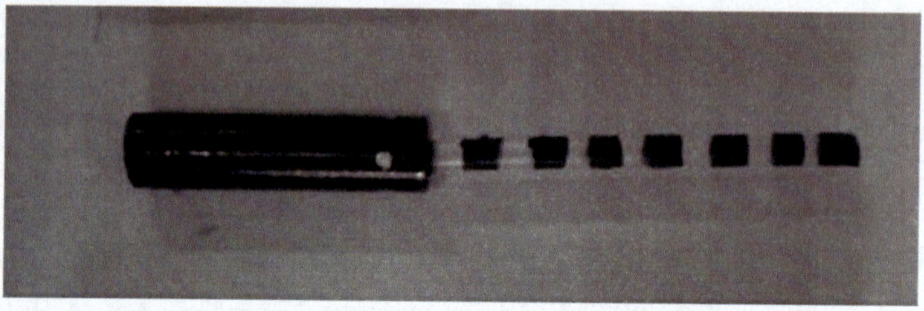

Fig. 13.4 Light plastic tube attached to the cylindrical rod with black marks

13.2.3 Penetration Depth Measurement

To measure the penetration depth, a very light plastic tube is attached vertically atop the cylindrical rod containing several black marks, as shown in Fig. 13.4. The images obtained with the high-speed digital camera (FastCAM SA50316010502, 1024 × 1024 pixels) at a frame rate of 1000 per second during the penetration test were analyzed to calculate the penetration depth. Based on the black mark position in each obtained image, the displacement and velocity of the cylindrical rod could be determined.

13.3 Experimental Results

13.3.1 Penetration into Dry Fused Quartz Sand

In total, twelve tests were conducted in the dry fused quartz sand model with twelve cylindrical rods at an initial velocity of approximately 5 m/s. A velocity of 5 m/s was selected considering that the penetration velocity in offshore geotechnical engineering is not high. If the velocity is very low, the penetration depth is very shallow.

Penetration depth and time curve

By tracing the black marks on the plastic tube attached to the cylindrical steel rod in the obtained images during penetration, the penetration depth at different times can be obtained. Since the camera exhibits a very high frame rate, the entry time of the cylindrical rod into the soil can be determined relatively accurately based on the captured images.

For the dry fused quartz sand model, the obtained penetration depth and time data are shown in Fig. 13.5. The penetration depth ranges from 30 to 100 mm, which is greater than D (the diameter of the cylindrical rod) but less than L (the length of the cylindrical rod) since the initial penetration velocity is low. For all the cylindrical rods, the penetration depth first linearly increases over time, then quickly nonlinearly increases with the penetration time and finally slowly increases until stopping. The penetration process can be divided into two distinct stages: impact, when the projectile first contacts the granular medium; penetration, when a transient crater is formed and grains ahead of the projectile are fluidized. Since no cylindrical rods are fully immersed in the soil, no granular medium collapse is observed, as described by previous researchers [26].

As shown in Fig. 13.5a, the penetration depth of S15-C is greater than that of S15-Sp, and the penetration depth of S15-Sp is greater than that of S15-F. These results indicate that the cone-head cylindrical rod generates the least resistance during penetration, followed by the sphere-head cylindrical rod. This observation indicated that the penetration depth depends on the tip shape, which is consistent with previous research results whereby sharper tips yielded deeper penetration [22].

In regard to S15-F, S15-Sp, and S15-C, all these cylindrical rods have a length of 60 cm and a diameter of 15 mm with different head shapes (Fig. 13.5a), but the collision times are the same. The three cylindrical rods of S10-F, S10-Sp, and S10-C, as shown in Fig. 13.5b, also yield the same collision time. This indicates that the collision time is independent of the tip shape.

Regarding the long cylindrical rods, similar conclusions could be obtained as those for the short cylindrical rods, namely, the cone-head cylindrical rods penetrated deeper than did the other cylindrical rods. Among the long cylindrical rods, the collision time could be expected to remain the same even though the final stop time was not determined due to operation error.

The above test results indicate that the time required for a cylindrical rod to stop penetrating into dry granular sand media is independent of the mass (size) and head shape. Similar findings have also been reported by Ciamarra et al. [27] whereby the time required for a projectile to slow and finally stop penetrating a granular medium is independent of the velocity at impact.

Size effects

Regarding the cylindrical rods with the same head shape but different lengths and diameters, the penetration depth variation over time is separately shown in Fig. 13.6. The results reveal that for the same head shape, the long cylindrical rod achieves a larger penetration depth than does the short cylindrical rod. Regardless of the head

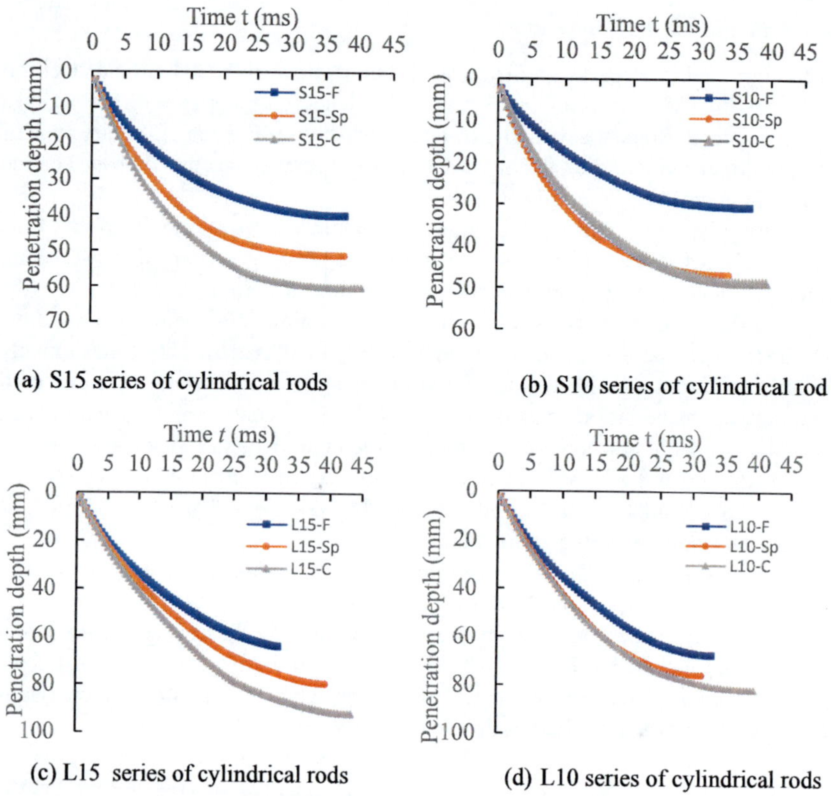

Fig. 13.5 Penetration depth variation over time in dry fused quartz sand: **a** S15 cylindrical rods; **b** S10 cylindrical rods; **c** L15 cylindrical rods; **d** L10 cylindrical rods

shape, the short and larger-diameter cylindrical rods attain a larger penetration depth than that attained by the short and smaller-diameter cylindrical rods, and the long and larger-diameter cylindrical rods also achieve a larger penetration depth than do the long and smaller-diameter cylindrical rods, except for L15-F and L10-F, which accomplish very similar penetration depths.

To observe the size effects, the normalized penetration depth by the soil volume expelled by the cylindrical rod $d/(V_s\, P_0)$ was plotted versus the kinetic energy (E_k) of the cylindrical rod in Fig. 13.7, where P_0 is the atmospheric pressure. For all the tested cylindrical rods, $d/(V_s\, P_0)$ occurs within a very narrow range from 0.11 to 0.14, which indicates that the normalized penetration depth is independent of the kinetic energy, but the penetration depth is highly related to the volume of the expelled soil or the diameter of the cylindrical rod.

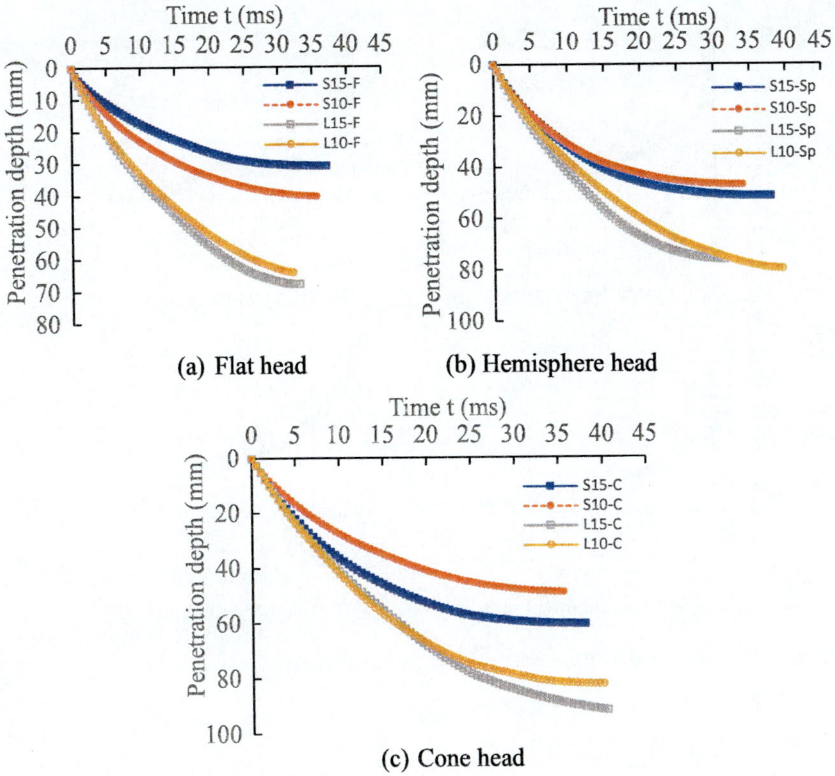

Fig. 13.6 Penetration depth variation over time: **a** flat head; **b** hemisphere head; **c** cone head

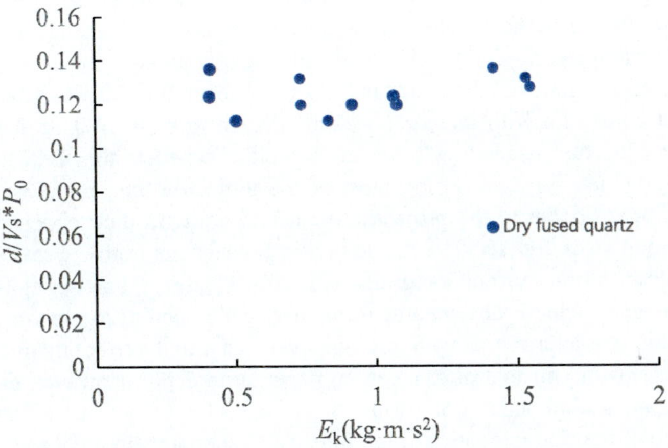

Fig. 13.7 Normalized depth versus kinetic energy in dry fused quartz

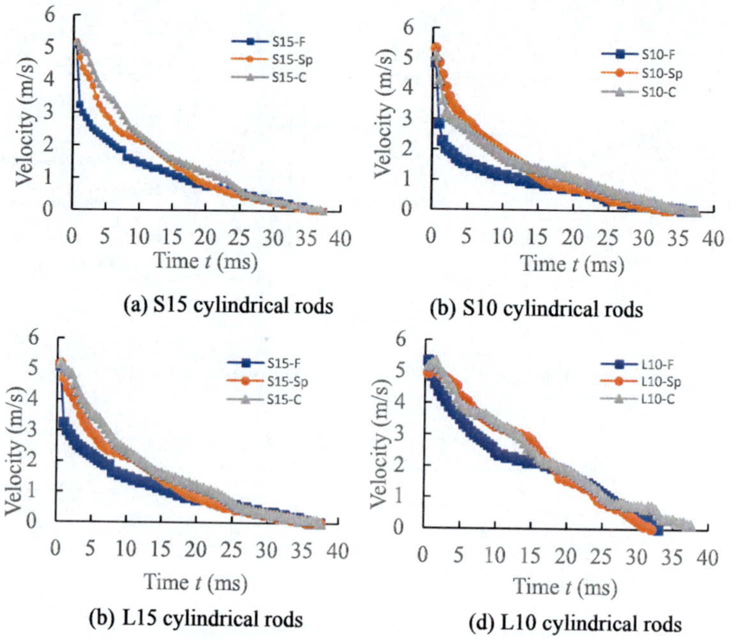

Fig. 13.8 Velocity curve for the cylindrical rods (dry fused quartz)

Penetration velocity

By differentiating the above penetration depth and time curves, penetration velocity and time curves can be obtained. Figure 13.8 shows velocity–time curves for the dry fused quartz model. There occurs a first linear reduction phase in the velocity corresponding to the impact stage. The second phase involves penetration, in which the velocity nonlinearly decreases. The third phase is another linear stage, where the acceleration again stabilizes beyond a certain penetration depth. This observation is not consistent with those reported by Ciamarra et al. [27]. In their experimental results, the mean drag force on a projectile penetrating into a granular medium remained constant during most of the projectile trajectory, and the drag force was proportional to the projectile impact velocity. In their experiments, the impact velocity was high (>50 m/s), and inertia played a major role. Previous experiments at low constant velocities (grain velocity $\leq$ 1.5 mm/s) and negligible inertia effects have also yielded constant drag forces in granular media [28]. In this study, the impact velocity is approximately 5 m/s, and the mean drag force acting on the cylindrical rod remains constant after a certain penetration depth. However, this occurs in the late penetration phase, not the middle phase, as found by Ciamarra et al. [27]. Knowledge of the characteristics of the forces acting on a cylindrical rod is helpful for better design and installation of offshore foundations.

A motion equation for a solid projectile impacting granular matter was established by Katsuragi and Durian [29], as follows [29]:

$$m\frac{d^2z}{dt^2} = mg - kz - m\frac{v^2}{d_1} \quad (13.1)$$

where m, t, z and v are the mass of the projectile, the elapsed time from the impact moment, the instantaneous penetration depth, and the instantaneous velocity of the projectile, respectively, g is the gravitational acceleration, k and d_I are two material parameters, kz denotes the friction force, and $m\frac{v^2}{d_1}$ is related to the mass inertia force.

At the impact moment, the maximum acceleration is reached, denoted as a_{peak}, which mostly results from the inertial force, as expressed in Eq. (13.1). During the penetration process, the friction force usually dominates at a low velocity, while the inertial force dominates at a high velocity. Since the velocity is not very high in this study, we can assume that both frictional and hydrostatic forces are important factors.

As shown in Fig. 13.8, the flat-head cylindrical rods first enter the nonlinear phase. Figure 13.8a shows that during the last several microseconds before stopping, the velocity tends to exhibit a similar slope value among the same-sized cylindrical rods with different head shapes, which indicates that the acceleration or resistance force remains the same during the last few microseconds. Similarly, Fig. 13.8b–d reveal that the slope of the velocity curve during the last few seconds is almost the same among the cylindrical rods of the same size. These results indicate that during the initial impact phase, the head shape plays an important role. During the last several microseconds, the resistance is attributed to friction at a very low velocity, and the aforementioned frictional force is not related to the depth, as analyzed from Eq. (13.1).

Figure 13.9 shows velocity–time curves for the cylindrical rods with the same head shape. Among the cylindrical rods with the same head shape, the first reduction in velocity is the most notable for the short and small-diameter cylindrical rods (S10-F, S10-Sp, and S10-C), followed by the short and large-diameter cylindrical rods and the long and large-diameter cylindrical rods. The possible reason is that the short and small-diameter cylindrical rods generate less energy, which is consequently easier to dissipate by the impact action. Among the flat-head cylindrical rods, the deflection in the velocity curve is significant for S15-F and S10-F but not significant for L15-F and L10-F. This occurs because the inertia resistance is high at impact. After impact, the inertia resistance decreases, and the exerted friction force is low, so there occurs a clear deflection in the velocity curve. However, the long cylindrical rod is subjected to a higher friction force after impact, so the deflection in the velocity curve is not significant. Similar conclusions can be obtained for the cone- and hemisphere-head cylindrical rods. However, the conclusions are less notable than those for the flat-head cylindrical rods. In regard to the short cylindrical rods with the same head shape, the velocity curves almost coincided during the later phase until stopping, independent of the diameter. This indicates that the acceleration becomes consistent during the last phase. During the last phase, the velocity is very low, and the friction force is the major resistance force, which is not related to the depth after a certain critical depth. The unit resistance per mass reaches the same value, as indicated by the test results.

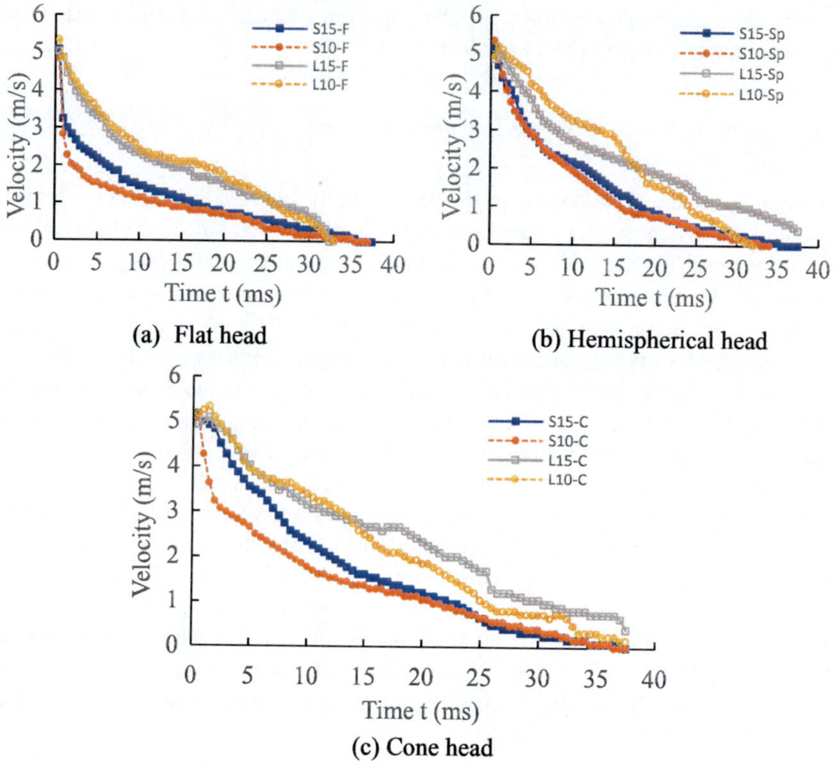

Fig. 13.9 Velocity–time curves: **a** flat head; **b** hemispherical head; **c** cone head Size effects

13.3.2 Penetration into Transparent Soil (Fully Saturated)

Penetration depth and time curve

In the transparent soil model, the obtained penetration depth and time data are shown in Fig. 13.10. The variation in penetration depth over time is similar to that observed for the dry fused quartz model. However, the nonlinear portion is very short. This indicates that the resistance during the first phase is relatively high, and a large amount of energy is therefore dissipated. This resistance includes the inertial force and fluid resistance. The penetration depth in the transparent soil ranges from approximately 8 mm to 20 mm, which is smaller than that in the dry fused quartz model. The collision time differs among the cylindrical rods with different head shapes. The flat-head cylindrical rod attains a short penetration time, the cone-head cylindrical rod achieves a longer penetration time, and the hemispherical head cylindrical rod yields a moderate penetration time. Similarly, the linear phase of the penetration curves also exhibits the above variation characteristics. The head shape of the cylindrical rod greatly affects the penetration process in the transparent soil model (saturated fused quartz). Because a proportion of the fluid resistance is exerted perpendicular to the front area of the rod head, the flat-head cylindrical rod exhibits the largest area and is subjected to the highest resistance. The resistance due to granular solids also follows similar rules.

Among the cylindrical rods with the same head shape, the penetration depth with the penetration time is plotted in Fig. 13.11. As observed, for the cylindrical rods with the same dimensions but different head shapes, the cone-head cylindrical rod yields the largest penetration depth, followed by the hemispherical head cylindrical rod, and the flat-head cylindrical rod attains the smallest penetration depth, since the cone-head cylindrical rod experiences the lowest resistance from the transparent soil. This is similar to what is observed for the dry fused quartz model.

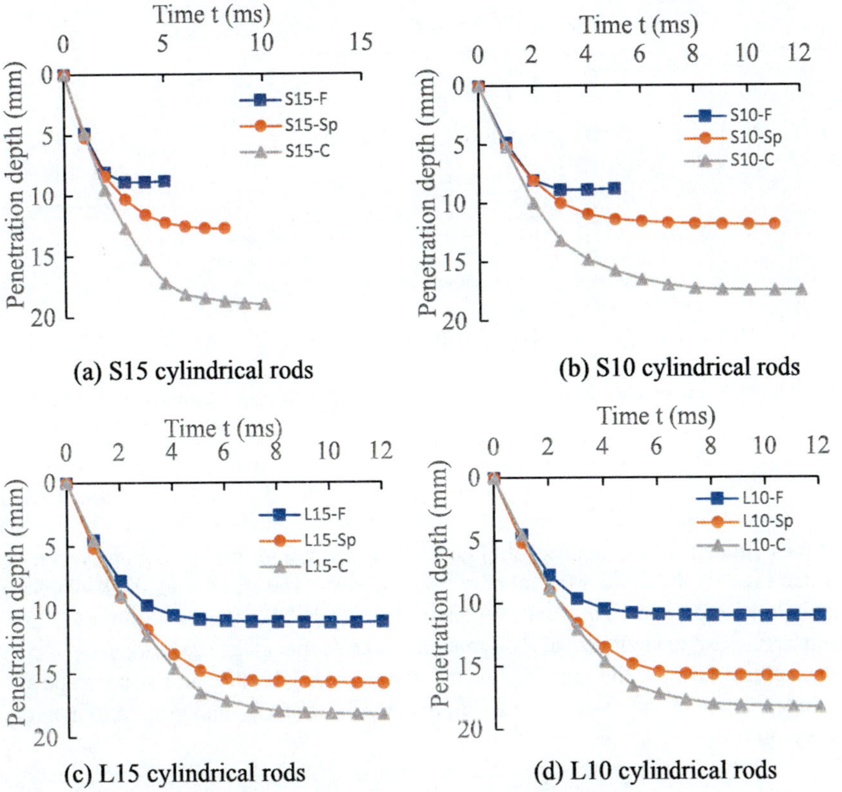

Fig. 13.10 Penetration depth variation over time in the transparent soil model: **a** S15 cylindrical rods; **b** S10 cylindrical rods; **c** L15 cylindrical rods; **d** L10 cylindrical rods

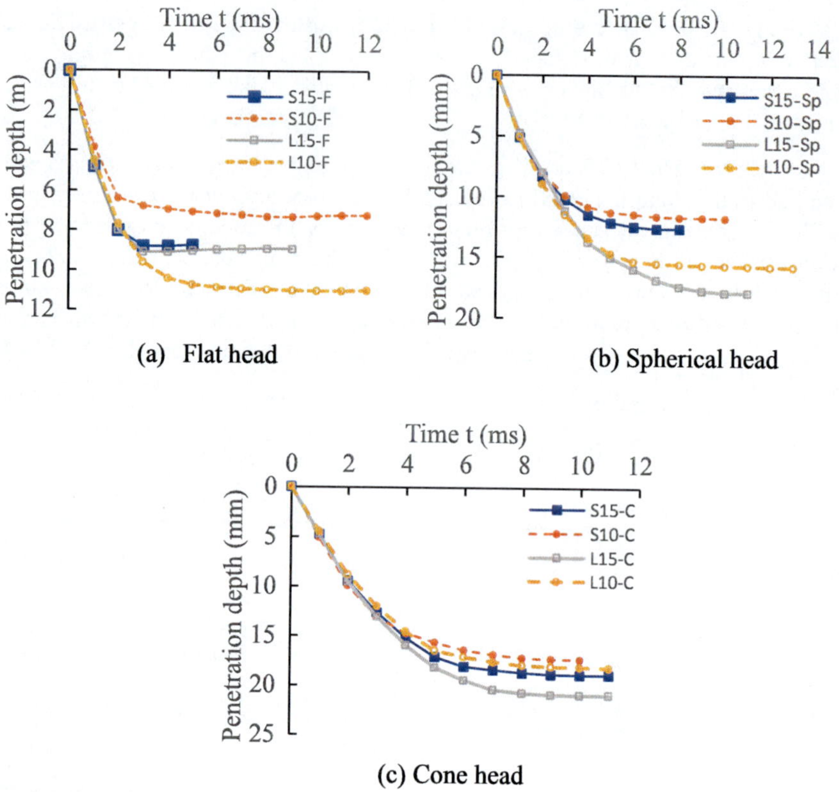

Fig. 13.11 Penetration depth change over time: **a** flat head; **b** spherical head; **c** cone head

Size effects

As shown in Fig. 13.11, it is clear that the penetration depth depends on the dimension of the cylindrical rod. The long and large-diameter cylindrical rod achieves a larger penetration depth than do the other cylindrical rods. The short- and small-diameter cylindrical rod attains the smallest penetration depth. This occurs because the long and large-diameter cylindrical rods generate more energy at the same impact velocity. The collision time is the same for the cylindrical rods with the same head shape and is independent of the diameter and length. This is similar to the observations in dry fused quartz.

Similarly, the normalized penetration depth $d/(Vs*P_0)$ versus the kinetic energy E_k is plotted in Fig. 13.12 for transparent soil. It can be observed that for the cylindrical rods with the same radius, the normalized depth is independent of E_k, the length and the head shape. However, the normalized depth of the larger-diameter cylindrical rod is greater than that of the smaller-diameter cylindrical rod. This is different from the observations in dry fused quartz, which might be caused by the

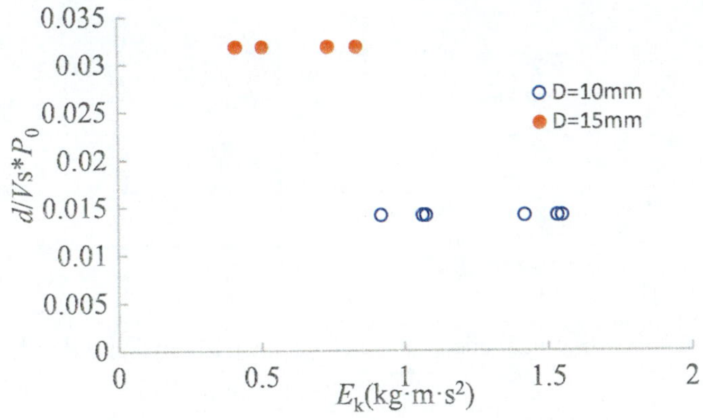

Fig. 13.12 Normalized depth versus kinetic energy in transparent soil

hydrodynamic drag exerted by the pore fluid. The normalized depth of $d/(Vs*P_0)$ depends on the diameter of the cylindrical rod in transparent soil.

Penetration velocity and time curve

The momentum of the cylindrical rod can be reflected by the velocity. Velocity–time curves obtained for the penetration tests are plotted in Fig. 13.13. The velocity–time curves for the transparent soil model are different from those for the dry fused quartz model. It is observed that the velocity exhibits a large linear decrease phase, followed by a small nonlinear decrease phase to a zero velocity. The velocity data for all the cylindrical rods exhibit a very sudden decrease after the impact moment.

Basically, the linear phase dominates the velocity curve, which indicates that the acceleration is constant during a long penetration period, i.e., the cylindrical rods are subjected to a constant resistance. This constant resistance mostly results from the saturated transparent soil, which includes the fluid resistance and granular resistance in addition to the inertia resistance. This is different from the velocity curves for the dry fused quartz model, where the linear velocity reduction phase is short and the resistance mostly results from the granular solid particles. The additional resistance from the pore fluid results in a smaller penetration depth for the same cylindrical rods in the transparent soil model than that in the dry fused quartz model.

For the cylindrical rods with the same head shape, the velocity is plotted over time in Fig. 13.14. As shown in Fig. 13.14a, the velocity of the flat-head cylindrical rods quickly linearly decreased to zero. Figure 13.14b shows that all the hemisphere-head cylindrical rods exhibited a similar variation in velocity over time. The velocity decreased to zero following a logarithmic function of time. As shown in Fig. 13.14c, all the cone-head cylindrical rods yielded similar velocity time curves except for the S15-C cylindrical rod. First, there is a slow decrease in speed at the beginning of penetration, after which the velocity quickly decreases in a nonlinear manner to

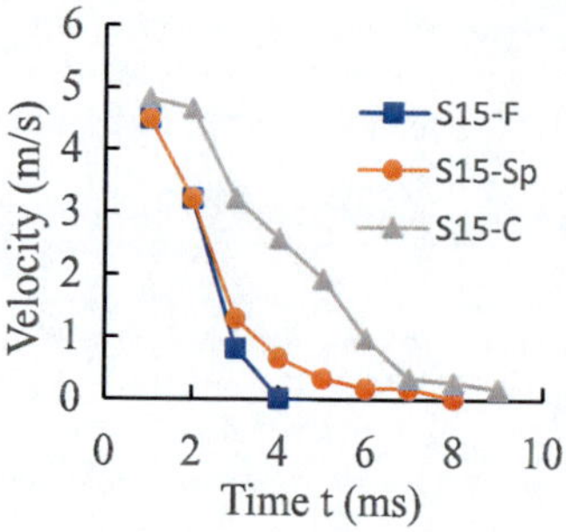

(a) S15 cylindrical rods

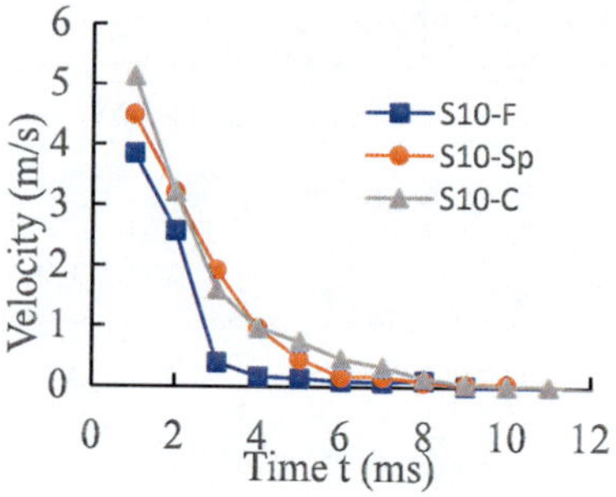

(b) S10 cylindrical rods

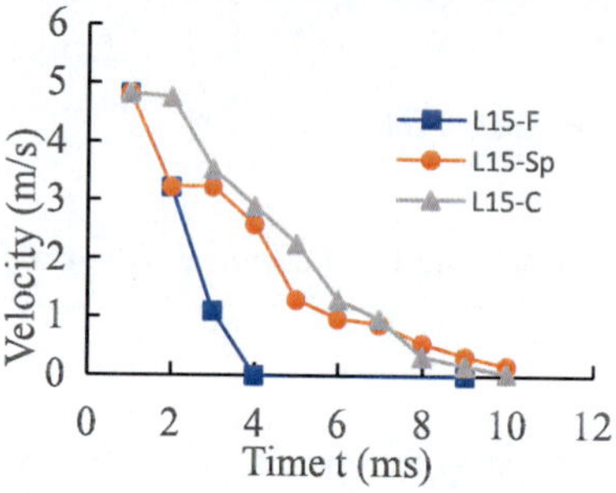

(c) L15 cylindrical rods

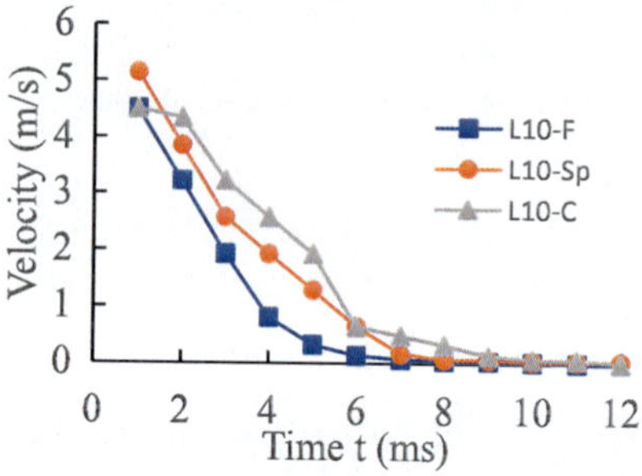

(d) L10 cylindrical rods

Fig. 13.13 Velocity curves (transparent soil model): **a** S15 cylindrical rods; **b** S10 cylindrical rods; **c** L15 cylindrical rods; **d** L10 cylindrical rods

zero. The first gradual decrease in velocity occurs because the resistance acting on the cone head is low at impact with only a portion of the cone head immersed in the saturated soil. After the cone head has completely entered the saturated soil, the resistance increases, and the velocity quickly declines. However, the variation for the short and small-diameter cylindrical rod (S10-C) is significantly different from that for the other cylindrical rods. The former rod attains a short linear reduction phase that quickly changes into a nonlinear reduction phase. This can be explained by the fact that the size of the S10-C rod is small, and both the inertia resistance and the frictional force play an important role during most of the penetration process. Hence, we observe a rather large nonlinear phase. All the cone-head cylindrical rods require approximately the same time to fully dissipate the kinetic energy, as shown in Fig. 13.14c.

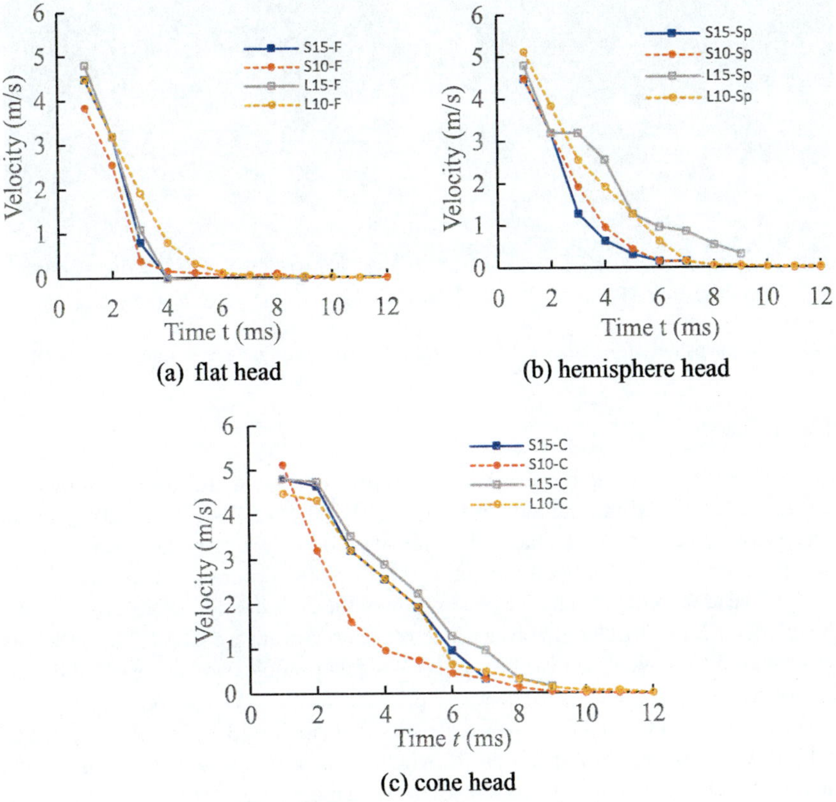

Fig. 13.14 Velocity versus penetration time for the cylindrical rods; **a** flat head; **b** hemisphere head; **c** cone head

13.4 Visualization of the Penetration Event

Typical snap shots of the penetration process into the transparent soil model along the center plane are shown in Fig. 13.15 for the flat-, cone- and hemisphere-head projectiles at a penetration depth of 1R (R is the radius of the cylindrical rod) at an impact velocity of 5 m/s. The impact and the initial penetration of the cylindrical rods resulted in cavity formation under air entrainment, and the laser beam was reflected by the unsaturated solid medium as a white region. The long and slim cylindrical rod with a cone head (L10-C) generated a larger contact area in the pore fluid-saturated fused quartz model, and more granular particles were displaced, which led to a larger white zone. The unsaturated fused quartz particles mainly occurred in front of the hemisphere- and flat-head cylindrical rods. Hence, the white region was smaller.

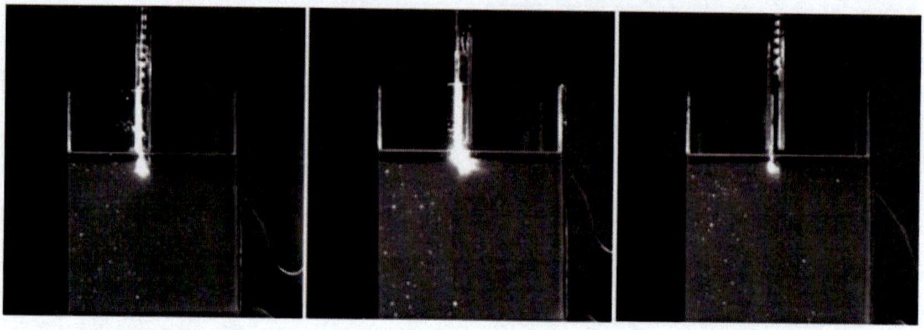

(a) L10-F at 1R (b) L10-C at 1R (c) L10-Sp at 1R

Fig. 13.15 Snapshots of penetration for the three cylindrical rods at a penetration depth of 1R

Velocity profiles

Velocity fields resulting from PIV analysis of successive images are shown in Fig. 13.16 for the cylindrical rods of L10-F, L10-Sp, and L10-C from initial impact to a penetration depth of 1R. Due to the page limitation, the other cylindrical rods are not described here. The velocity magnitude follows the order of L10-F > L10-Sp > L10-C. The area influenced by the penetration of the cylindrical rods also follows the same order. More particles are observed to move downward for L10-F. Few particles are observed to move downward for L10-C. A few particles are also ejected due to the impact.

Distributions of the velocity magnitude are shown in Fig. 13.17 for L10-F, L10-Sp, and L10-C at a penetration depth of 1R, which exhibits a bulb shape. L10-C attains the lowest velocity values. This is consistent with the results of the velocity vectors. The maximum velocity (v_{max}) of the transparent soil particles occurs approximately 3–5 diameters ahead of the initial impact location at a penetration depth of 1R. With

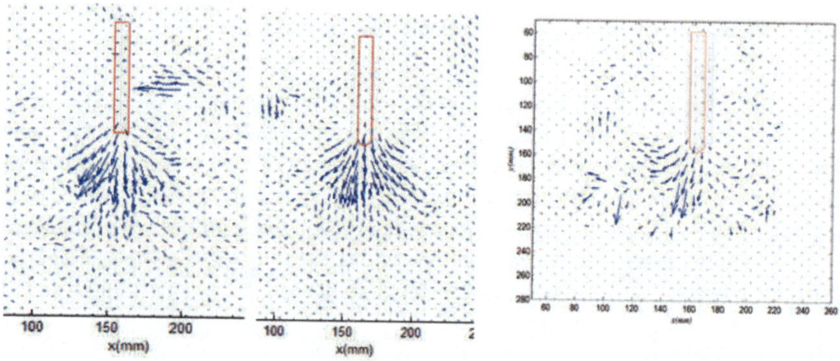

(a) L10-F (b) L10-S (c) L10-C

Fig. 13.16 Velocity vector field for the three cylindrical rods at a penetration depth of $z = 1R$ ($0 \rightarrow 1R$)

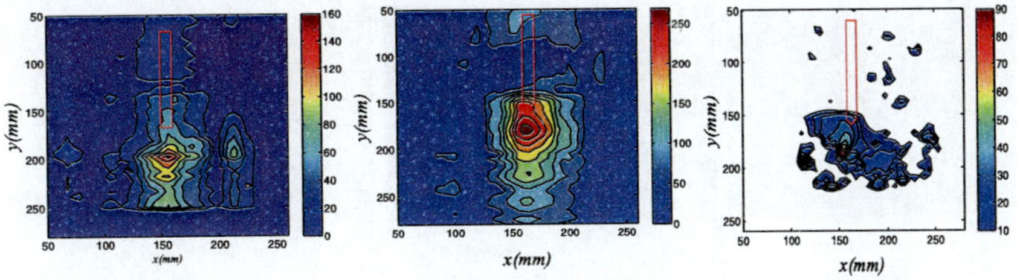

(a) L10-F (b) L10-Sp (c) L10-C

Fig. 13.17 Velocity profile for the three cylindrical rods at $z = 1R$ ($0 \rightarrow 1R$), unit: mm/s

increasing distance, the velocity magnitude is reduced. At a similar impact velocity, v_{max} of L10-F is approximately 160 mm/s, v_{max} of L10-Sp is approximately 260 mm/s, and v_{max} of L10-C is approximately 90 mm/s. This indicates that the hemispherical head cylindrical rod results in a higher maximum velocity of the soil particles in front, while the cone-head cylindrical rod cannot impart most of its energy to the soil particles in front because of its sharp tip. There are more soil particles in front of the flat-head rod, and these particles do not obtain the same velocity as those in front of the hemispherical head rod.

To quantify the soil-projectile interactions, the vertical and horizontal velocities (v_v and v_h, respectively) were extracted via image analysis of the penetration of the cylindrical rods, as shown in Figs. 13.18 and 13.19, respectively, for L10-F, L10-Sp, and L10-C. A comparison of the vertical velocity magnitude indicated that the distribution pattern was similar for the three cylindrical rods, while the depth of disturbance was deeper for L-10-F, followed by L10-Sp and finally L10-C. L10-C also attained a narrower zone of vertically moving soil particles, and L10-F and L10-Sp attained a similar zone of vertically mobilized soil particles during impact and penetration. The L10-F cylindrical rod achieved a large contact area with the fused quartz particles. Hence, the mobilized depth was larger. The cone-head cylindrical rod attained a smaller contact area with the fused quartz particles; hence, the mobilized depth was smaller. The front contact surface plays an important role in mobilizing soil particles to vertically move downward. The maximum v_v value was the largest for L10-Sp among the three cylindrical rods with the different head shapes, as explained above.

As shown in Fig. 13.19, v_h is much lower than v_v. The distribution of v_h indicated that L10-F generated a smaller horizontal zone of influence. L10-Sp and L10-C generated a larger zone of influence due to v_h. This revealed that most soil particles were pushed downward by the flat-head cylindrical rod, while the spherical head and cone-head cylindrical rods tended to push soil particles to move laterally.

Displacement profile

The displacement profiles of the soil particles at the final penetration depth of the cylindrical rods of L10-F, L10-Sp and L10-C are shown in Fig. 13.20. Among the

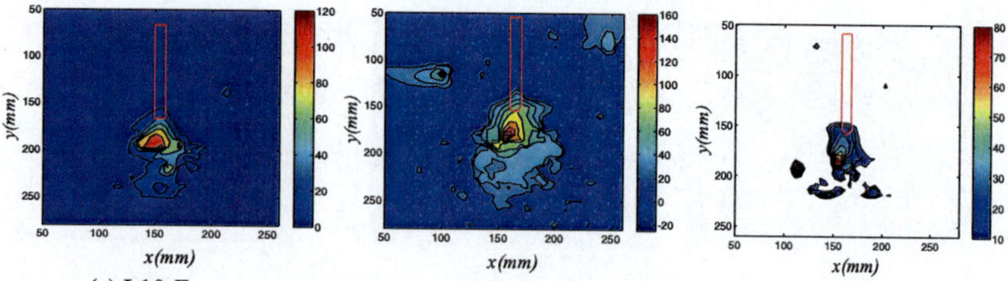

(a) L10-F (b) L10-Sp (c) L10-C

Fig. 13.18 v_v profile for L10-F, L10-S, and L10-C at $z = 1R$ ($0 \rightarrow 1R$), unit: mm/s

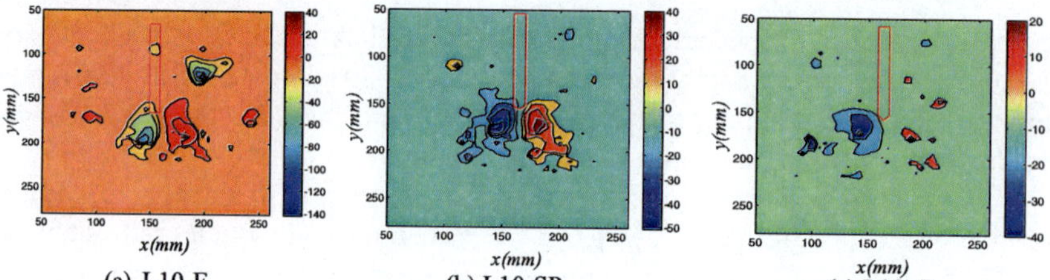

(a) L10-F (b) L10-SP (c) L10-C

Fig. 13.19 v_h profile for L10-F, L10-Sp, and L10-C at $z = 1R$ ($0 \rightarrow 1R$), unit: mm/s

three cylindrical rods with the different head shapes, the flat-head rod induced a larger displacement in the soil mass, while the cone-head rod induced the smallest displacement values. There were two larger displacement zones located around the two corners of the flat head. For the cylindrical rod with the hemisphere head, one larger deformation zone was located beneath the head. The mobilized soil particles occurred in a larger zone for L10-F, followed by L10-Sp. Because the soil particles beneath the flat head tended to move together with the cylindrical rod after initial impact, they did not exhibit large deformation, while the soil particles around the edge corners were pushed aside during the penetration of the cylindrical rod and exhibited large displacement. Since the hemisphere head does not have sharp edges, the mobilized soil particles were more centered beneath the head.

To obtain further details on soil particle movement during the penetration of the cylindrical rods, horizontal displacement profiles obtained after PIV analysis are shown in Fig. 13.21. The horizontal profiles for the three different cylindrical rods are similar. The lateral displacement of the soil particles was basically distributed symmetrically along the left and right sides of the cylindrical rods. The horizontal displacement or lateral spreading was the greatest for L10-F, followed by L10-Sp and finally L10-C. Collins et al. [7] pointed out that lateral splitting of the material ahead of a projectile is an important process during ballistic penetration into sand at an impact velocity of 200 m/s. Lateral splitting was also found to be important in our

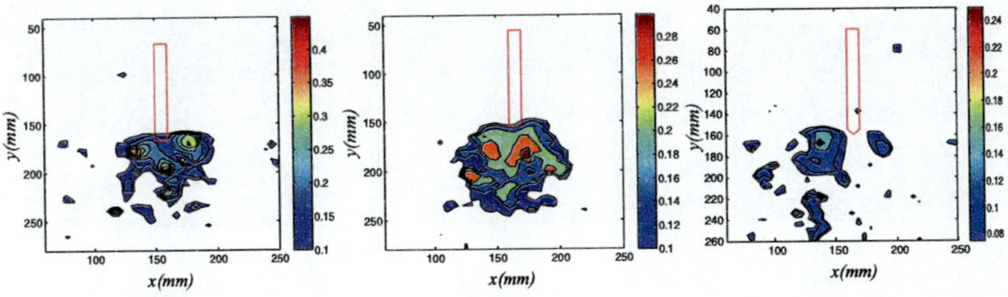

(a) L10-F (b) L10-Sp (c) L10-C

Fig. 13.20 Displacement profiles at the final penetration depth for L10-F, L10-Sp, and L10-C

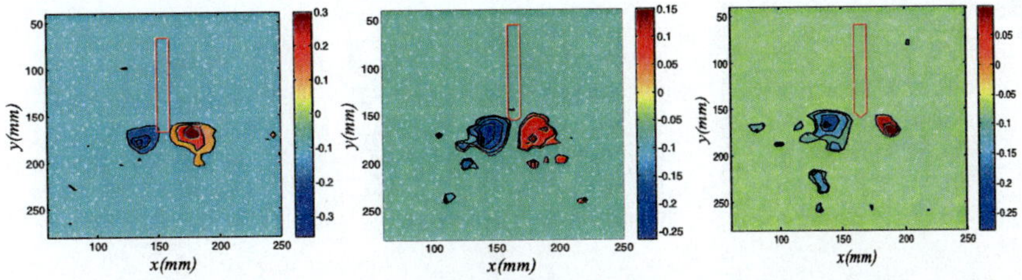

(a) L10-F (b) L10-Sp (c) L10-C

Fig. 13.21 Horizontal displacement at the final penetration depth for L10-F, L10-Sp, and L10-C

experiments at an impact velocity of 5 m/s. The head shape of the cylindrical rods greatly influenced the resulting lateral splitting, namely, the flat-head rod exerted the least influence and the cone-head rod exerted the greatest influence. This occurred because the lateral soil-projectile contact surface of the cone-head rod was the largest.

The vertical displacement profiles for the three cylindrical rods are different, as shown in Fig. 13.22. There is a small zone of large vertical displacement for L10-C since most soil particles are pushed toward the two sides of the cylindrical rod. There are few soil particles directly beneath the cone head, while L10-Sp attains a larger vertical displacement zone because the hemispherical head pushes more particles to move downward. Where the largest displacement occurs, the soil particles are also compacted. This is referred to as the compaction front, which can be observed in the case of L10-F and L10-Sp. However, for L10-C, no obvious compaction front is observed. A traveling compaction front was also observed by Collins et al. [12] during high-speed penetration. In offshore foundation design, this compaction front might increase the soil bearing capacity.

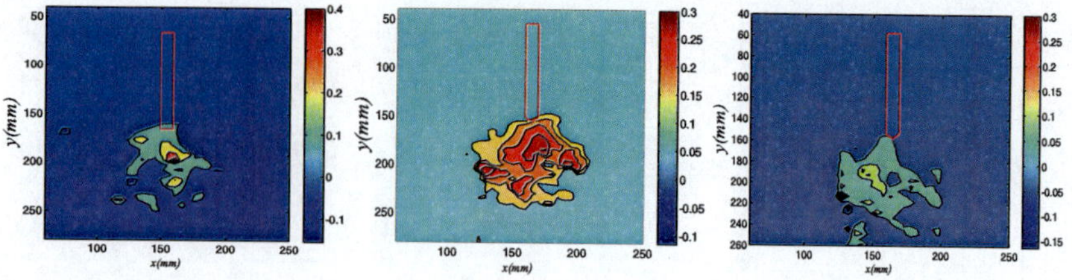

(a) L10-F (b) L10-Sp (c) L10-C

Fig. 13.22 Vertical displacement at the final penetration depth for L10-F, L10-Sp, and L10-C

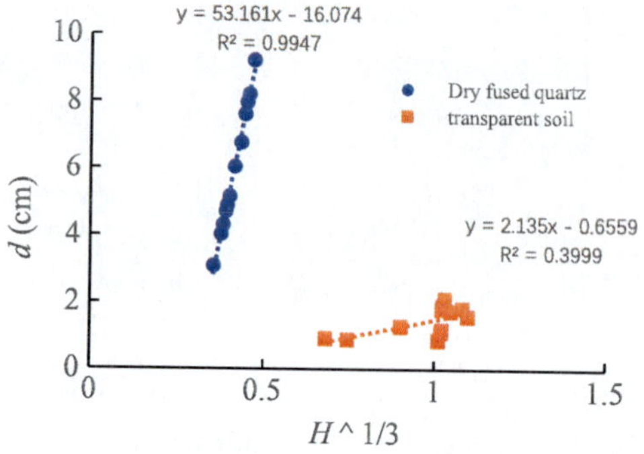

Fig. 13.23 Penetration depth *d* versus drop height *H*

13.5 Discussions

13.5.1 Penetration Depth Scaling

Several penetration depth scaling relationships are available [4, 6, 13, 16, 29]. The power law relationship of $d \sim H^{1/3}$ ($H = h + d$, with $h = v_0^2/2/g$) is mentioned the most [4, 13], but this relationship is mostly employed for spherical projectiles, not cylindrical rods. Therefore, $d \sim H^{1/3}$ is shown in Fig. 13.23 for the tested cylindrical rods in this study. It can be seen that $d \sim H^{1/3}$ performs well for the dry fused quartz model but not as well for the transparent soil model.

13.5.2 Collision Duration t_c

Several studies have investigated the collision duration. Ambroso et al. [8] found that spheres all come to rest at approximately $t_c = 3d/v_0$. However, this conclusion does not apply in this study. Ciamarra et al. [27] found that the collision time t_c was independent of v_0 for a disk impacting smaller disks. In their study, the velocity was approximately 5 m/s. Previously, Goldman and Umbanhowar [14] also revealed that t_c was nearly independent of v_0 for $v_0 \geq 1.5$ m/s. Seguin et al. [24] found that the collision time decreased with increasing impact velocity at a sufficiently low impact velocity v_0 ($v_0/V <= 2$, with $V = (\rho g d/\rho_g)^{1/2}$, where ρ_g is the density of the granular material). At high impact velocities ($v_0/V >= 2$), a plateau was reached, and the stopping time did not depend on the impact velocity and tended to reach a constant value. Katsuragi [29] presented a study involving impact velocities ranging from 0 to 4 m/s. The stopping time remained constant during high-speed impact ($v_0 > 2$ m/s) but increased with the velocity during low-speed impact ($v_0 <$ 2 m/s). Goldman and Umbanhowar [14] found that the collision duration remained constant at a sufficiently high impact velocity ($v_0 > 1$ m/s). Based on these studies, it can be concluded that during high-velocity impact ($v_0 > 1.5$ m/s), the duration is independent of the impact velocity.

This study also revealed that t_c is independent of v_0, as shown in Fig. 13.24, in both the dry fused quartz and transparent soil models even though the change in $v_0$is small. The observation in this study is consistent with previous conclusions. The collision time in the transparent soil model is much shorter than that in the dry fused quartz model because the pore fluid generates additional resistance toward the cylindrical rod. Furthermore, this study found that the collision time $t_c$is independent of the geometric dimension and head shape of the cylindrical rods in dry fused quartz but not in transparent soil.

Goldman and Umbanhowar [14] proposed a scaling equation of $t_c \sim (\rho_s/\rho_g)(R/g)^{1/2}$ for a sphere impacting granular media, where R is the radius of the projectile. The experimental data in this study are shown in Fig. 13.25a. This scaling rule performs well but is not linear.

A good linear relationship can be obtained, as shown in Fig. 13.25b, for $T_c = t_c[\rho_g R^2/(g^2\rho)]^{-1/4}$. Goldman and Umbanhowar [14] also obtained the same conclusion in their studies.

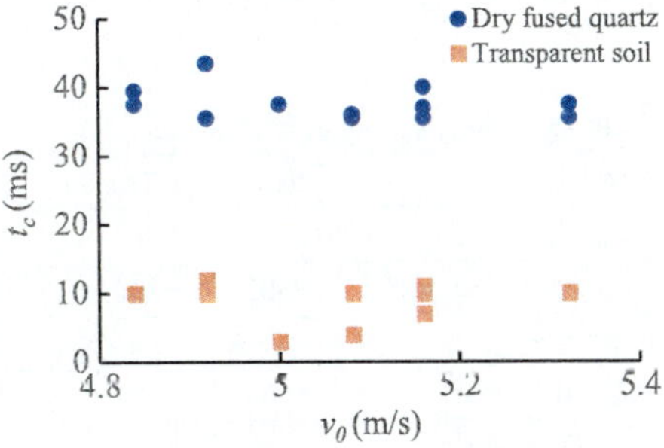

Fig. 13.24 Collision time for all the cylindrical rods tested

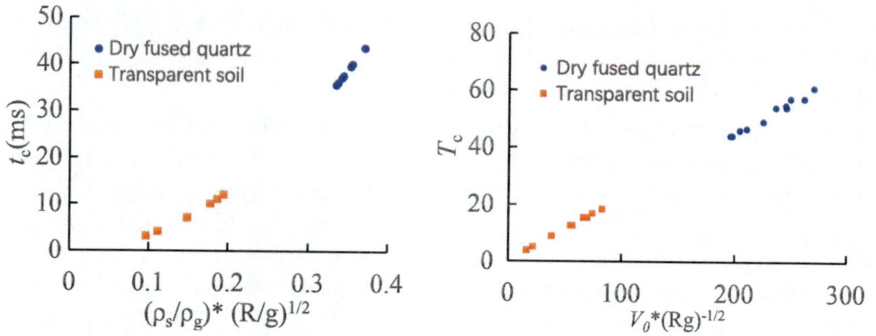

(a) t_c~$(\rho_s/\rho_g)(R/g)^{1/2}$ (b) T_c=$t_c[\rho_s R^2/(g^2\rho_p)]^{-1/4}$~$V_0(Rg)^{-1/2}$

Fig. 13.25 Collision time: **a** $t_c \sim (\rho_s/\rho_g)(R/g)^{1/2}$; **b** $T_c = t_c[\rho_s R^2/(g^2\rho_p)]^{-1/4} \sim V_0(Rg)^{-1/2}$

13.5.3 Peak Acceleration

The peak acceleration is an important variable since it indicates the maximum resistance forces acting on the projectile at impact. The peak acceleration a_{peak} obtained is provided in Table 13.2. In general, a_{peak} is the highest for the flat-head cylindrical rod, followed by the hemisphere-head cylindrical rod and finally the cone-head cylindrical rod in both dry and saturated fused quartz. The a_{peak} value in the saturated fused quartz model is larger than that in the dry fused quartz model, which is possibly attributable to the resistance force generated by the pore fluid during impact. The abnormally large value for S10-C in dry fused quartz might be due to measurement error. The results indicate that the resistance at impact depends on the head shape with the lowest resistance for the cone-head cylindrical rod. Kasuragi and Blum [30] proposed an equation to relate a_{peak} with the penetration depth and impact velocity. This equation is examined with our testing data and is found to not apply.

Table 13.2 Peak acceleration at impact

Specimen	S15-F	S15-Sp	S15-C	S10-F	S10-Sp	S10-C
a_{peak} (dry) m/s^2	3710	968	645	4032	968	1613
a_{peak} (saturated) m/s^2	2403	1923	1442	2179	2244	1602
Specimen	L15-F	L15-Sp	L15-C	L10-F	L10-Sp	L10-C
a_{peak} (dry) m/s^2	806	484	484	968	323	484
a_{peak} (saturated) m/s^2	2115	1602	1218	1282	1282	1122

13.5.4 Kinetic Energy

Clark and Behringer [6] proposed a kinetic energy-based equation (Eq. 13.2) to describe the dynamics of a projectile impacting granular materials. This equation performs well for two-dimensional objects but has not been examined in three-dimensional cases. It is applied in this study to examine the dynamics of cylindrical rods penetrating dry fused quartz and transparent soil in a three-dimensional case. The fitted equation for the hemispherical head cylindrical rod is shown in Fig. 13.26.

$$E_k(z) = (K_0 - C_1)\exp(-C_2 z) + C_1 - C_3 z \tag{13.2}$$

$$C_1 = \left[\left(m_g - f_0\right)C_2 + k_f\right]/C_2^2;\ C_2 = 2b/m;$$
$$C_3 = k_f/C_2 = k_f m/(2b)$$

where K_0 is the initial impact energy, b is a coefficient for the inertial force, f_0 is a constant for the friction force, k_f is a coefficient for the friction force, and m is the mass of the cylindrical rod.

The solid line in Fig. 13.26 indicates the fitted kinetic equation to the data (Eq. 13.2). The kinetic energy equation fits all the experimental data fairly well.

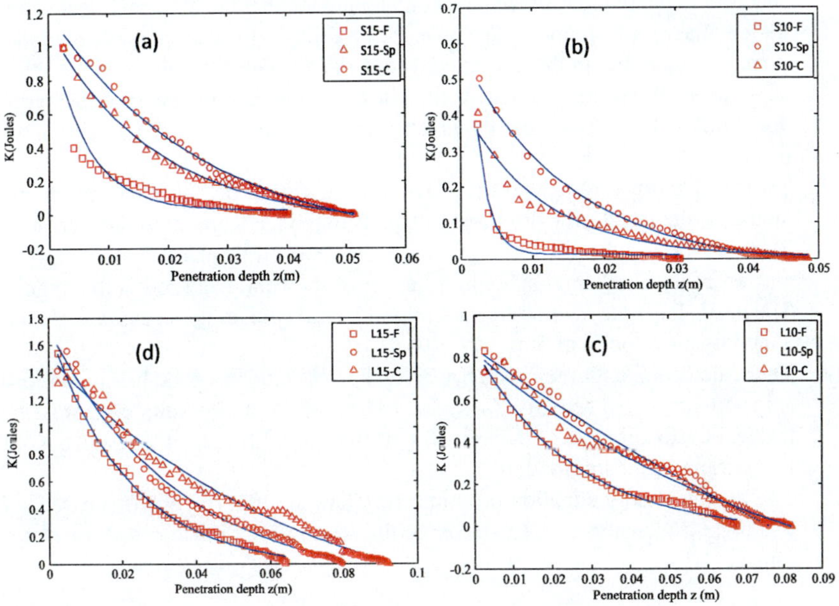

Fig. 13.26 Kinetic energy during cylindrical rod penetration: **a** S15-F, S15-Sp, and S15-C; **b** S10-F, S10-Sp, and S10-C; **c** L15-F, L15-Sp, and L15-C; **d** L10-F, L10-Sp, and L10-C

When the transition phase indicates a high-gradient change, there is a slight deviation in the fitting line from the experimental data, such as S15-F, S10-F and S10-C. This fitting equation can be applied to cylindrical rods with different head shapes in general. Therefore, the kinetic energy-based equation remains applicable to the three-dimensional case of cylindrical rods regardless of the head shape.

13.6 Summary and Conclusions

This research investigated the impact and penetration of cylindrical rods into granular media at an impact velocity of approximately 5 m/s. Dry and saturated fused quartz (transparent soil) models were used in this study. The penetration depth and collision time were examined. Visualization of the penetration process was achieved with the transparent soil model and PIV technique. The conclusions obtained in this study are as follows:

(1) An accelerator is constructed for a cylindrical steel rod by using the electromagnetic principle. A laser sheet is employed to generate speckle images of the transparent soil model, the penetration phenomena are visualized through images captured by a high-speed digital camera, and particle movement is determined via PIV analysis.
(2) Penetration can be divided into three phases for the cylindrical rods: a first linear penetration phase after impact, a second nonlinear phase and a final linear penetration phase until stopping. The penetration depth in the transparent soil model is smaller than that in the dry fused quartz model. The normalized penetration depth $d/(V_s\ \mathrm{P}_0)$ is independent of the kinetic energy and diameter of the cylindrical rod in dry fused quartz but is related to the cylindrical rod diameter in the transparent soil model.
(3) For the cylindrical rods with the different head shapes but the same length and diameter, the flat-head cylindrical rod tends to induce larger soil displacement. The cone-head cylindrical rod gives the smallest soil displacement among the various rods. The lateral spreading impact of the cone-head rod is the largest. The head shape plays an important role in lateral spreading during penetration at an impact velocity of 5 m/s.
(4) The scaled collision time $T_c = t_c[\rho_s \mathrm{R}^2/(\mathrm{g}^2 \rho_\mathrm{p})]^{-1/4}$ attains a good linear relationship with the scaled velocity $V_0(\mathrm{Rg})^{-1/2}$. The power function suitably represents the penetration process of cylindrical rods in the dry fused quartz model but not in the transparent soil model.
(5) The kinetic energy equation proposed by Clark and Behringer [6] is extended to a three-dimensional case to describe the penetration of cylindrical rods after impact.

References

1. Wang W, Wu Y, Wu H, Yang C (2021) Numerical analysis of dynamic compaction using FEM-SPH coupling method. Soil Dyn Earthq Eng 140:106420
2. Ahmed M, Al Qadhi S, Mallick J (2020) Advances in projectile penetration mechanism in soil media. Appl Sci 10(19):6810. https://doi.org/10.3390/app10196810
3. Allen WA, Mayfield EB, Morrison HL (1957) Dynamics of a projectile penetrating into sand. J Applied Physics 28(3):370–376
4. Uehara SJ, Ambroso AM, Ojha PR, Durian JD (2003) Low-Speed Impact Craters in Loose Granular Media. Phys Rev Lett 90:194301
5. Katsuragi H, Durian JD (2013) Drag force scaling for penetration into granular media. Phys Rev E 87:052208
6. Clark HA, Behringer PR (2013) Granular impact model as an energy-depth relation. Phys Rev Lett 101:64001
7. Forrestal MJ, Grady DE (1982) Penetration experiments for normal impact into geological targets. Int J Solids Struct 18(1):229–234
8. Forrestal MJ, Lee LM, Jenrette BD (1986) Laboratory-scale penetration experiments into geological targets to impact velocities of 2.1 km/s. J Appl Mech 53(2):317–320. https://doi.org/10.1115/1.3171758
9. Omidvar M, Malioche JD, Bless S, Iskander M (2015) Phenomenology of rapid projectile penetration into granular soils. Int J Impact Eng 85:146–160
10. Grantham SG, Goldrein HT, Proud WG, Field JE (2003) Digital speckle radiography: a new ballistic measurement technique. Imaging Sci J 51:175–186
11. Guzman LI, Iskander M, Bless S (2015) Observation of projectile penetration into a transparent soil. Mech Res Commun 70:4–11
12. Collins LA, Addiss WJ, Walley MS, Promratana K, Bobaru F, Proud GW, Williamson MD (2011) The effect of rod nose shape on the internal flow fields during the ballistic penetration of sand. Int J Impact Eng 38:951–963
13. Ambroso AM, Kamien DR, Durian JD (2005) Dynamics of shallow impact cratering. Phys Rev E 72:041305
14. Goldman ID, Umbanhowar P (2008) Scaling and dynamics of sphere and disk impact into granular media. Phys Rev E 77:021308
15. Guzman LI, Iskander M, Bless S, Qi C (2014) Terminal depth of penetration of spherical projectiles in transparent granular media. Granular Matter 16:829–842. https://doi.org/10.1007/s10035-014-0528-y
16. Hinch J (2014) Particles impacting on a granular bed. J Eng Math 84:41–48
17. Guo J (2018) Exact solution for depth of impact crater into granular bed. J Eng Mech 144(1):06017018
18. Bless SJ, Berry DT, Pedersen B, Lawhorn W (2009) Sand penetration by high speed projectiles. American Institute of Physics Conference Series. American Institute of Physics, 1361–1364
19. Chian CS, Tan CVB, Sarma A (2017) Projectile penetration into sand: Relative density of sand and projectile nose shape and mass. Int J Impact Eng 103:29–37
20. Glößner C, Moser S, Külls R (2017) Instrumented projectile penetration testing of granular materials. Exp Mech 57(2):261–272
21. Omidvar M, Iskander M, Bless S (2016) Soil-projectile interactions during low velocity penetration. Int J Impact Eng 93:211–221
22. Newhall KA, Durian DJ, Projectile-shape dependence of impact craters in loose granular media. Phys Rev E 68, 060301(R)
23. Takita H, Sumita I (2013) Low-velocity impact cratering experiments in a wet sand target. Phys Rev E 88(2):022203
24. Seguin A, Bertho Y, Gondret P, Crassouos J (2009) Sphere penetration by impact in a granular medium: A collision process. EPL (Europhysics Letters) 88(4):44002

25. Bragov MA, Blandin VV, Igumnov AL, Kotov LV, Kruszka, L.: Lomunov K.A., (2018) Impact and penetration of cylindrical bodies into dry and water-saturated sand. Int J Impact Eng 122:197–208
26. Dunatunga S, Kamrin K (2017) Continuum modeling of projectile impact and penetration in dry granular media. J Mech Phys Solids 100:45–60
27. Ciamarra PM, Lara HA, Lee TA, Goldman ID, Vishik I, Swinney LH (2004) Dynamics of drag and force distributions for projectile impact in a granular medium. Phys Rev Lett 92:194301
28. Cao Z, Kong G, Liu H (2014) Model tests on 3-D soil deformation during pile penetration using transparent soils. Chinese J Geotech Eng 36(2):395–400
29. Katsuragi H, Durian JD (2007) Unified force law for granular impact cratering. Nat Phys 3:420–423
30. Kasuragi H, Blum J (2017) The physics of protoplanetesimal dust agglomerates.x. mechanical properties of dust aggregates probed by a solid-projectile impact. The Astrophysical J 851:23